Bioinformatics and Machine Learning for Cancer Biology

Bioinformatics and Machine Learning for Cancer Biology

Editors

Shibiao Wan
Yiping Fan
Chunjie Jiang
Shengli Li

MDPI • Basel • Beijing • Wuhan • Barcelona • Belgrade • Manchester • Tokyo • Cluj • Tianjin

Editors

Shibiao Wan
Center for Applied
Bioinformatics
St. Jude Children's Research
Hospital
Memphis
United States

Yiping Fan
Center for Applied
Bioinformatics
St. Jude Children's Research
Hospital
Memphis
United States

Chunjie Jiang
Department of Medicine
Division of Diabetes,
Endocrinology and
Metabolism
Baylor College of Medicine
Houston
United States

Shengli Li
Precision Research Center for
Refractory Disease
Shanghai Jiao Tong
University
Shanghai
China

Editorial Office
MDPI
St. Alban-Anlage 66
4052 Basel, Switzerland

This is a reprint of articles from the Special Issue published online in the open access journal *Biology* (ISSN 2079-7737) (available at: www.mdpi.com/journal/biology/special_issues/Bioinformatics_Machine_Learning).

For citation purposes, cite each article independently as indicated on the article page online and as indicated below:

LastName, A.A.; LastName, B.B.; LastName, C.C. Article Title. *Journal Name* **Year**, *Volume Number*, Page Range.

ISBN 978-3-0365-4814-2 (Hbk)
ISBN 978-3-0365-4813-5 (PDF)

Contents

About the Editors

Shibiao Wan

Dr. Shibiao Wan is currently a Bioinformatics Research Scientist at St. Jude Children's Research Hospital, one of the top pediatric cancer hospitals in the US. With more than 12 years of experience in machine learning, bioinformatics, and computational biology, Dr. Wan has published >40 articles in prestigious journals such as *Genome Research* (impact factor: 11.093), *Nature Communications* (impact factor: 14.919), *Science Advances* (impact factor: 14.136), *Circulation Research* (impact factor: 15.862), *Briefings in Bioinformatics* (impact factor: 11.622), and *Bioinformatics* (impact factor: 6.937). Dr. Wan is an academic editor for *BioMed Research International* (impact factor: 3.411) and a guest associate editor for a series of high-impact journals including *Frontiers in Cell and Developmental Biology* (impact factor: 6.684), *Frontiers in Pharmacology* (impact factor: 5.810), *Biology* (impact factor: 5.079), and *Frontiers in Genetics* (impact factor: 4.599). He is a TPC member for >20 machine learning-related international conferences including IEEE ICTAI, IEEE IAICT, IEEE IoTaIS, and IEEE SOFTT. Dr. Wan is also a reviewer for >50 prestigious journals including *IEEE TNNLS* (impact factor: 10.451), *Nucleic Acids Research* (impact factor: 16.971), *Genomic Medicine* (impact factor: 11.117) and *Briefings in Bioinformatics* (impact factor: 11.622). He was a recipient of the global peer review awards (top 1%) in "Cross-Field" and "Biology and Biochemistry" in 2019, awarded by Clarivate. Dr. Wan was also an Outstanding Young Alumni Awardee in 2021–2022. Dr. Wan is a member of IEEE, ACM and ISCB.

Yiping Fan

Dr. Yiping Fan is currently the Associated Director of Center for Applied Bioinformatics at St. Jude Children's Research Hospital. He obtained his Bioengineering PhD in 2002 at the University of California at San Diego and MS degree in Biophysics at University of Illinois Urbana Champaign in 1999. His research interests include the development and optimization of bioinformatics methods, biomedical applications of machine learning, the epigenetic regulation of T-cell functions and cancer immunotherapy.

Chunjie Jiang

Dr. Chunjie Jiang's research focuses on the integrative analysis of multi-omics next-generation sequencing data to reveal the underlying molecular mechanism, identify clinical biomarkers for diagnosis and treatment, and make contributions to precision medicine in the fields of metabolism and cancer biology.

Until now, Dr. Jiang has published 30 papers in *Science, Cell, Nature Neuroscience, Cell Metabolism, Cell Stem Cell,* etc., including two cover stories (Hu and Jiang et al., Cell Stem Cell, 2019; Hu and Jiang et al., Genes & Development, 2022) as co-first author. The study Precision Medicine Approach to Glucocorticoid Therapy (Hu and Jiang et al., Cell Metabolism, 2021, co-first author) has been selected as the 2022 Top 10 Clinical Research Finalists.

Additionally, Dr. Jiang serves as a Guest/Associate/Youth Editor for 8 journals and a reviewer for more than 20 journals.

Shengli Li

Dr. Shengli Li received his Ph.D. in Medical Systems Biology from Fudan University in 2018 and Postdoctoral training from the University of Texas Health Science center in Houston in 2018–2020. He is a Principal Investigator with the Precision Research Center for Refractory Disease at Shanghai General Hospital affiliated to Shanghai Jiao Tong University School of Medicine. Shengli is a computational biologist with expertise in human disease genetics, including cancer and cardiovascular diseases. His research focuses on the integrative utilization of cutting-edge computational biology algorithms and molecular experimental technologies to reveal systematical changes and functional factors in human disease progression, disease microenvironment and response to drug treatment.

Preface to "Bioinformatics and Machine Learning for Cancer Biology"

Cancer is a leading cause of death worldwide, claiming millions of lives each year. Cancer biology is an essential research field to understand how cancer develops, evolves, and responds to therapy. By taking advantage of a series of "omics" technologies (e.g., genomics, transcriptomics, and epigenomics), computational methods in bioinformatics and machine learning can help scientists and researchers to decipher the complexity of cancer heterogeneity, tumorigenesis, and anticancer drug discovery. Particularly, bioinformatics enables the systematic interrogation and analysis of cancer from various perspectives, including genetics, epigenetics, signaling networks, cellular behavior, clinical manifestation, and epidemiology. Moreover, thanks to the influx of next-generation sequencing (NGS) data in the postgenomic era and multiple landmark cancer-focused projects, such as The Cancer Genome Atlas (TCGA) and Clinical Proteomic Tumor Analysis Consortium (CPTAC), machine learning has a uniquely advantageous role in boosting data-driven cancer research and unraveling novel methods for the prognosis, prediction, and treatment of cancer.

This book presents some of the latest progresses on leveraging bioinformatics and machine learning for cancer biology, which is particularly useful and attractive for cancer biologists, bioinformaticians, machine learning experts, computational biologists and other scientists or researchers in life sciences and biology. We would like to thank all the authors contributing to this book who made significant contributions to the better understanding of various cancers as well as cancer-related analysis method improvements. We are eager to see more exciting discoveries in cancer biology with the help of bioinformatics analysis and machine learning in the near future.

Shibiao Wan, Yiping Fan, Chunjie Jiang, and Shengli Li
Editors

biology

MDPI

Editorial

Special Issue on Bioinformatics and Machine Learning for Cancer Biology

Shibiao Wan [1,*], Chunjie Jiang [2], Shengli Li [3] and Yiping Fan [1]

[1] Center for Applied Bioinformatics, St. Jude Children's Research Hospital, Memphis, TN 38105, USA; yiping.fan@stjude.org
[2] Division of Diabetes, Endocrinology and Metabolism, Department of Medicine, Baylor College of Medicine, Houston, TX 77030, USA; chunjie.jiang917@outlook.com
[3] School of Medicine, Shanghai Jiao Tong University, Shanghai 201620, China; shengli.li@shsmu.edu.cn
* Correspondence: shibiao.wan@stjude.org; Tel.: +1-901-595-1905

Citation: Wan, S.; Jiang, C.; Li, S.; Fan, Y. Special Issue on Bioinformatics and Machine Learning for Cancer Biology. *Biology* **2022**, *11*, 361. https://doi.org/10.3390/biology11030361

Received: 15 February 2022
Accepted: 23 February 2022
Published: 24 February 2022

Publisher's Note: MDPI stays neutral with regard to jurisdictional claims in published maps and institutional affiliations.

Cancer is a leading cause of death worldwide, claiming millions of lives each year. Cancer biology is an essential research field to understand how cancer develops, evolves, and responds to therapy. By taking advantage of a series of "omics" technologies (e.g., genomics, transcriptomics, and epigenomics), computational methods in bioinformatics and machine learning can help scientists and researchers decipher the complexity of cancer heterogeneity, tumorigenesis, and anticancer drug discovery. Particularly, bioinformatics enables the systematic interrogation and analysis of cancer from various perspectives, including genetics, epigenetics, signaling networks, cellular behavior, clinical manifestation, and epidemiology. Moreover, thanks to the influx of next-generation sequencing (NGS) data in the postgenomic era and multiple landmark cancer-focused projects, such as The Cancer Genome Atlas (TCGA) and Clinical Proteomic Tumor Analysis Consortium (CPTAC), machine learning has a uniquely advantageous role in boosting data-driven cancer research and unraveling novel methods for the prognosis, prediction, and treatment of cancer.

This special issue aims to leverage bioinformatics analysis and machine learning to further our understanding of cancer biology in different perspectives. Specifically, Yao et al. [1] identified and validated an Annexin-related prognostic signature and therapeutic targets for bladder cancer. Furthermore, for bladder cancer, Wei et al. [2] demonstrated CPA4 to be a poor prognostic biomarker correlated with immune cells infiltration, and for ovarian cancer, Li et al. [3] identified the RNA modification gene PUS7 as a potential biomarker. Another interesting development is that Serna-Blasco et al. [4] proposed a new measurement called R-score to assess the quality of variants' calls using liquid biopsies for non-small cell lung cancer. Additionally, for breast cancer, Zainab et al. [5] used a drug–drug interaction network approach to identify estrogen receptor alpha inhibitors, and simultaneously, they revealed the role of persistent organic pollutants in the progression of the breast cancer. Furthermore, Chiu et al. [6] identified a DNA damage repair gene set as a potential biomarker to stratifying patients with high tumor mutational burden. Methodically, Rehman et al. [7] proposed a depth-wise convolutional neural network for architecture distortion-based digital mammograms classification. Obermayer et al. [8] proposed a web-based framework called DRPPM-EASY for integrative analysis of multiomics cancer datasets.

It is exciting to know that, with the help of integrative bioinformatics analyses, our understanding of multiple cancers such as bladder cancer, ovarian cancer, breast cancer, and non-small cell lung cancer has been remarkably enhanced. Additionally, with the application of machine learning approaches (especially deep learning) and webtool development, our capabilities of extending the analysis and understanding of other less-studied cancers are expected to be consolidated. On the other hand, we are also aware that with heterogeneity and complexity properties, even the internal mechanisms of tumorigenesis for those well-studied cancers [9,10] remain to be further unraveled, let alone the discovery of anticancer drugs and treatment. More integrative, genome-wide, and global-scale studies are required to further shed light on the driving forces behind the tumorigenesis and the development of anticancer drugs and treatments.

In summary, we would like to thank all the authors for the articles published within this special issue who made significant contributions to this special issue, and more importantly, to the better understanding of various cancers as well as cancer-related analysis method improvements. We are eager to see more exciting discoveries in cancer biology with the help of bioinformatics analysis and machine learning in the near future.

Author Contributions: S.W. wrote the main manuscript. S.W., C.J., S.L. and Y.F. participated in revising the manuscript. All authors have read and agreed to the published version of the manuscript.

Funding: This research was funded by in part by the National Cancer Institute grant P30 CA021765. The content is solely the responsibility of the authors and does not necessarily represent the official views of the National Institutes of Health. This work was in part supported by National Natural Science Foundation of China (32100517) and Shanghai General Hospital Startup Funding (02.06.01.20.06).

Conflicts of Interest: The authors declare no conflict of interest. The funders had no role in the design of the study; in the collection, analyses, or interpretation of data; in the writing of the manuscript, or in the decision to publish the results.

References

1. Yao, X.; Qi, X.; Wang, Y.; Zhang, B.; He, T.; Yan, T.; Zhang, L.; Wang, Y.; Zheng, H.; Zhang, G.; et al. Identification and Validation of an Annexin-Related Prognostic Signature and Therapeutic Targets for Bladder Cancer: Integrative Analysis. *Biology* **2022**, *11*, 259. [CrossRef]
2. Wei, C.; Zhou, Y.; Xiong, Q.; Xiong, M.; Hou, Y.; Yang, X.; Chen, Z. Comprehensive Analysis of CPA4 as a Poor Prognostic Biomarker Correlated with Immune Cells Infiltration in Bladder Cancer. *Biology* **2021**, *10*, 1143. [CrossRef] [PubMed]
3. Li, H.; Chen, L.; Han, Y.; Zhang, F.; Wang, Y.; Han, Y.; Wang, Y.; Wang, Q.; Guo, X. The Identification of RNA Modification Gene PUS7 as a Potential Biomarker of Ovarian Cancer. *Biology* **2021**, *10*, 1130. [CrossRef] [PubMed]
4. Serna-Blasco, R.; Sánchez-Herrero, E.; Berrocal Renedo, M.; Calabuig-Fariñas, S.; Molina-Vila, M.Á.; Provencio, M.; Romero, A. R-Score: A New Parameter to Assess the Quality of Variants' Calls Assessed by NGS Using Liquid Biopsies. *Biology* **2021**, *10*, 954. [CrossRef] [PubMed]
5. Zainab, B.; Ayaz, Z.; Rashid, U.; Al Farraj, D.A.; Alkufeidy, R.M.; AlQahtany, F.S.; Aljowaie, R.M.; Abbasi, A.M. Role of Persistent Organic Pollutants in Breast Cancer Progression and Identification of Estrogen Receptor Alpha Inhibitors Using In-Silico Mining and Drug-Drug Interaction Network Approaches. *Biology* **2021**, *10*, 681. [CrossRef] [PubMed]
6. Chiu, T.-Y.; Lin, R.W.; Huang, C.-J.; Yeh, D.-W.; Wang, Y.-C. DNA Damage Repair Gene Set as a Potential Biomarker for Stratifying Patients with High Tumor Mutational Burden. *Biology* **2021**, *10*, 528. [CrossRef] [PubMed]
7. Rehman, K.U.; Li, J.; Pei, Y.; Yasin, A.; Ali, S.; Saeed, Y. Architectural Distortion-Based Digital Mammograms Classification Using Depth Wise Convolutional Neural Network. *Biology* **2022**, *11*, 15. [CrossRef] [PubMed]
8. Obermayer, A.; Dong, L.; Hu, Q.; Golden, M.; Noble, J.D.; Rodriguez, P.; Robinson, T.J.; Teng, M.; Tan, A.-C.; Shaw, T.I. DRPPM-EASY: A Web-Based Framework for Integrative Analysis of Multi-Omics Cancer Datasets. *Biology* **2022**, *11*, 260. [CrossRef]
9. Singh, S.; Quarni, W.; Goralski, M.; Wan, S.; Jin, H.; Van de Velde, L.A.; Fang, J.; Sing, R.; Fan, Y.; Johnson, M.; et al. Targeting the Spliceosome through RBM39 Degradation Results in Exceptional Responses in High-Risk Neuroblastoma Models. *Sci. Adv.* **2021**, *7*, 47, eabj5405. [CrossRef]
10. Wang, R.; Zheng, X.; Wang, J.; Wan, S.; Song, F.; Wong, M.H.; Leung, K.S.; Cheng, L. Improving Bulk RNA-seq Classification by Transferring Gene Signature from Single Cells in Acute Myelemia Leukemia. *Brief. Bioinform.* **2022**, bbac002. [CrossRef] [PubMed]

 *biology*

Article

Machine Learning-Based Identification of Colon Cancer Candidate Diagnostics Genes

Saraswati Koppad [1], Annappa Basava [1], Katrina Nash [2], Georgios V. Gkoutos [3,4,5,6,7,8] and Animesh Acharjee [3,4,5,*]

1 Department of Computer Science and Engineering, National Institute of Technology Karnataka, Mangalore 575025, India; saraswatikoppad@gmail.com (S.K.); annappa@ieee.org (A.B.)
2 College of Medical and Dental Sciences, University of Birmingham, Birmingham B15 2TT, UK; katrinanash649@outlook.com
3 Institute of Cancer and Genomic Sciences, University of Birmingham, Birmingham B15 2TT, UK; g.gkoutos@bham.ac.uk
4 Institute of Translational Medicine, University of Birmingham, Birmingham B15 2TT, UK
5 NIHR Surgical Reconstruction and Microbiology Research Centre, University Hospital Birmingham, Birmingham B15 2WB, UK
6 MRC Health Data Research UK (HDR UK), Midlands Site, Birmingham B15 2TT, UK
7 NIHR Experimental Cancer Medicine Centre, Birmingham B15 2TT, UK
8 NIHR Biomedical Research Centre, University Hospital Birmingham, Birmingham B15 2TT, UK
* Correspondence: a.acharjee@bham.ac.uk; Tel.: +44-07403642022

Simple Summary: We developed a predictive approach using different machine learning methods to identify a number of genes that can potentially serve as novel diagnostic colon cancer biomarkers.

Abstract: Background: Colorectal cancer (CRC) is the third leading cause of cancer-related death and the fourth most commonly diagnosed cancer worldwide. Due to a lack of diagnostic biomarkers and understanding of the underlying molecular mechanisms, CRC's mortality rate continues to grow. CRC occurrence and progression are dynamic processes. The expression levels of specific molecules vary at various stages of CRC, rendering its early detection and diagnosis challenging and the need for identifying accurate and meaningful CRC biomarkers more pressing. The advances in high-throughput sequencing technologies have been used to explore novel gene expression, targeted treatments, and colon cancer pathogenesis. Such approaches are routinely being applied and result in large datasets whose analysis is increasingly becoming dependent on machine learning (ML) algorithms that have been demonstrated to be computationally efficient platforms for the identification of variables across such high-dimensional datasets. Methods: We developed a novel ML-based experimental design to study CRC gene associations. Six different machine learning methods were employed as classifiers to identify genes that can be used as diagnostics for CRC using gene expression and clinical datasets. The accuracy, sensitivity, specificity, F1 score, and area under receiver operating characteristic (AUROC) curve were derived to explore the differentially expressed genes (DEGs) for CRC diagnosis. Gene ontology enrichment analyses of these DEGs were performed and predicted gene signatures were linked with miRNAs. Results: We evaluated six machine learning classification methods (Adaboost, ExtraTrees, logistic regression, naïve Bayes classifier, random forest, and XGBoost) across different combinations of training and test datasets over GEO datasets. The accuracy and the AUROC of each combination of training and test data with different algorithms were used as comparison metrics. Random forest (RF) models consistently performed better than other models. In total, 34 genes were identified and used for pathway and gene set enrichment analysis. Further mapping of the 34 genes with miRNA identified interesting miRNA hubs genes. Conclusions: We identified 34 genes with high accuracy that can be used as a diagnostics panel for CRC.

Keywords: biomarker identification; transcriptomics; machine learning; prediction; variable selection

Citation: Koppad, S.; Basava, A.; Nash, K.; Gkoutos, G.V.; Acharjee, A. Machine Learning-Based Identification of Colon Cancer Candidate Diagnostics Genes. *Biology* **2022**, *11*, 365. https://doi.org/10.3390/biology11030365

Academic Editors: Shibiao Wan, Yiping Fan, Chunjie Jiang, Shengli Li and Lucia Mangone

Received: 17 December 2021
Accepted: 23 February 2022
Published: 25 February 2022

Publisher's Note: MDPI stays neutral with regard to jurisdictional claims in published maps and institutional affiliations.

1. Introduction

Colorectal cancer (CRC) is the third most common cause of death due to cancer and the fourth most commonly diagnosed cancer worldwide [1,2]. Considering demographic estimates, nearly 2.2 million new cases and about 1.1 million deaths are expected by 2030, and the global burden of CRC is estimated to increase by 60% [3]. CRC cancer is a genotype and phenotype heterogeneous disease, characterized by a display of distinct molecular signatures [4]. Around 1.4 million new cases and nearly 700,000 deaths were recorded in 2012 due to colorectal cancer [5].

Advancements in omics technologies, such as microarrays, RNAseq [6], next-generation sequencing (NGS) [7], and mass spectrometry [8], have enabled employing molecular markers for the diagnosis of CRC [9]. For example, recent studies have used gene microarrays, as well as high-throughput sequencing technologies, to explore differential expressing novel genes in colon cancer [10]. Fang-Ze et al. [11] reported that *CLCA1* may be a candidate diagnostic and prognostic differentially expressed gene or biomarker for colon cancer. Li et al. [12] identified *CDK1* and *CDC20* genes as candidate targets for diagnosis of CRC. Most studies reported individual markers such as the *CEA*, *CK19*, and *CK20* genes [13]. However, the resulting specificity (89%) and sensitivity (78%) of those biomarkers have rendered them unsuitable for the development of a noninvasive diagnostic method for the detection of colon cancer [14]. Dasi et al. [15] and Schiedeck et al. [16] investigated *TERT*, *GCC*, *MAGEA*, *TS*, *CGM2*, and *L6* as biomarkers for detecting colon cancer, reporting a sensitivity and specificity of around 85% and 95%. Furthermore, Liu et al. [17] identified seven prognostic genes, namely, *TIMP1*, *LZTS3*, *AXIN2*, *CXCL1*, *ITLN1*, *CPT2*, and *CLDN23*, for the application of novel diagnostic and prognostic biomarkers for the treatment of colon cancer.

Torres et al. [18] investigated the proteome profiling of human and mouse tissue which revealed a novel association of cancer-associated fibroblasts with cancer progression. This study further unveiled the role of the *LTBP2*, *CDH11*, *OLFML3*, *CALU*, *CDH11*, and *FSTL1* proteins in migration and invasion of CRC and, hence, their use as a biomarker. Moreover, Kim et al. (2019) [19] identified abnormal concentrations of the taurine, alanine, 3-aminoisobutyrate, and citrate metabolites from urine samples in CRC patients.

Although the various molecular characteristics, biological markers, and therapeutic targets of colon cancer previously discovered have contributed significantly to its diagnosis and treatment, the biological complexity, outcome severity, and high metastasis of this complex disease necessitate further predictive and prognostic biomarker identification [20,21]. Currently, CRC prognosis is based on a classification of clinicopathological features, including, tumor, node, metastasis (TNM) stage, cancer numbers, histologic type (mucinous carcinoma or signet ring-cell carcinoma), histology type, tumor grade, tumor size, number of lymph nodes, and tumor location [22]. Furthermore, the right and left localization and the excision of lymph nodes are included in the histological type and grading in the prognosis of colorectal cancer [23].

This study aimed to design and develop novel ML-based, computationally efficient platforms to study CRC gene associations and identify signature genes used as diagnostics markers across transcriptomics datasets.

2. Methods

In this study, we used three gene expression datasets (GSE44861, GSE20916, GSE113513), available from the GEO database [24], and applied six different machine learning methods (Adaboost, ExtraTrees, logistic regression, naïve Bayes, random forest, and XGBoost) to identify genes that can be used as diagnostics markers. We used different combinations of the GSE44861, GSE20916, and GSE113513 datasets for training and validation. We then performed an enrichment analysis and associated the resulting gene signatures with miRNA. Lastly, we estimated the number of samples required for the markers selected for the future validation experiments.

2.1. Data

The gene expression matrixes and clinical data were downloaded from the GEO database repository (https://www.ncbi.nlm.nih.gov/geo/) accessed on 1 October 2020. The details of the datasets used in this study are summarized in Table 1. The detailed workflow of the methods and process used in this study is presented in Figure 1.

Table 1. List of the datasets and platforms used in this study.

GEO Dataset	No. of Samples			Platform ID	References
	Normal	CRC	Total		
GSE44861	55	56	111	GPL3921	[25]
GSE20916	44	46	90	GPL570	[26]
GSE113513	14	14	28	GPL15207	[27]

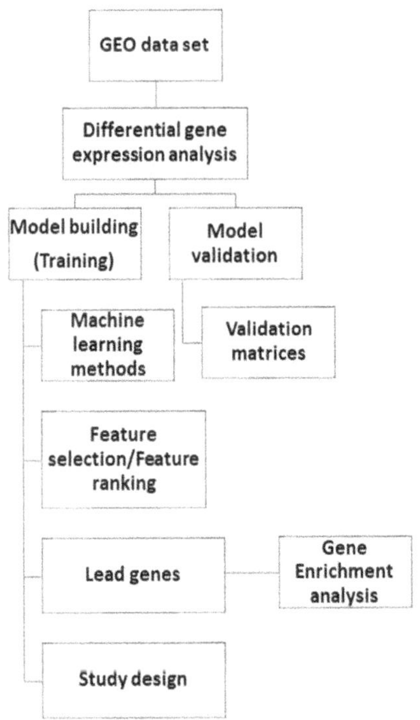

Figure 1. A schematic representation of the biomarker identification workflow.

Differentially Expressed Genes (DEGs) Identified by GEO2R

GEO2R (http://www.ncbi.nlm.nih.gov/geo/geo2r, accessed on 5 January 2021), an online data analysis tool, was used to identify differentially expressed genes (DEGs) between colon cancer patients and healthy controls. We used three GEO series, namely GSE20916, GSE44861, and GSE113513, and identified differential expressed genes. Genes without a corresponding gene symbol and genes with more than one probe set were removed. Adjusted p-values ≤ 0.0001 were considered statistically significant. Subsequently, the top 500 most statistically significant DEG genes from each dataset were selected for further analysis.

2.2. Machine Learning Algorithms and Predictive Analytics

Six different machine learning algorithms, namely, Adaboost [28], ExtraTrees [29], logistic regression [30,31], naïve Bayes (NB) classifier [32], random forest [33], and XG-Boost [34], were employed to develop models using the selected GEO datasets (GSE44861, GSE20916, and GSE113513). These datasets were employed to generate different combinations of training and test data to assess the derived models' performance.

The python Scikit-learn libraries [35] were employed for the implementation of the different classifiers and feature selection methods.

2.2.1. Hyperparameter Optimization

We used the GridSearchCV [35] function to find the optimal values for each model hyperparameter. GridSearchCV is a function, available as part of the Scikit-learn's library, that caters the looping through predefined hyperparameters and the fitting of the model on the training set. GridSearchCV uses all the combinations of the predefined parameter values and evaluates a model's performance for each combination using cross-validation. The accuracy results obtained for every hyperparameter combination can then be used to identify the best-performing model.

2.2.2. Machine Learning Model Evaluation

The analysis was carried out using three different GEO datasets (GSE44861, GSE20916, and GSE113513) as training and testing data for performance comparison in a combinatorial way with six different machine learning models including logistic regression [36], naïve Bayes [37], random forest [38], ExtraTrees [39], Adaboost [40], and XGBoost [41]. Each model was evaluated with different evaluation metrics such as precision, recall [42], specificity, sensitivity [43], F1 score, AUROC [44], and accuracy.

We also included multiple validation strategies to validate the performance of the model. The most commonly used k-fold cross-validation technique was applied in our experimental work. In the k-fold (here, k = 5) cross-validation technique, the dataset is randomly split into k subsets, whereby k − 1 subsets are used for training, and the remaining subset is used for testing; the is process repeated k times. In addition to this, we used resampling with the bootstrap method and leave-one-out cross-validation (LOOCV) in our experimental work for validation of the model performance. The model performance was evaluated for the mean value of performance metrics over 100 iterations. In the LOOCV method, the dataset is split into training data considering all data samples, excluding one data sample used as the test dataset. The model developed with training data finally measures the mean performance value for the repeated process. The experimental results in this method for different models are also provided in Supplementary Table S1.

2.2.3. Feature Selection

We performed feature selection using two methods, mean decrease in impurity (MDI) [45] and Boruta [46], for the selection of important genes. MDI or Gini importance [47] computes the total reduction in loss or impurity contributed by all splits for a given feature. This method evaluates the importance of a variable X_m for predicting Y by adding up the weighted impurity decreases $p(t) \Delta i(s_t, t)$ for all nodes t where X_m is used, averaged over all N_T trees in the forest as shown in the equation below.

$$\mathrm{Imp}(X_m) = \frac{1}{N_T} \sum_T \sum_{t \in T_{iv(s_t)} = X_m} p(t) \Delta i(s_t, t),$$

where $p(t)$ is the proportion N_t/N of samples reaching t, and $v(s_t)$ is the variable used in split s_t. When using the Gini index as an impurity function, this measure is known as the Gini importance or mean decrease Gini. MDI is computationally very efficient and has been widely used in a variety of applications. Gini importance represents the total decrease in node impurity, i.e., how much the model fit or accuracy decreases when dropping a variable. A larger decrease in node impurity results in a more significant variable. The top 15 genes across 10 iterations were selected with the MDI technique.

In addition to MDI, we also used Boruta which is a feature selection algorithm and works as a wrapper algorithm around random forest [48]. It attempts to capture all the important, interesting features from a dataset with respect to an outcome variable and can be used in combination with tree-based ensemble learning algorithms.

2.3. Gene Enrichment Analysis

Gene ontology (GO) enrichment analysis of DEGs was carried out using the FunRich (functional enrichment analysis tool) (http://www.funrich.org/, accessed on 25 January 2021). DEGs were classified according to the biological process and cellular component GO collections. Biological terms with an FDR *p*-value lower than 0.05 were considered significantly enriched. Correction for multiple hypothesis testing was carried out by the Benjamini–Hochberg method.

2.3.1. Association of the Gene Markers with miRNA

We used the NetworkAnalyst (www.networkanalyst.ca, accessed on 28 January 2021) [49] tool and more specifically the gene–miRNA module that employs the miRTarBase v8 database to calculate the number of connections or links for each gene, also termed degrees.

2.3.2. Sample Size Estimates for Future Validation Experiments

We then used PowerTools (https://joelarkman.shinyapps.io/PowerTools/, accessed on 10 February 2021) [50] to estimate the number of samples required for future experiments.

3. Results

3.1. Differential Expressed Genes (DEGs)

We identified the top 500 DEGS across each of the GEO datasets examined. For the GSE44861 dataset, 324 genes were found to be upregulated and 176 genes were downregulated, while, for the GSE20916 and the GSE113513 datasets, 171 and 223 genes were upregulated and 329 and 277 genes were downregulated, respectively. The identified differentially expressed genes and their respective *p*-values, as well as the fold changes, are listed in Supplementary Table S2.

Performance Evaluation

For each of the three GEO datasets examined, their respective DEGs were used as features across six different classification models, namely, Adaboost, ExtraTrees, logistic regression, naïve Bayes classifier, random forest, and XGBoost. The performance of these models was evaluated against different combination of training and test datasets.

The results of the different performance metrics for each classifier are presented in Supplementary Table S1. With GSE44861 as training data and GSE20916 as test data, the random forest model achieved better performance with an accuracy of 98.2% and 90% using the bootstrap and LOOCV methods, respectively. With GSE44861 as training and GSE113513 as testing data, the logistic regression model achieved an accuracy of 96.4% and 84% using bootstrap and LOOCV, respectively. When we used GSE20916 as training data and GSE44861 as testing data, the naïve Bayes classifier achieved an accuracy of 90.1% and 96% using bootstrap and LOOCV, respectively. With GSE20916 as training data and GSE113513 as testing data, logistic regression resulted in better performance. With GSE113513 as training and GSE44861 as testing data, the ExtraTree classifier model achieved better performance. With GSE113513 as training data and GSE20916 as testing data, none of the models achieved good performance.

A comparison of the accuracy and AUROC results for each model evaluations is presented in Figure 2. When using GSE44861 as training data and GSE20916 as test data, the random forest classifier achieved the best performance across all classifiers with an accuracy of 98.2% and an AUROC of 99.9% (Figure 2A). With GSE44861 as training data and GSE113513 as test data, a logistic regression model achieved an accuracy of 96.4% and an AUROC of 99% (Figure 2B). When using GSE20916 as training data and GSE44861 as test data, the naïve Bayes classifier exhibited the best performance with an accuracy of 90.1% and AUROC of 90%, as shown in Figure 2C. Using GSE20916 as the training data and GSE113513 as the test data, the logistic regression model achieved the best performance (Figure 2D). Lastly, with GSE113513 as the training data and GSE44861 as the test data, as well as with GSE113513 as the training data and GSE20916 as the test data, all classifiers achieved an accuracy of 50% to 51% and an AUROC of 50% to 51%, apart from logistic regression, which resulted in an AUROC of 99% (Figure 2E,F).

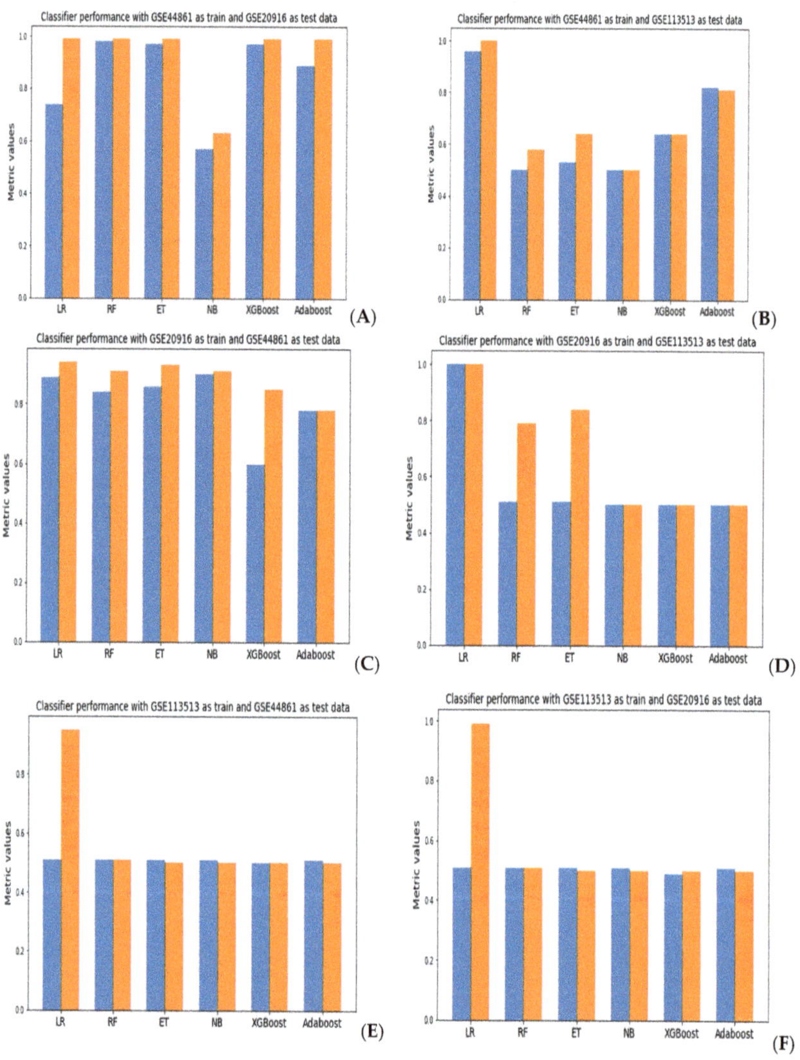

Figure 2. A comparison of accuracy (blue) and AUROC (orange) values obtained across the different classifiers using combinations of the GEO datasets as training and test datasets. (**A**) GSE44861 (training) and GSE20916 (test); (**B**) GSE44861 (training) and GSE20916 (test); (**C**) GSE20916 (training) and GSE44861 (test); (**D**) GSE20916 (training) and GSE113513 (test); (**E**) GSE113513 (training) and GSE44861 (test); (**F**) GSE113513 (training) and GSE20916 (test).

The AUROC plots for the models that had the best performance across the different training and test data combinations are presented in Figure 3. Across the three datasets tested, random forest and logistic regression achieved the best performance when we combined GSE44861 and GSE20916 datasets as training and test data. However, none of the classifiers assessed achieved a good performance using the GSE113513 dataset. The best performances of each classification model are represented as AUROC plots. Overall, the random forest models exhibited consistently better performance across all classification models tested.

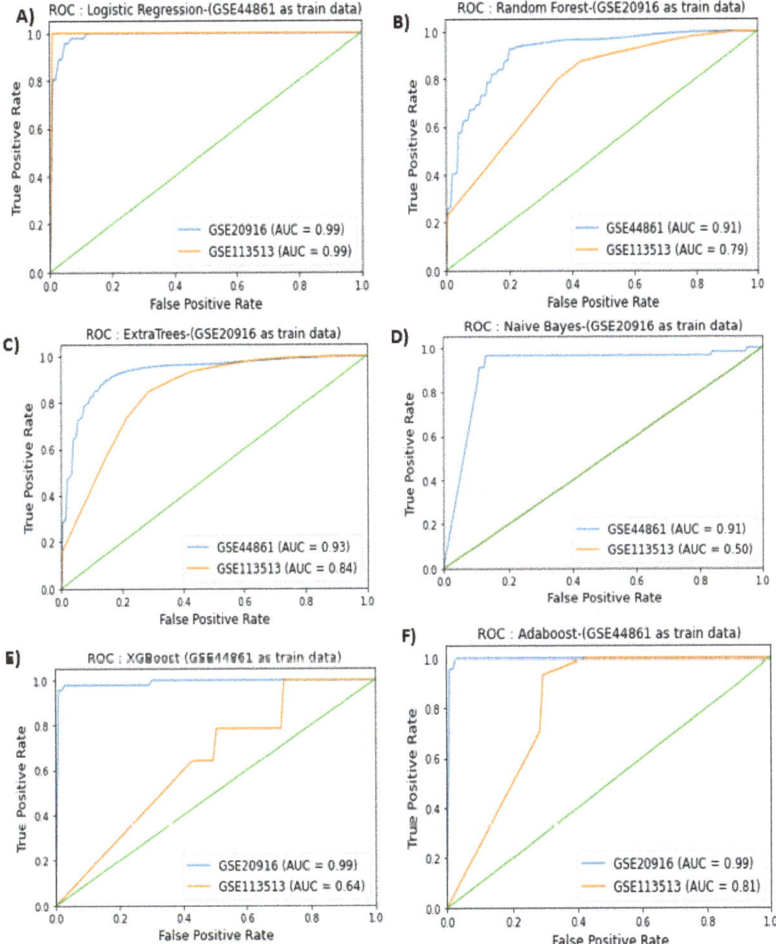

Figure 3. ROC curves for the different classifiers. (**A**) Performance of logistic regression model with GSE44861 as training and GSE20916, GSE113513 as test data; (**B**) performance of random forest model with GSE20916 as training and GSE44861, GSE113513 as test data; (**C**) performance of ExtraTrees model with GSE20916 as training and GSE44861, GSE113513 as test data; (**D**) performance of naïve Bayes model with GSE20916 as training and GSE44861, GSE113513 as test data; (**E**) performance of XGBoost model with GSE44861 as training and GSE20916, GSE113513 as test data; (**F**) performance of Adaboost model with GSE44861 as training and GSE20916, GSE113513 as test data.

3.2. Gene Selection

Random forest classification, on the basis of the performance previously reported, was applied in combination with MDI to select the top 15 genes with the highest importance score in 10 different iterations. We then identified the union of all the genes selected from all 10 iterations. Figure 4 shows the important genes selected using the mean decrease in impurity (MDI) technique in combination with the random forest classifier. Figure 4A depicts the important genes selected using the GSE44861 dataset, while Figure 4B presents the important genes selected using the GSE20916 dataset.

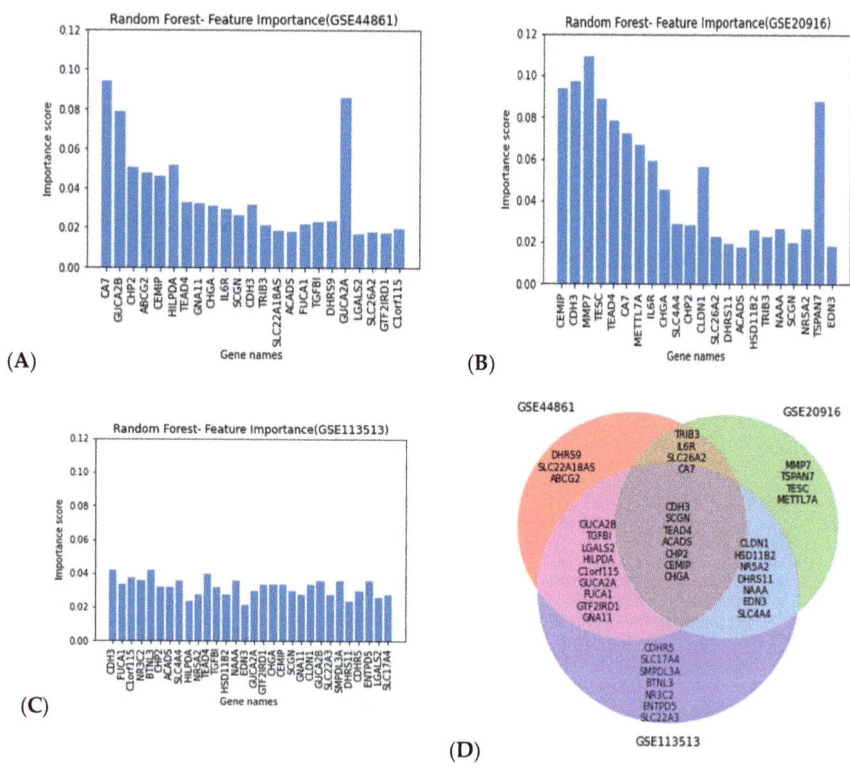

Figure 4. Important genes selected using the mean decrease in impurity (MDI) technique in combination with random forest classifier. The *x*-axis represents the gene names, and the *y*-axis represents importance score values across the GSE44861 (**A**), GSE20916 (**B**), and GSE113513 datasets (**C**). The common genes from all three datasets (**D**).

Gene Ontology (GO) Enrichment Analysis

MDI in combination with the random forest classifier for feature selection resulted in the selection of 34 genes that were used for the pathway and gene set enrichment analysis. These genes were found to be associated with a number of molecular functions including cell adhesion molecule activity (*CDH3* and *CLDN*), transporter activity (*ABCG2*, *SLC22A18AS*, and *SLC26A2*), catalytic activity (*CA7*, *DHRS9*, and *HSD11B2*) and oxidoreductase activity (*ACADS* and *DHRS11*). The pathways for which 34 genes were found to be enriched are presented in Figure 5A.

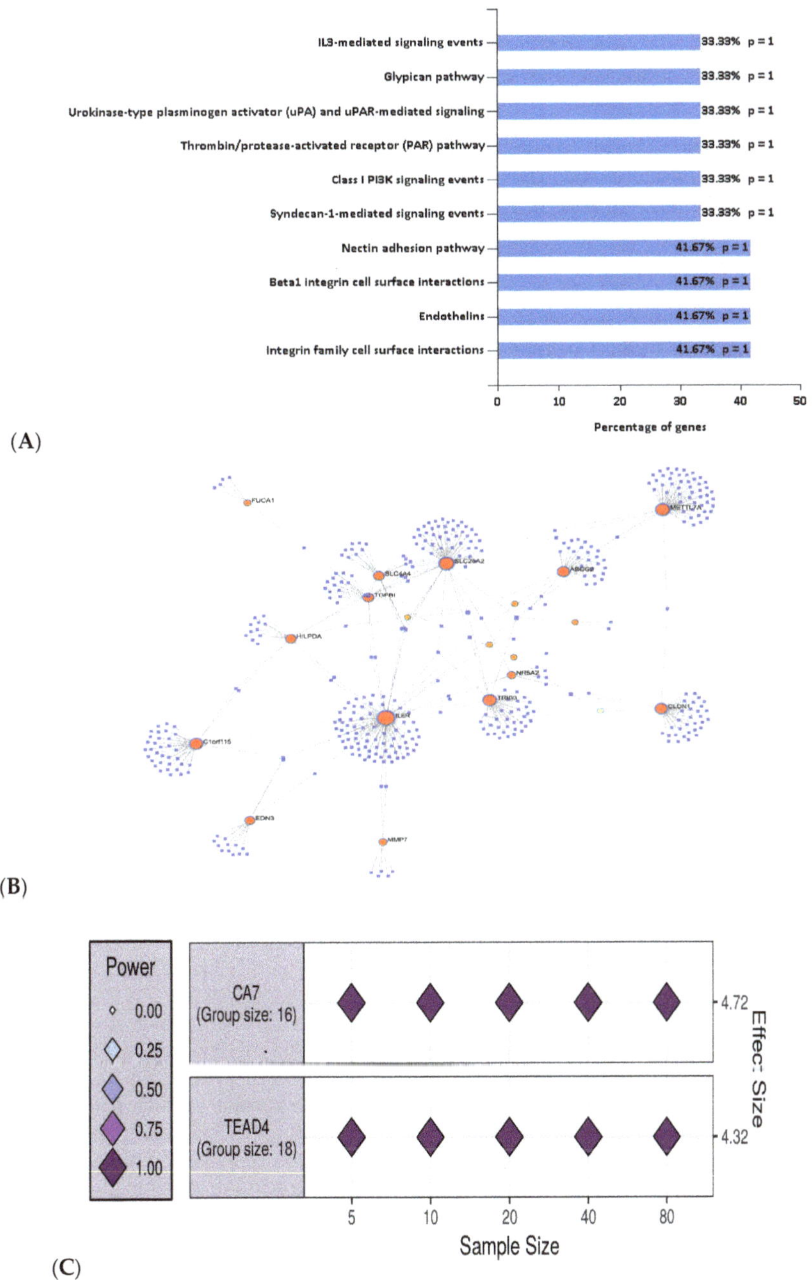

(A)

(B)

(C)

Figure 5. (**A**) Pathway enrichment analysis with the genes selected using the MDI method; (**B**) mapping of the 19 most interacting genes out of 34 genes with miRNAs and their interconnections; (**C**) the two clusters of genes with representative genes *CA7* and *TEAD4* for the GSE44861 are visualized by the largest effect size. The effect size of each assessed variable is shown along the *y*-axis, with a series of sample sizes along the *x*-axis.

3.3. Associating Selected Genes with miRNA Using NetworkAnalyst

We mapped the 34 identified genes using the NetworkAnalyst tool and found that 19 genes out of 34 genes formed hub genes (Figure 5B). For example, *IL6R* had the highest number of miRNA interactions (degree, 94). A list of the identified genes and their miRNA associations is provided in the Supplementary Table S3.

Lastly, we also performed a power analysis over the GSE44861 dataset. For this purpose, we used the 34 genes that were identified and ranked by the random forest algorithm. We then applied hierarchical clustering over these 34 genes and identified two clusters. We selected the genes that presented the highest correlation across normal vs. cancer samples.

CA7 and *TEAD4* were selected as representative genes across the two clusters as they had the highest correlation with the normal vs. CRC samples (i.e., lowest *p*-values). For both clusters of genes including *CA7* and *TEAD4*, we estimated $N = 5$ samples, required for both control and CRC samples. Figure 5C represents the number of the estimated samples required for genes from each cluster.

4. Discussion

The three GEO datasets used in our experimental work with six different machine learning methods were validated across different combinations of training and test datasets. The performance of each model was reported and compared using a number of performance metrics, such as accuracy, sensitivity, specificity, AUC, etc. The random forest method showed the best performance against the GSE44861 and GSE20916 datasets when used as a combination of training and test data. It was less prone to overfitting when compared to the other methods used. This method has also been applied successfully in other diseases such as NAFLD [51], obesity [52], and IBD [53]; therefore, we applied the random forest method to select the important features from these two datasets.

The GSE113513 dataset had a lower number of samples or observations compared to the GSE44861 and GSE20916 datasets, which resulted in lower performance compared to the other datasets, thus indicating an overfitting problem. We used multiple approaches to protect against the overfitting problem, such as the widely used fivefold cross-validation, LOOCV, and bootstrapping. Compared with k-fold cross validation and LOOCV, the bootstrap method could use the entire sample in model development and validation, thus helping to estimate optimism and measure overfitting. The optimism-corrected estimated performance by the bootstrap method is relatively stable because it uses the full sample size and the bootstrap samples vary in composition [54]. We incorporated 100 iterations with the bootstrap method for the experimental work, and each of these evaluation metrics were averaged over these 100 iterations. Datasets GSE44861 and GSE20961 were observed to perform better, and the random forest method was chosen for the feature selection process.

The gene ontology enrichment analysis identified several genes and their associated pathways, most notably, cell adhesion molecule activity, transporter activity, catalytic activity, and oxidoreductase activity. *CDH3*, a gene encoding P-cadherin that forms a major component of the adherens junctions that are essential for cell adhesion, has been identified as being upregulated in CRC in multiple studies and as a diagnostic or prognostic marker [55,56]. Conversely, *CLDN*, encoding for the claudin protein forming tight junctions, has been found to be a potential diagnostic marker with downregulation in CRC patients [56,57]. Furthermore, previous research has postulated that the *HDS11B2* gene, involved in catalytic activity pathways, plays a vital role in migration, invasion, and metastasis of CRC [58]. Other genes identified to be involved in catalytic activity (*CA7* and *DHRS9*) have been found to be downregulated in CRC cells, and have been proposed as promising diagnostic and/or prognostic markers [59,60]. Genes associated with transporter activity have also been identified in existing studies. Of particular note is the upregulation of the *ABCG2* gene, which has been postulated to play a protective role against oxidative stress through cell signaling pathways, which may explain why it has been found to be upregulated in CRC [61–63]. Similarly, genes involved in oxidoreductase activity (*ACADS* and *DHRS11*) have been found to be downregulated in previous studies [64,65]. These genes are involved in fatty-acid metabolism and energy production within mitochondria; thus, their downregulation may partially explain the changes in metabolism often observed in cancer cells [66]. Many of the identified genes have been previously associated with colon cancer via miRNA interactions. Multiple studies, including Bian et al., Hua et al., and Xu et al., have reported that serum *IL-6* may be a potential biomarker for CRC diagnosis and a miR-34a target [67]. IL6R has also been implicated in other cancer types, including prostate cancer [68]. Another gene, *SLC4A4*, was found to be significantly correlated with shorter survival of CRC patients and a marker of poorer progression for patients with breast cancer, lung cancer, gastric cancer, and ovarian cancer. This suggests a potential role *of SLC4A4* in tumor suppression, as well as in prognostic prediction in multiple malignancies, including CRC, thus representing a potential novel therapeutic CRC target [69]. Yang et al. (2019) [70] identified a similar *SLC4A4* expression association and proposed the expression of six further genes, namely, *SGCG, CLDN23, CCDC78, SLC17A7, OTOP3,* and *SMPDL3A*, as novel colon cancer prognostic biomarkers. Zhang et al. (2020) [71] reported that hsa_circRNA_001587 upregulates *SLC4A4* expression to inhibit migration, invasion, and angiogenesis of pancreatic cancer cells via binding to microRNA-223. Furthermore, Mencia et al. (2011) reported miR-224 to be one of the most differentially expressed miRNAs associated with *SLC4A4* [72]. Andersen et al. (2015) [73] reported changes in gene expression levels (high *ABCC2* and low *ABCG2*) as early events in the colon adenoma–carcinoma sequence. Moreover, miR-132 has been reported to regulate the *SIRT1/CREB/ABCG2* signaling pathway, contributing to cisplatin resistance and serving as a novel therapeutic target against gastric cancer [74]. Cherradi et al. found *CLDN1* to be significantly overexpressed ($p < 0.001$) in CRC samples, and they proposed it as a new potential therapeutic target of miR-7-2 [75]. Lastly, Miwa et al. (2011) [76] reported *CLDN1* as a target of TCF/LEF signaling, while Singh et al. (2011) [77] suggested the involvement of *CLDN1* in the regulation of the WNT signaling pathway.

Our approach utilized a limited number of public datasets, and the potential causal relationships identified necessitate experimental validation. We did not consider the effect of multiple factors, such as age, gender, ethnicity, and tumor grade and stage, on gene expression patterns since we focused only on genes that have been previously reported as having significant variation between control and cancer samples. In the context of translational medicine [78], further research is required to investigate the selected prognostic/diagnostic signature's clinical utility in predicting clinical outcomes in various tumor types.

In CRC diagnostics, colonoscopy is the current gold-standard screening method. However, this approach has some limitations that include internal hemorrhage, colonic perforation, and cardiorespiratory problems [79].

Another approach is the guaiac fecal occult blood test (gFOBT) [80], which detects hemoglobin peroxidase activity in the feces, and it is the most often used noninvasive screening procedure. Although FOBT is a simple and inexpensive way to screen for CRC, it has a high percentage of false positives and false negatives.

As a result, alternative CRC screening approaches that are cost-effective, noninvasive, easily quantifiable, and accurate are urgently needed. Thus, gene signature-based biomarkers in the clinical applications in CRC are required for early cancer detection, prognostic stratification, and surveillance [80]. Genes identified in this study will need to go through targeted validation experiments using qPCR. A new trial needs to be set up to replicate the gene signature's effect. This step will ensure the clinical efficacy of those markers identified and will allow a better clinical decision on CRC [81].

5. Conclusions

This study aimed to identify novel genes associations with CRC that can potentially be used as diagnostic markers in translational research. To achieve this, we applied a predictive analytics approach that employed a variety of machine learning methods. In addition, we estimated the required number of samples for future validation experiments.

Supplementary Materials: The following supporting information can be downloaded at https://www.mdpi.com/article/10.3390/biology11030365/s1: Table S1. The experimental results from different models and performance comparison of all methods across training and test datasets; Table S2. List of identified differentially expressed genes and their respective *p*-values, and the fold changes; Table S3. A list of the identified genes and their miRNA associations.

Author Contributions: S.K. and A.A. performed the machine learning analysis; A.A. conceptualized and designed the data analytics strategy; S.K., A.B., K.N., G.V.G. and A.A. wrote the first draft; A.B., G.V.G. and A.A. supervised the project; K.N. provided clinical interpretation of the results. All authors co-wrote, edited, and reviewed the manuscript. All authors have read and agreed to the published version of the manuscript.

Funding: The authors acknowledge support from the Ministry of Electronics and Information Technology (MeitY), Government of India, as well as from the NIHR Birmingham ECMC, NIHR Birmingham SRMRC, Nanocommons H2020-EU (731032), NIHR Birmingham Biomedical Research Center, MRC Health Data Research UK (HDRUK/CFC/01), an initiative funded by UK Research and Innovation, Department of Health and Social Care (England) and the devolved administrations, and leading medical research charities. The views expressed in this publication are those of the authors and not necessarily those of the NHS, the National Institute for Health Research, the Medical Research Council, or the Department of Health.

Institutional Review Board Statement: Not applicable.

Informed Consent Statement: This study did not involve an animal and/or human tissue/individual data/participants; thus, there were no ethics-related issues. No permission was required to use any repository data involved in the present study.

Data Availability Statement: Availability of data and materials: GEO datasets used in this study can be obtained from the GEO database (https://www.ncbi.nlm.nih.gov/geo/, accessed on 5 January 2021).

Conflicts of Interest: The authors declare no conflict of interest.

Abbreviations

AUROC	Area under the receiver operating characteristic curve
CRC	Colon cancer
DEGs	Differential expressed genes
GEO	Gene Expression Omnibus
GO	Gene ontology
miRNA	microRNA
MDI	Mean decrease in impurity
RF	Random forest

References

1. Siegel, R.; DeSantis, C.; Jemal, A. Colorectal cancer statistics, 2014. *CA A Cancer J. Clin.* **2014**, *64*, 104–117. [CrossRef] [PubMed]
2. Worldwide incidence and mortality of colorectal cancer and human development index (HDI): An ecological study. *WCRJ* **2019**, *6*, 1433.
3. Rawla, P.; Sunkara, T.; Barsouk, A. Epidemiology of colorectal cancer: Incidence, mortality, survival, and risk factors. *Prz. Gastroenterol.* **2019**, *14*, 89–103. [CrossRef] [PubMed]
4. Bogaert, J.; Prenen, H. Molecular genetics of colorectal cancer. *Ann. Gastroenterol.* **2014**, *27*, 9–14.
5. Torre, L.A.; Siegel, R.L.; Ward, E.M.; Jemal, A. Global Cancer Incidence and Mortality Rates and Trends—An Update. *Cancer Epidemiol. Biomark. Prev.* **2016**, *25*, 16–27. [CrossRef]
6. Stefano, G.B.; Mantione, K.J.; Kream, R.M.; Kuzelova, H.; Ptacek, R.; Raboch, J.; Samuel, J.M. Comparing Bioinformatic Gene Expression Profiling Methods: Microarray and RNA-Seq. *Med. Sci. Monit. Basic Res.* **2014**, *20*, 138–142. [CrossRef]
7. Metzker, M.L. Sequencing technologies—The next generation. *Nat. Rev. Genet.* **2010**, *11*, 31–46. [CrossRef]
8. Kim, H.-Y.; Lee, S.-G.; Oh, T.-J.; Lim, S.R.; Kim, S.-H.; Lee, H.J.; Kim, Y.-S.; Choi, H.-K. Antiproliferative and Apoptotic Activity of Chamaecyparis obtusa Leaf Extract against the HCT116 Human Colorectal Cancer Cell Line and Investigation of the Bioactive Compound by Gas Chromatography-Mass Spectrometry-Based Metabolomics. *Molecules* **2015**, *20*, 18066–18082. [CrossRef]
9. Dalal, N.; Jalandra, R.; Sharma, M.; Prakash, H.; Makharia, G.K.; Solanki, P.R.; Singh, R.; Kumar, A. Omics technologies for improved diagnosis and treatment of colorectal cancer: Technical advancement and major perspectives. *Biomed. Pharmacother.* **2020**, *131*, 110648. [CrossRef]

10. Chen, M.; Yang, X.; Yang, M.; Zhang, W.; Li, L.; Sun, Q. Identification of a novel biomarker-CCL5 using antibody microarray for colorectal cancer. *Pathol. Res. Pract.* **2019**, *215*, 1033–1037. [CrossRef]
11. Wei, F.-Z.; Mei, S.-W.; Wang, Z.-J.; Chen, J.-N.; Shen, H.-Y.; Zhao, F.-Q.; Li, J.; Liu, Z.; Liu, Q. Differential Expression Analysis Revealing CLCA1 to Be a Prognostic and Diagnostic Biomarker for Colorectal Cancer. *Front. Oncol.* **2020**, *10*, 573295. [CrossRef] [PubMed]
12. Li, J.; Wang, Y.; Wang, X.; Yang, Q. CDK1 and CDC20 overexpression in patients with colorectal cancer are associated with poor prognosis: Evidence from integrated bioinformatics analysis. *World J. Surg. Oncol.* **2020**, *18*, 1–11. [CrossRef] [PubMed]
13. Gonzalez-Pons, M.; Cruz-Correa, M. Colorectal Cancer Biomarkers: Where Are We Now? *BioMed. Res. Int.* **2015**, *2015*, 1–14. [CrossRef] [PubMed]
14. Lin, S.-R.; Huang, M.-Y.; Chang, H.-J. Molecular Detection of Circulating Tumor Cells With Multiple mRNA Markers by Genechip for Colorectal Cancer Early Diagnosis and Prognosis Prediction. *Genom. Med. Biomark. Health Sci.* **2011**, *3*, 9–16. [CrossRef]
15. Dasí, F.; Lledó, S.; García-Granero, E.; Ripoll, R.; Marugán, M.; Tormo, M.; García-Conde, J.; Aliño, S.F. Real-time quantification in plasma of human telomerase reverse transcriptase (hTERT) mRNA: A simple blood test to monitor disease in cancer patients. *Lab. Investig.* **2001**, *81*, 767–769. [CrossRef] [PubMed]
16. Schiedeck, T.H.K.; Wellm, C.; Roblick, U.J.; Broll, R.; Bruch, H.-P. Diagnosis and Monitoring of Colorectal Cancer by L6 Blood Serum Polymerase Chain Reaction Is Superior to Carcinoembryonic Antigen-Enzyme-Linked Immunosorbent Assay. *Dis. Colon Rectum* **2003**, *46*, 818–825. [CrossRef]
17. Liu, X.; Bing, Z.; Wu, J.; Zhang, J.; Zhou, W.; Ni, M.; Meng, Z.; Liu, S.; Tian, J.; Zhang, X.; et al. Integrative Gene Expression Profiling Analysis to Investigate Potential Prognostic Biomarkers for Colorectal Cancer. *Med. Sci. Monit.* **2020**, *26*, e918906. [CrossRef]
18. Torres, S.; Bartolome, R.A.; Mendes, M.; Barderas, R.; Fernández-Aceñerp, M.J.; Peláez-García, A.; Peña, C.; Lopez-Lucendo, M.; Villar-Vázquez, R.; De Herreros, A.G.; et al. Proteome Profiling of Cancer-Associated Fibroblasts Identifies Novel Proinflammatory Signatures and Prognostic Markers for Colorectal Cancer. *Clin. Cancer Res.* **2013**, *19*, 6006–6019. [CrossRef]
19. Kim, E.R.; Kwon, H.N.; Nam, H.; Kim, J.J.; Park, S.; Kim, Y.-H. Urine-NMR metabolomics for screening of advanced colorectal adenoma and early stage colorectal cancer. *Sci. Rep.* **2019**, *9*, 1–10. [CrossRef]
20. Schirripa, M.; Lenz, H.-J. Biomarker in Colorectal Cancer. *Cancer J.* **2016**, *22*, 156–164. [CrossRef]
21. Shi, K.; Lin, W.; Zhao, X.-M. Identifying Molecular Biomarkers for Diseases with Machine Learning Based on Integrative Omics. *IEEE/ACM Trans. Comput. Biol. Bioinform.* **2020**, *18*, 2514–2525. [CrossRef] [PubMed]
22. Wang, R.; Wang, M.-J.; Ping, J. Clinicopathological Features and Survival Outcomes of Colorectal Cancer in Young Versus Elderly: A Population-Based Cohort Study of SEER 9 Registries Data (1988–2011). *Medicine* **2015**, *94*, e1402. [CrossRef] [PubMed]
23. Mangone, L.; Pinto, C.; Mancuso, P.; Ottone, M.; Bisceglia, I.; Chiaranda, G.; Michiara, M.; Vicentini, M.; Carrozzi, G.; Ferretti, S.; et al. Colon cancer survival differs from right side to left side and lymph node harvest number matter. *BMC Public Health* **2021**, *21*, 1–10. [CrossRef]
24. Edgar, R.; Domrachev, M.; Lash, A.E. Gene Expression Omnibus: NCBI gene expression and hybridization array data repository. *Nucleic Acids Res.* **2002**, *30*, 207–210. [CrossRef] [PubMed]
25. Ryan, B.M.; Zanetti, K.A.; Robles, A.; Schetter, A.J.; Goodman, J.; Hayes, R.; Huang, W.-Y.; Gunter, M.J.; Yeager, M.; Burdette, L.; et al. Germline variation inNCF4, an innate immunity gene, is associated with an increased risk of colorectal cancer. *Int. J. Cancer* **2014**, *134*, 1399–1407. [CrossRef]
26. Skrzypczak, M.; Goryca, K.; Rubel, T.; Paziewska, A.; Mikula, M.; Jarosz, D.; Pachlewski, J.; Oledzki, J.; Ostrowsk, J. Modeling oncogenic signaling in colon tumors by multidirectional analyses of microarray data directed for maximization of analytical reliability. *PLoS ONE* **2010**, *5*, e13091. [CrossRef]
27. Barrett, T.; Wilhite, S.E.; Ledoux, P.; Evangelista, C.; Kim, I.F.; Tomashevsky, M.; Marshall, K.A.; Phillippy, K.H.; Sherman, P.M.; Holko, M.; et al. NCBI GEO: Archive for functional genomics data sets—Update. *Nucleic Acids Res.* **2013**, *41*, D991–D9955. [CrossRef]
28. Friedman, J.; Hastie, T.; Tibshirani, R. Additive logistic regression: A statistical view of boosting. *Ann. Stat.* **2000**, *28*, 337–407. [CrossRef]
29. Huynh-Thu, V.A.; Irrthum, A.; Wehenkel, L.; Geurts, P. Inferring Regulatory Networks from Expression Data Using Tree-Based Methods. *PLoS ONE* **2010**, *5*, e12776. [CrossRef]
30. Yuan, Z.; Ghosh, D. Combining Multiple Biomarker Models in Logistic Regression. *Biometrics* **2008**, *64*, 431–439. [CrossRef]
31. Tolles, J.; Meurer, W.J. Logistic Regression: Relating Patient Characteristics to Outcomes. *JAMA* **2016**, *316*, 533–534. [CrossRef] [PubMed]
32. Sambo, F.; Trifoglio, E.; Di Camillo, B.; Toffolo, G.M.; Cobelli, C. Bag of Naïve Bayes: Biomarker selection and classification from genome-wide SNP data. *BMC Bioinform.* **2012**, *13*, S2. [CrossRef] [PubMed]
33. Chen, X.; Ishwaran, H. Random forests for genomic data analysis. *Genomics* **2012**, *99*, 323–329. [CrossRef] [PubMed]
34. Li, W.; Yin, Y.; Quan, X.; Zhang, H. Gene Expression Value Prediction Based on XGBoost Algorithm. *Front. Genet.* **2019**, *10*, 1077. [CrossRef] [PubMed]
35. Pedregosa, F.; Varoquaux, G.; Gramfort, A.; Michel, V.; Thirion, B.; Grisel, O.; Blondel, M.; Prettenhofer, P.; Weiss, R.; Dubourg, V.; et al. Scikit-learn: Machine Learning in Python. *J. Mach. Learn. Res.* **2011**, *12*, 2825–2830.

36. Dreiseitl, S.; Ohno-Machado, L. Logistic regression and artificial neural network classification models: A methodology review. *J. Biomed. Inform.* **2002**, *35*, 352–359. [CrossRef]

37. Bauer, E.; Kohavi, R. An Empirical Comparison of Voting Classification Algorithms: Bagging, Boosting, and Variants. *Mach. Learn.* **1999**, *36*, 105–139. [CrossRef]

38. Breiman, L. Random Forests. *Mach. Learn.* **2001**, *45*, 5–32. [CrossRef]

39. Geurts, P.; Maree, R.; Wehenkel, L. Extremely Randomized Trees and Random Subwindows for Image Classification, Annotation, and Retrieval. *Mach. Learn.* **2013**, *63*, 3–42. [CrossRef]

40. Schapire, R.E. Explaining AdaBoost. In *Empirical Inference*; Springer: Berlin/Heidelberg, Germany, 2013; pp. 37–52.

41. Chen, T.; Guestrin, C. Xgboost: A scalable tree boosting system. In Proceedings of the 22nd ACM SIGKDD International Conference on Knowledge Discovery and Data Mining; KDD '16, San Francisco, CA, USA, 13–17 August 2016; pp. 785–794.

42. Davis, J.; Goadrich, M. The Relationship Between Precision-Recall and ROC Curves. In *Proceedings of the 23rd International Conference on Machine Learning*; Association for Computing Machinery: New York, NY, USA, 2006; pp. 233–240.

43. Hand, D.J. Assessing the Performance of Classification Methods. *Int. Stat. Rev.* **2012**, *80*, 400–414. [CrossRef]

44. Sokolova, M.; Japkowicz, N.; Szpakowicz, S. Beyond Accuracy, F-Score and ROC: A Family of Discriminant Measures for Performance Evaluation. In *AI 2006: Advances in Artificial Intelligence*; Lecture Notes in Computer Science; Sattar, A., Kang, B., Eds.; Springer: Berlin/Heidelberg, Germany, 2006; Volume 4304, pp. 1015–1021, ISBN 978-3-540-49787-5.

45. Gilles, L.; Wehenkel, L.; Sutera, A.; Geurts, P. Understanding variable importances in forests of randomized trees. In Proceedings of the Twenty-Seventh Conference on Neural Information Processing Systems—NIPS, Lake Tahoe, CA, USA, 5–10 December 2013.

46. Kursa, M.B.; Jankowski, A.; Rudnicki, W.R. Boruta—A System for Feature Selection. *Fundam. Inform.* **2010**, *101*, 271–285. [CrossRef]

47. Sandri, M.; Zuccolotto, P. A Bias Correction Algorithm for the Gini Variable Importance Measure in Classification Trees. *J. Comput. Graph. Stat.* **2008**, *17*, 611–628. [CrossRef]

48. Chen, R.-C.; Dewi, C.; Huang, S.-W.; Caraka, R.E. Selecting critical features for data classification based on machine learning methods. *J. Big Data* **2020**, *7*, 1–26. [CrossRef]

49. Zhou, G.; Soufan, O.; Ewald, J.; Hancock, R.E.W.; Basu, N.; Xia, J. NetworkAnalyst 3.0: A visual analytics platform for comprehensive gene expression profiling and meta-analysis. *Nucleic Acids Res.* **2019**, *47*, W234–W241. [CrossRef]

50. Acharjee, A.; Larkman, J.; Xu, Y.; Cardoso, V.R.; Gkoutos, G.V. A random forest based biomarker discovery and power analysis framework for diagnostics research. *BMC Med. Genom.* **2020**, *13*, 1–14. [CrossRef]

51. Shafiha, R.; Bahcivanci, B.; Gkoutos, G.V.; Acharjee, A. Machine Learning-Based Identification of Potentially Novel Non-Alcoholic Fatty Liver Disease Biomarkers. *Biomedicines* **2021**, *9*, 1636. [CrossRef]

52. Acharjee, A.; Ament, Z.; West, J.A.; Stanley, E.; Griffin, J.L. Integration of metabolomics, lipidomics and clinical data using a machine learning method. *BMC Bioinform.* **2016**, *17* (Suppl. S15), 440. [CrossRef]

53. Quraishi, M.N.; Acharjee, A.; Beggs, A.D.; Horniblow, R.; Tselepis, C.; Gkoutos, G.; Ghosh, S.; Rossiter, A.E.; Loman, N.; van Schaik, W.; et al. A Pilot Integrative Analysis of Colonic Gene Expression, Gut Microbiota, and Immune Infiltration in Primary Sclerosing Cholangitis-Inflammatory Bowel Disease: Association of Disease With Bile Acid Pathways. *J. Crohn's Colitis* **2020**, *14*, 935–947. [CrossRef]

54. Frank, H. *Regression Modeling Strategies: With Applications to Linear Models, Logistic Regression, and Survival Analysis*, 2nd ed.; Springer: New York, NY, USA, 2015.

55. Kumara, H.S.; Bellini, G.A.; Caballero, O.L.; Herath, S.A.; Su, T.; Ahmed, A.; Njoh, L.; Cekic, V.; Whelan, R.L. P-Cadherin (CDH3) is overexpressed in colorectal tumors and has potential as a serum marker for colorectal cancer monitoring. *Oncoscience* **2017**, *4*, 139–147. [CrossRef]

56. Xu, Y.; Zhao, J.; Dai, X.; Xie, Y.; Dong, M. High expression of CDH3 predicts a good prognosis for colon adenocarcinoma patients. *Exp. Ther. Med.* **2019**, *18*, 841–847. [CrossRef]

57. Hahn-Strömberg, V.; Askari, S.; Ahmad, A.; Befekadu, R.; Nilsson, T.K. Expression of claudin 1, claudin 4, and claudin 7 in colorectal cancer and its relation with CLDN DNA methylation patterns. *Tumor Biol.* **2017**, *39*, 1010428317697569. [CrossRef] [PubMed]

58. Chen, J.; Liu, Q.-M.; Du, P.-C.; Ning, D.; Mo, J.; Zhu, H.-D.; Wang, C.; Ge, Q.-Y.; Cheng, Q.; Zhang, X.-W.; et al. Type-2 11β-hydroxysteroid dehydrogenase promotes the metastasis of colorectal cancer via the Fgfbp1-AKT pathway. *Am. J. Cancer Res.* **2020**, *10*, 662–673. [PubMed]

59. Yang, G.-Z.; Hu, L.; Cai, J.; Chen, H.-Y.; Zhang, Y.; Feng, D.; Qi, C.-Y.; Zhai, Y.-X.; Gong, H.; Fu, H.; et al. Prognostic value of carbonic anhydrase VII expression in colorectal carcinoma. *BMC Cancer* **2015**, *15*, 209. [CrossRef] [PubMed]

60. Hu, L.; Chen, H.-Y.; Han, T.; Yang, G.-Z.; Feng, D.; Qi, C.-Y.; Gong, H.; Zhai, Y.-X.; Cai, Q.-P.; Gao, C.-F. Downregulation of DHRS9 expression in colorectal cancer tissues and its prognostic significance. *Tumor Biol.* **2015**, *37*, 837–845. [CrossRef] [PubMed]

61. Nie, S.; Huang, Y.; Shi, M.; Qian, X.; Li, H.; Peng, C.; Kong, B.; Zou, X.; Shen, S. Protective role of ABCG2 against oxidative stress in colorectal cancer and its potential underlying mechanism. *Oncol. Rep.* **2018**, *40*, 2137–2146. [CrossRef] [PubMed]

62. Expression of ABCG2 and its Significance in Colorectal Cancer. *Asian Pac. J. Cancer Prev.* **2010**, *11*, 845–848.

63. Tuy, H.D.; Shiomi, H.; Mukaisho, K.I.; Naka, S.; Shimizu, T.; Sonoda, H.; Mekata, E.; Endo, Y.; Kurumi, Y.; Sugihara, H.; et al. ABCG2 expression in colorectal adenocarcinomas may predict resistance to irinotecan. *Oncol. Lett.* **2016**, *12*, 2752–2760. [CrossRef]

64. Yang, W.; Ma, J.; Zhou, W.; Li, Z.; Zhou, X.; Cao, B.; Zhang, Y.; Liu, J.; Yang, Z.; Zhang, H.; et al. Identification of hub genes and outcome in colon cancer based on bioinformatics analysis. *Cancer Manag. Res.* **2018**, *11*, 323–338. [CrossRef]
65. Pira, G.; Uva, P.; Scanu, A.M.; Rocca, P.C.; Murgia, L.; Uleri, E.; Piu, C.; Porcu, A.; Carru, C.; Manca, A.; et al. Landscape of transcriptome variations uncovering known and novel driver events in colorectal carcinoma. *Sci. Rep.* **2020**, *10*, 1–12. [CrossRef]
66. Coller, H.A. Is Cancer a Metabolic Disease? *Am. J. Pathol.* **2014**, *184*, 4–17. [CrossRef]
67. Li, H.; Rokavec, M.; Hermeking, H. Soluble IL6R represents a miR-34a target: Potential implications for the recently identified IL-6R/STAT3/miR-34a feed-back loop. *Oncotarget* **2015**, *6*, 14026–14032. [CrossRef] [PubMed]
68. Vainer, N.; Dehlendorff, C.; Johansen, J.S. Systematic literature review of IL-6 as a biomarker or treatment target in patients with gastric, bile duct, pancreatic and colorectal cancer. *Oncotarget* **2018**, *9*, 29820–29841. [CrossRef] [PubMed]
69. Dai, G.; Wang, L.; Wen, Y.; Ren, X.; Zuo, S. Identification of key genes for predicting colorectal cancer prognosis by integrated bioinformatics analysis. *Oncol. Lett.* **2019**, *19*, 388–398. [CrossRef]
70. Yang, H.; Liu, H.; Lin, H.-C.; Gan, D.; Jin, W.; Cui, C.; Yan, Y.; Qian, Y.; Han, C.; Wang, Z. Association of a novel seven-gene expression signature with the disease prognosis in colon cancer patients. *Aging* **2019**, *11*, 8710–8727. [CrossRef] [PubMed]
71. Zhang, X.; Tan, P.; Zhuang, Y.; Du, L. hsa_circRNA_001587 upregulates SLC4A4 expression to inhibit migration, invasion, and angiogenesis of pancreatic cancer cells via binding to microRNA-223. *Am. J. Physiol. Liver Physiol.* **2020**, *319*, G703–G717. [CrossRef] [PubMed]
72. Mencia, N.; Selga, E.; Noe, V.; Ciudad, C.J. Underexpression of miR-224 in methotrexate resistant human colon cancer cells. *Biochem. Pharmacol.* **2011**, *82*, 1572–1582. [CrossRef]
73. Andersen, V.; Vogel, L.K.; Kopp, T.I.; Sæbø, M.; Nonboe, A.W.; Hamfjord, J.; Kure, E.H.; Vogel, U. High ABCC2 and Low ABCG2 Gene Expression Are Early Events in the Colorectal Adenoma-Carcinoma Sequence. *PLoS ONE* **2015**, *10*, e0119255. [CrossRef]
74. Zhang, L.; Guo, X.; Zhang, D.; Fan, Y.; Qin, L.; Dong, S. Upregulated miR-132 in Lgr5+gastric cancer stem cell-like cells contributes to cisplatin-resistance via SIRT1/CREB/ABCG2 signaling pathway. *Mol. Carcinog.* **2017**, *56*, 2022–2034. [CrossRef]
75. Cherradi, S.; Ayrolles-Torro, A.; Vezzo-Vié, N.; Gueguinou, N.; Denis, V.; Combes, E.; Boissière, F.; Busson, M.; Canterel-Thouennon, L.; Mollevi, C.; et al. Antibody targeting of claudin-1 as a potential colorectal cancer therapy. *J. Exp. Clin. Cancer Res.* **2017**, *36*, 89. [CrossRef]
76. Miwa, N.; Furuse, M.; Tsukita, S.; Niikawa, N.; Nakamura, Y.; Furukawa, Y. Involvement of claudin-1 in the beta-catenin/Tcf signaling pathway and its frequent upregulation in human colorectal cancers. *Oncol. Res.* **2001**, *12*, 469–476. [CrossRef]
77. Singh, A.B.; Sharma, A.; Smith, J.J.; Krishnan, M.; Chen, X.; Eschrich, S.; Washington, M.K.; Yeatman, T.J.; Beauchamp, R.D.; Dhawan, P. Claudin-1 Up-regulates the Repressor ZEB-1 to Inhibit E-Cadherin Expression in Colon Cancer Cells. *Gastroenterology* **2011**, *141*, 2140–2153. [CrossRef]
78. Bravo-Merodio, L.; Acharjee, A.; Russ, D.; Bisht, V.; Williams, J.A.; Tsaprouni, L.G.; Gkoutos, G.V. Translational biomarkers in the era of precision medicine. *Int. Rev. Cytol.* **2021**, *102*, 191–232. [CrossRef]
79. Bailey, J.R.; Aggarwal, A.; Imperiale, T.F. Colorectal Cancer Screening: Stool DNA and Other Noninvasive Modalities. *Gut Liver* **2016**, *10*, 204–211. [CrossRef] [PubMed]
80. de Wit, M.; Fijneman, R.J.; Verheul, H.M.; Meijer, G.A.; Jimenez, C.R. Proteomics in colorectal cancer translational research: Biomarker discovery for clinical applications. *Clin. Biochem.* **2013**, *46*, 466–479. [CrossRef] [PubMed]
81. Alvarez-Chaver, P.; Otero-Estévez, O.; Páez de la Cadena, M.; Rodríguez-Berrocal, F.J.; Martínez-Zorzano, V.S. Proteomics for discovery of candidate colorectal cancer biomarkers. *World J. Gastroenterol.* **2014**, *20*, 3804–3824. [CrossRef]

 *biology*

MDPI

Article

A Novel Approach to Modeling and Forecasting Cancer Incidence and Mortality Rates through Web Queries and Automated Forecasting Algorithms: Evidence from Romania

Cristiana Tudor 

International Business and Economics Department, The Bucharest University of Economic Studies, 010374 Bucharest, Romania; cristiana.tudor@net.ase.ro

Simple Summary: Cancer remains a global burden, currently causing nearly one in six deaths worldwide. Accurate projections of cancer incidence and mortality are needed for effective and efficient policymaking, accurate resource allocation, and to assess the impact of newly introduced policies and measures. However, the COVID-19 pandemic disrupted public health systems and caused a significant number of cancers to remain undiagnosed, thus affecting the quality of official statistics and their usefulness for health studies. This paper addresses this issue by proposing novel cancer incidence/cancer mortality models based on population web-search habits and historical links with official health variables. The models are empirically estimated using data from one of the most vulnerable European Union (EU) members, Romania, a country that consistently reports lower survival rates than the EU average, and are further used to forecast cancer incidence and mortality rates in the country. Research findings have important policy implications, and the novel framework, owing to its generalizability, can be applied to the same task in other countries. Overall, the results indicate a continuation of the increasing trends in cancer incidence and mortality in Romania and thus underline the urgency to change the status quo in the Romanian public-health system.

Citation: Tudor, C. A Novel Approach to Modeling and Forecasting Cancer Incidence and Mortality Rates through Web Queries and Automated Forecasting Algorithms: Evidence from Romania. *Biology* **2022**, *11*, 857. https://doi.org/10.3390/biology11060857

Academic Editors: Shibiao Wan, Yiping Fan, Chunjie Jiang and Shengli Li

Received: 9 April 2022
Accepted: 30 May 2022
Published: 3 June 2022

Publisher's Note: MDPI stays neutral with regard to jurisdictional claims in published maps and institutional affiliations.

Abstract: Cancer remains a leading cause of worldwide mortality and is a growing, multifaceted global burden. As a result, cancer prevention and cancer mortality reduction are counted among the most pressing public health issues of the twenty-first century. In turn, accurate projections of cancer incidence and mortality rates are paramount for robust policymaking, aimed at creating efficient and inclusive public health systems and also for establishing a baseline to assess the impact of newly introduced public health measures. Within the European Union (EU), Romania consistently reports higher mortality from all types of cancer than the EU average, caused by an inefficient and underfinanced public health system and lower economic development that in turn have created the phenomenon of "oncotourism". This paper aims to develop novel cancer incidence/cancer mortality models based on historical links between incidence and mortality occurrence as reflected in official statistics and population web-search habits. Subsequently, it employs estimates of the web query index to produce forecasts of cancer incidence and mortality rates in Romania. Various statistical and machine-learning models—the autoregressive integrated moving average model (ARIMA), the Exponential Smoothing State Space Model with Box-Cox Transformation, ARMA Errors, Trend, and Seasonal Components (TBATS), and a feed-forward neural network nonlinear autoregression model, or NNAR—are estimated through automated algorithms to assess in-sample fit and out-of-sample forecasting accuracy for web-query volume data. Forecasts are produced with the overperforming model in the out-of-sample context (i.e., NNAR) and fed into the novel incidence/mortality models. Results indicate a continuation of the increasing trends in cancer incidence and mortality in Romania by 2026, with projected levels for the age-standardized total cancer incidence of 313.8 and the age-standardized mortality rate of 233.8 representing an increase of 2%, and, respectively, 3% relative to the 2019 levels. Research findings thus indicate that, under the no-change hypothesis, cancer will remain a significant burden in Romania and highlight the need and urgency to improve the status quo in the Romanian public health system.

Keywords: cancer; incidence; mortality; modeling; forecasting; Google Trends; Romania; ARIMA; TBATS; NNAR

---

1. Introduction

Cancer remains a primary cause of death worldwide [1] and acknowledged as a growing global burden [2]. Moreover, many healthcare systems in less developed countries are ill-equipped to adequately deal with this burden, and a huge percentage of cancer patients worldwide lack access to timely, high-quality diagnosis and treatment [3]. As of 2020, cancer accounted for approximately 10 million deaths worldwide, or nearly one in six deaths. Furthermore, cancer maintains its place as the second leading cause of death in many nations, trailing only cardiovascular disease [3–6]. Additionally, the number of cancer diagnoses and fatalities is expected to significantly increase over the next decade, with projections for 2030 indicating 26 million new cancer cases and 17 million cancer deaths per year [7].

Concurrently, cancer is one of the most critical economic and financial burdens that the globe faces today [8]. In the United States alone, national costs of cancer totaled USD 183 billion as of 2015, with projections that include only population growth indicating an increase of 34% by 2030, reaching USD 246 billion [9].

Consequently, with this escalating global burden, cancer prevention and cancer mortality reduction are counted among the most serious public health concerns of the twenty-first century [10]. In particular, the term "primary prevention" refers to measures to reduce the incidence of the disease, whereas "secondary prevention" refers to efforts to diagnose cancer early or to reduce second cancers among cancer survivors [11]. Accurate cancer projections for future time points are paramount for both primary and secondary prevention and are additionally critical for planning future services and resource allocation, as well as establishing and evaluating cancer control programs [12]. However, time series forecasting is a challenging task [13], whereas producing accurate estimates for the future rates of cancer incidence and mortality is additionally complicated due to the short time series available. For example, at the time of the study the Eurostat (i.e., the statistical office of the European Union) database provides statistics for cancer deaths at a European level spanning the period 2011–2018, whereas the World Development Indicators (WDI) database of the World Bank offers data on the mortality rate from cardiovascular disease, cancer, diabetes, or chronic respiratory disease, and thus does not individualize cancer. Additionally, with short series, out-of-sample forecasting accuracy is hard to assess, and time series cross-validation can be difficult to implement [14].

To solve such research obstacles, monitoring health-seeking behavior in the form of public interest indicated by online search queries has emerged as an essential technique for early identification of health problem occurrences throughout certain periods and geographies [15]. This in turn is based on the fact that the internet has grown in importance as a source of health information accessed by the world population [16,17]. As a direct result, Google Trends has become increasingly popular in health and medical research over the past decade [15,18].

Our data confirm the relevance of web searches for highlighting real occurrences of health problems. Thus, [1] indicates that the three most common types of cancer in 2020, in terms of new cases, were breast cancer with 2.26 million cases, lung cancer with 2.21 million cases, and colon and rectum cancer with 1.93 million cases. Concurrently, as reflected in Figure 1, these were the exact web queries related to the term "cancer" over recent years at the world level, confirming that Internet searches are an accurate reflection of health issue incidences.

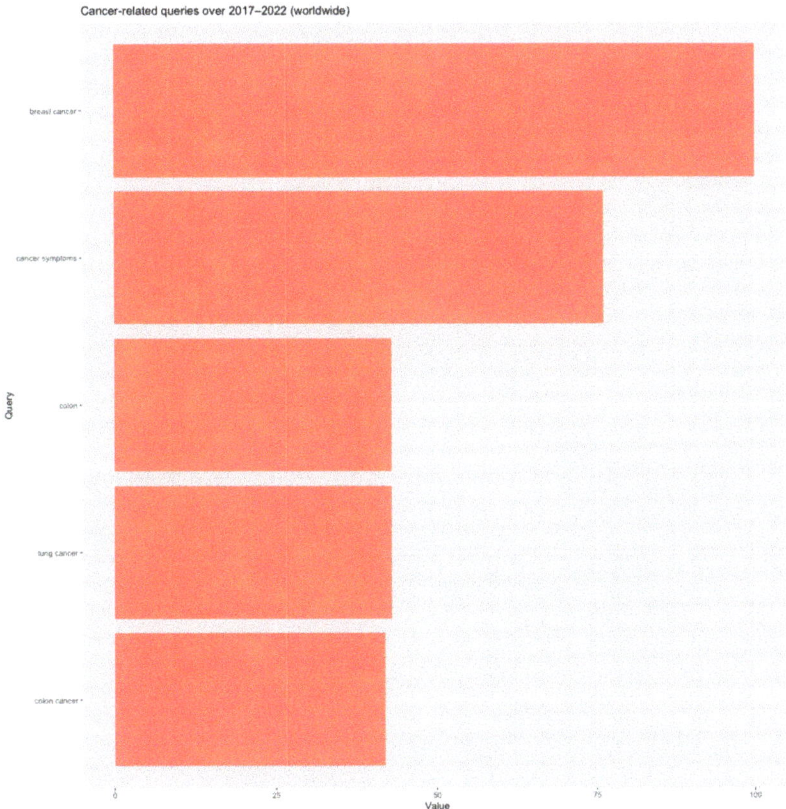

Figure 1. Most common queries related to the search term "cancer": worldwide (April 2017–March 2022). Source of data: Google Trends. Estimation results using the "gtrendsR" package [19] in R software.

Moreover, a visualization of the global web-search interest reveals that most normalized searches emerged in countries that also reported the highest age-standardized cancer rates. Of note, to accurately comprehend the geography of search interest for a given keyword, the term should be searched across all the world's languages. However, Google Trends provides a specific tool capable of dealing with this issue, i.e., "Topics," which collects all related words, variant spellings, and names in other languages under a single label to help with comprehending topics in a multilingual setting. Topics can thus be particularly effective in combining translations into multiple languages under a single subject [20]. As such, we specified the topic "cancer" when sourcing Google Trends data. Thus, Figure 2 reflects the normalized number of internet searches over the most recent five years, confirming that the highest population interest in the topic "cancer" was encountered in countries including Australia, the US, and Ireland. On the other hand, WHO data confirm that Australia registered the world's highest age-standardized cancer rate at 452.4 cases per 100,000 people in 2020, followed by New Zealand (422.9), Ireland (372.8), and the United States (362.2).

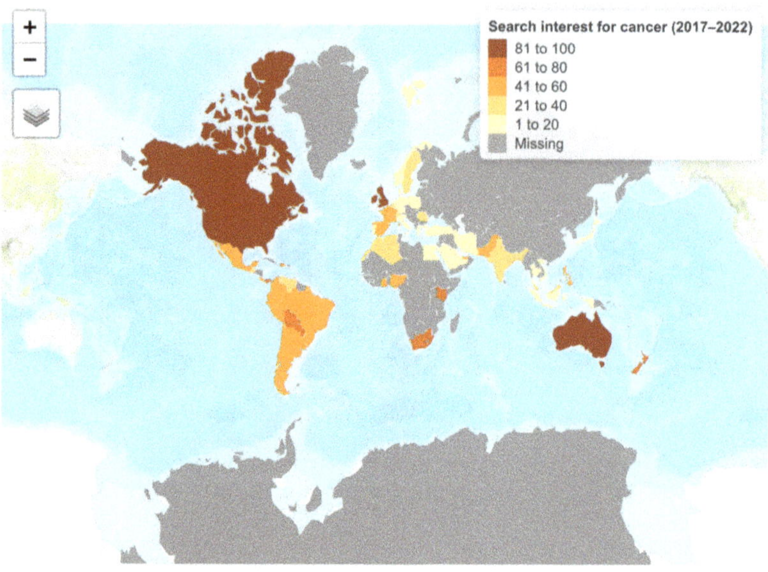

Leaflet | Tiles © Esri — Esri, DeLorme, NAVTEQ, TomTom, Intermap, iPC, USGS, FAO, NPS, NRCAN, GeoBase, Kadaster NL, Ordnance Survey, Esri Japan, METI, Esri China (Hong Kong), and the GIS User Community

Figure 2. Internet search interest for "cancer" at the world level: (April 2017–March 2022). Source of data: Google Trends. Map is based on estimation results and uses the packages "gtrendsR" [19] and "tmap" [21] in R software.

Additionally, studies increasingly confirm that many cancers remained undiagnosed as a result of healthcare system disruptions caused by the COVID-19 outbreak [22–24]. In this context, with official statistics failing to accurately capture the variation in incidence, people's search interest for specific symptomatology emerges as the most relevant indication of the health problem's occurrence.

In light of the above considerations, this study sourced Google Trends data to extract information on internet searches for the word "cancer" and employed it as a proxy to forecast cancer incidence rates. Google Trends (www.trends.google.com, accessed on 30 March 2022) is a web-based tool that shows the popularity of a search phrase in a certain location over time. It provides a time series index of the number of Google queries submitted in a given location. The query weight or share is calculated by dividing the overall query volume for a specific search term within a geographic region by the total number of searches in that region throughout the period in question. Following that, the result is scaled from 0 to 100. As a result, the maximum query share of a search phrase for a given period is normalized to 100, reflecting the point when the search was at its most popular. In conclusion, on a scale of 0–100, Google Trends calculates relative search interest (RSI), with 100 reflecting peak interest [25]. Ref. [26] explore the utility of Google Trends data to examine population web searches for cancer screening and conclude that web queries can capture awareness and interest in cancer screening. Thus, Google Trends data may complement traditional data collection and analysis about cancer screening and related interests, providing important scientific possibilities. However, given the aforementioned disruption of public health systems caused by COVID-19 that altered official statistics, we argue that Google Trends data can now be used as a substitute for traditional statistics, which further expands its scientific value.

Of note, among European countries, cancer survival is significantly lower in newer and less developed EU members from Central and Eastern Europe [27]. Higher death rates at the CEE level are caused by two main factors: delayed diagnosis and suboptimal treatment [28], which are in turn related to inefficient and underfinanced public health systems and lower overall economic development [29].The situation is particularly challenging in Romania, which continues to register a divergent trend in mortality rates relative to its EU counterparts, including in the CEE area [30]. Figure 3 shows that the age-standardized mortality rate from all cancers follows an increasing trend in Romania from 2011 to 2018, reflecting the inefficiency of the public healthcare system in the country, whereas most other CEE countries have managed to reverse the trend.

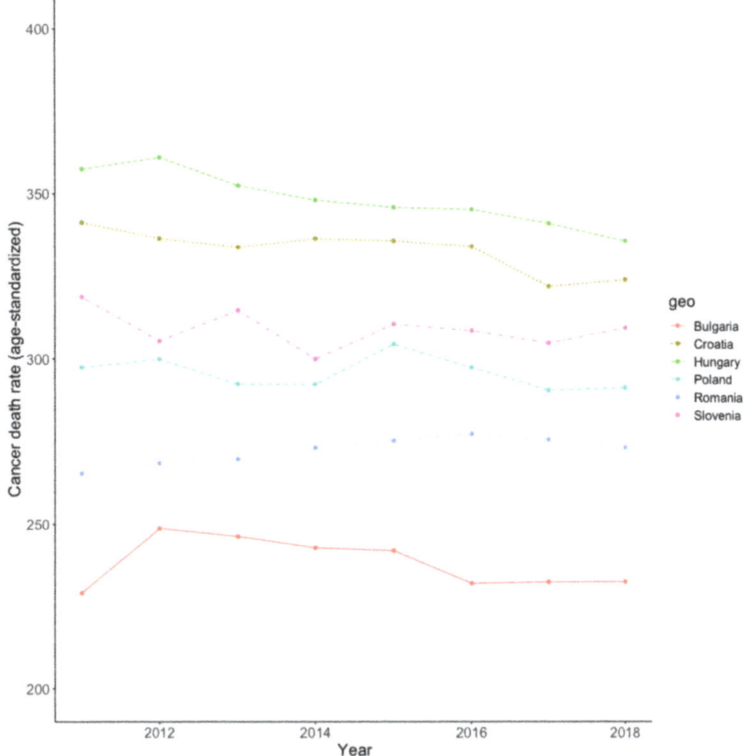

Figure 3. Trends in cancer mortality rates in selected CEE countries (2011–2018). Estimation results. Plot created in R software ("ggplots" function). Source of data: Eurostat.

Moreover, the excess mortality from the main types of cancer registered in Romania relative to the EU average is reflected by the difference in the five-year survival rates presented in Table 1. For example, whereas recent statistics show that the survival rate of breast cancer patients has rapidly increased over recent years, due to the availability of early diagnostic tools and treatment [31], Romania still reports significant health gaps, which is heavily influenced by the fact that there is no organized population screening for breast cancer in the country [30].

Table 1. The five-year survival rates from main types of cancer (Romania versus EU26).

Type of Cancer	5-Year Survival Rate	
	Romania	EU26
Lung	11%	15%
Breast	75%	83%
Prostate	77%	87%

Source of data: Romanian Ministry of Health (2021) [32].

A worrisome disaggregation also occurs between cancer incidence and mortality rates registered in Romania, which has not managed to reduce cancer mortality despite periods of decreased incidence and which further highlights the necessity and urgency of better policies aimed at providing an efficient and inclusive public health system (Figure 4, Panel a). Moreover, the implementation of join-point regression analysis, also known as change-point regression or segmented regression analysis [33] to detect changing trends in cancer incidence (see for example [34–39]) further confirms that the incidence rate in Romania presents two join points in 2014 and 2017, leading to three periods with a different trend over the analysis period, as follows: a positive trend with a slope coefficient of 14.68 until 2014, a negative trend with a slope coefficient of −8.83 during 2014–2017, and a slightly increasing trend with a slope of 2.05 after 2017 (Figure 4, Panel b). Hence, whereas the incidence follows an increasing trend until the first join-point (i.e., 2014), a reversal is detected thereafter, and a decreasing trend is confirmed over the second segment (i.e., 2014–2017). However, the decreasing trend is reversed thereafter, as the incidence rate presents a subsequent rise. On the other hand, the join-point regression analysis found no join-point in the mortality rate, confirming the disaggregation between the two series.

Patient migration from CEE countries has grown with the implementation of a European Union Directive issued in 2011. According to this directive, European nationals are eligible to use European healthcare services in any of the European member states, and their treatments are covered (at least partially) by their home country's health insurance system [40]. As a result, a phenomenon called "oncotourism" or "cancer tourism" has emerged, whereas diagnosed patients move away from inefficient Eastern and Central European public healthcare systems, particularly from Romania, toward the private system or the healthcare systems of more developed EU countries [41].

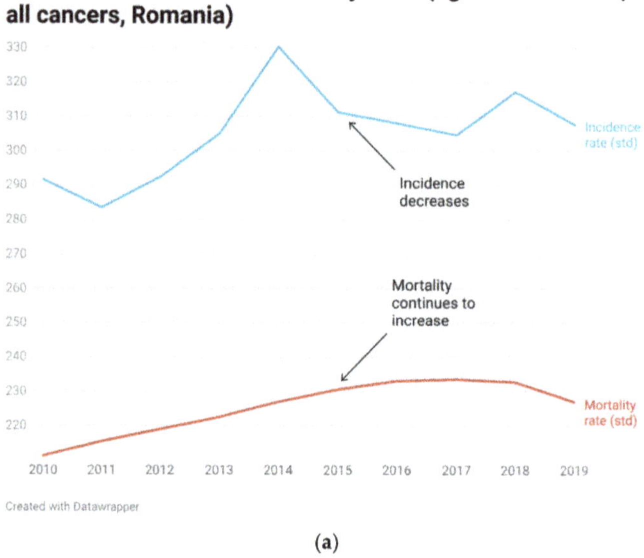

(a)

Figure 4. *Cont.*

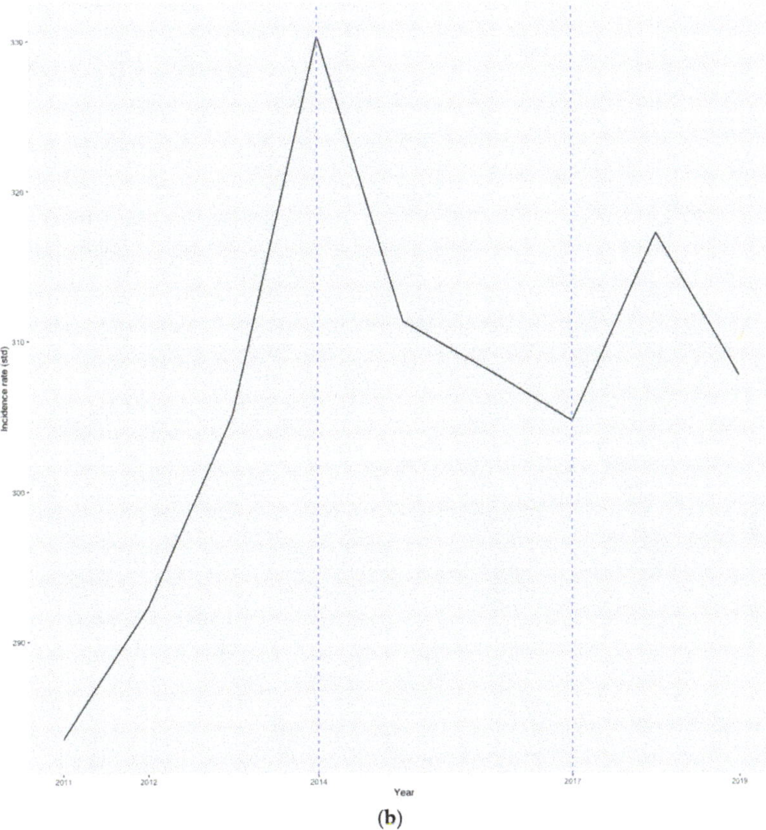

(b)

Figure 4. Trends in age-standardized cancer incidence and mortality rates in Romania (2010–2019) (panel **a**); join-points in cancer incidence rate (panel **b**). Source of data: Romanian Ministry of Health (2021) [32]. Chart in panel (**a**) is produced in Datawrapper. Chart in panel (**b**) is produced with the "ggplot" function in R software; join-point regression analysis is performed with the "segmented" package within R software.

Consequently, accurate predictions for cancer incidences are paramount for early detection and for issuing effective and more inclusive public health policies, especially in the most vulnerable EU members in the CEE area. Additionally, cancer incidence projections are also useful for planning health services and establishing a baseline for evaluating the impact of public health measures [42]. Thus, the main goal of this study is to develop cancer incidence/cancer mortality models, and subsequently to make use of web-search data extracted from Google Trends and its point estimates issued through an array of statistical and machine-learning models to ultimately produce accurate forecasts of cancer incidence and mortality rates, while taking a special focus on a vulnerable CEE country significantly plagued by this disease, i.e., Romania. From a methodological perspective, the robustness of results is assured through various approaches, such as: (i) the estimation of alternative predictive models (statistical and machine-learning); (ii) the assessment of the relative out-of-sample forecasting accuracy through the hold-out forecasting technique; (iii) the estimation of the Diebold-Mariano test for superior forecasting accuracy, and (iv) resampling Google Trends data and employing the sampling average for the web-query index.

Of note, the vast majority of previous research either employs one forecasting method or assesses the predictive ability of concurrent methods by estimating forecasting accuracy metrics. This study implements alternative predictive models, both statistical and machine learning, through automated forecasting algorithms. Moreover, the random sampling issue that arises from using Google Trends data is mitigated through resampling and averaging. Additionally, the forecasting results are defended against the Diebold-Mariano (DM) predictive accuracy test. Hence, various robustness checks confirm the reliability of current findings. Furthermore, the strand of literature on cancer research, particularly with a focus on Central and Eastern Europe, remains thin. Hence, whereas most related studies focus on developed countries, the current research contributes to filling the literature void and is thus concerned with a rather under-investigated EU member, Romania, a country that constitutes an interesting playing field for cancer research due to divergent trends relative to its EU counterparts and plagued by the worrisome phenomenon of "oncotourism". The proposed method is novel and carries the generalizability advantage, being suitable to further investigate other countries for which official statistics have been heavily affected by the coronavirus pandemic.

Thus, compared to previous studies, the contributions of the current research are threefold: (i) we develop two novel models to explain cancer incidence and cancer mortality rates that embed both official statistics and data on population health-seeking behavior as reflected in internet search habits, whereas most previous studies employ some version of the age-period-cohort model (APC) for the same task; (ii) we propose a robust and integrated approach for web query volume forecasting that includes statistical and machine-learning forecasting methods and assures the robustness of results through multiple model calibration on training and test datasets and estimation of multiple accuracy metrics; and (iii) we apply this novel framework to data from one of the most vulnerable EU members, Romania, a country increasingly defined by the phenomenon of "oncotourism", whereby diagnosed patients avoid the inefficient national public health system. We additionally provide evidence on the link between internet-seeking behavior and the incidence and mortality of the disease in Romania, thus contributing to the extent of infodemiological literature. Research findings have important policy implications, and the framework, owing to its generalizability, can be applied to the same task in other countries. The novel approach is particularly relevant in the aftermath of the COVID-19 pandemic, which has disrupted public health systems and caused a significant number of cancers to remain undiagnosed, thus affecting the quality of official statistics and their usefulness for health studies.

Results overall indicate a continuation of the increasing trends in cancer incidence and mortality in Romania, with a standardized cancer incidence rate of 313.8 by 2026 and a standardized cancer mortality rate of 233.8 by the same horizon, and thus underline the urgency to change the status quo in the Romanian public health system.

The paper continues as follows. Section 2 presents the data used in model development and explains the integrated method. Section 3 describes the empirical findings that emerge from implementing the novel-forecasting framework on the Romanian data. Section 4 discusses the main findings and, finally, Section 5 concludes the study.

2. Materials and Methods

In this study, we sourced annual data on Romanian cancer incidence and mortality rates spanning 2010–2019 from the Romanian Ministry of Health. Next, to develop a model capable of explaining and forecasting these relevant health indicators in the absence of reliable official statistics (which is a worldwide issue caused by the significant number of undetected cancer cases after the onset of the COVID-19 pandemic), we relied on previous infodemiological studies that acknowledge the population's internet search habits as a reliable proxy for the incidence of a health problem.

The Google Trends platform is a handy tool for determining the popularity of a specific search keyword among a particular demographic. In this study, we extracted the monthly volume of Google queries issued from Romania for "cancer" for the period spanning

January 2005–March 2022. It should be acknowledged that Google Trends implements random sampling and uses only a fraction of the entire search data to construct a search index [43] Thus, to overcome the sample instability issue, multiple samples (i.e., 12) were sourced and the average of samples was used to construct the web-query index, instead of only one sample (see [44] for relevant details on the sample bias and its correction). However, it has also been recognized that the Google Trends sampling procedure produces reasonably precise estimates, and consequently, there is often no need to employ more than a single sample [45]. The web query time series contained 207 monthly observations.

The relationship between web searches and the health indicators of interest for Romania was assessed through the linear model given by Equation (1).

$$\hat{y} = bX + a \tag{1}$$

where $\hat{y}$ is alternatively the cancer incidence rate, and subsequently the cancer mortality rate, and the independent variable is the web-query index.

Additionally, we used both linear and nonlinear statistical and machine-learning techniques, which allowed us to capture most of the properties of the web-query time series and further contributed to avoiding unreliable forecasts. Predictive models can be delineated into two main categories [46–48]: statistical and machine learning methods (self-learning systems that can learn from data and continuously increase performance), respectively. Thus, in this study, the autoregressive integrated moving average (ARIMA) model (Equation (2)), the Exponential Smoothing State Space Model with Box-Cox Transformation, ARMA Errors, Trend, and Seasonal Components (TBATS) given by Equation (3), and the neural network autoregression (NNAR) model reflected in Equation (4) were alternatively fitted.

An ARIMA(p,d,q)(P,D,Q)s model, first developed by [49], is given by:

$$(1 - \varphi_1 B - \ldots - \varphi_p B^p)(1 - \Phi_1 B^s - \ldots - \Phi_P B^{sP})(1 - B)^d (1 - B^s)^D Y_t = \\ (1 - \theta_1 B - \ldots - \theta_q B^q)(1 - \Theta_1 B^s - \ldots - \Theta_P B^{sQ})\varepsilon_t \tag{2}$$

where s is the seasonal period, the lowercase and the capital letters represent nonseasonal and seasonal parameters, and ε_t is a random variable with mean zero and the standard deviation σ.

A TBATS model [50] can accommodate complex seasonal behaviors of data [51] and is written as:

$$TBATS(\omega, p, q, \varphi, \{m_1, k_1\}, \{m_2, k_2\}, \ldots, \{m_T, k_T\}), \tag{3}$$

where ω is the Box-Cox transformation, k is the number of harmonics used for the seasonal trait, and φ is the dampening parameter.

Artificial neural networks (ANNs) are capable of simulating complicated real-world systems while properly accounting for nonlinearities [52]. Lagged values of time series are frequently utilized as inputs in an ANN structure when fitting time series data, which is then known as neural network autoregression (NNAR) [53] (Munim et al., 2019). As in [29,54], the NNAR model is written as:

$$Y = f(H) = f(W * X + B), X = [y(t-1), y(t-2), \ldots, y(t-p)] \tag{4}$$

where Y stands for the output vector, f is the activation function, H is the vector of n nodes in the hidden layer, W is the weight matrix between the input and hidden layers, X is the vector of inputs (i.e., the lagged values of the actual observations), and B is a bias vector.

All estimations are automated and performed in R software via dedicated algorithms included in the "forecast" package [55]. To implement the method robustly, the series of length N = 207 was first split into a training set (containing 187 observations) for in-sample fit purposes and a testing set (containing the last 20 observations) on which the models that reported the best fit on the training set were further estimated and their out-of-sample forecasting ability assessed. Lastly, we assessed the forecast accuracy of alternative predictive models by estimating both scale and scale-free accuracy metrics.

First, let us define the forecast error of a candidate model as:

$$e_{T+h} = y_{T+h} - \hat{y}_{T+h|T} \tag{5}$$

where $\{y_1, \ldots, y_T\}$ is the training set data and $\{y_{T+1}, y_{T+2}, \ldots\}$ is the test-set data.

Then, the following forecasting accuracy metrics are computed as:

Mean absolute error:

$$MAE = \sqrt{\frac{1}{N}\sum_{i=1}^{N}|y_i - \hat{y}_l|} \tag{6}$$

Root mean squared error:

$$RMSE = \sqrt{\frac{1}{N}\sum_{i=1}^{N}(y_i - \hat{y}_l)^2} \tag{7}$$

Mean absolute percentage error:

$$MAPE = mean(|p_t|) \tag{8}$$

where: $p_t = \frac{100e_t}{y_t}$

Mean absolute percentage error:

$$MASE = mean(|q_j|) \tag{9}$$

where qj is given by as: $q_j = \dfrac{e_t}{\frac{1}{N-1}\sum\limits_{i=2}^{N}|y_t - y_{t-1}|} f$ when the series is non-seasonal and by:

$q_j = \dfrac{e_t}{\frac{1}{N-m}\sum\limits_{i=m+1}^{N}|y_t - y_{t-m}|}$ when the time series is seasonal.

Lastly, all predictive models were fitted to the entire series of length N and point forecasts for the web query index for the following 4 years (i.e., a 48-month forecasting horizon) were produced by the overperforming method in the out-of-sample setting. Forecasted values were then fitted into the incidence/mortality models developed by estimating Equation (1), which then issued the expected values for standardized cancer incidence and mortality rates corresponding to the forecasting horizon. Figure 5 reflects the integrated method employed in this study and implemented in R software.

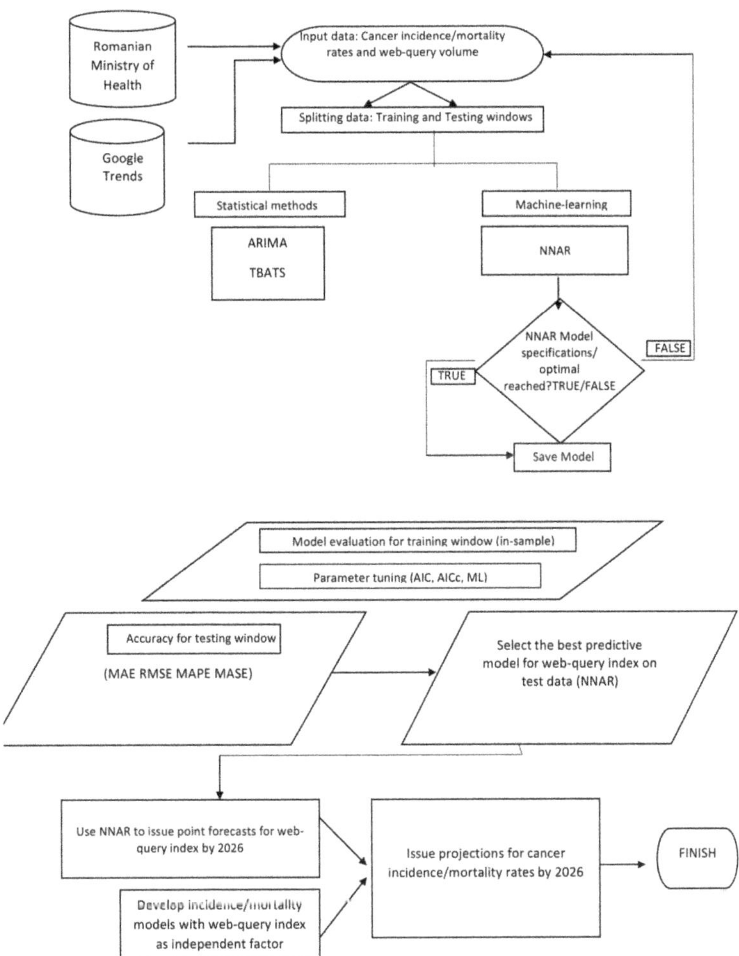

Figure 5. The integrated framework for modeling and forecasting cancer incidence and mortality rates.

3. Results

3.1. Relationship between Related Web Queries and the Age-Standardized Cancer Incidence/Cancer Mortality Rate in Romania

The best-fit linear model between the vector of web-query volume and the cancer incidence/cancer mortality rate is reflected in Figure 6, panels (a) and (b), respectively. Both representations highlight a positive link between the web-search interest and the variables reflecting the incidence and mortality of the disease. Additionally, both equations show a similar slope coefficient, equal to 0.47 in the incidence rate model, and equal to 0.46 in the mortality rate model.

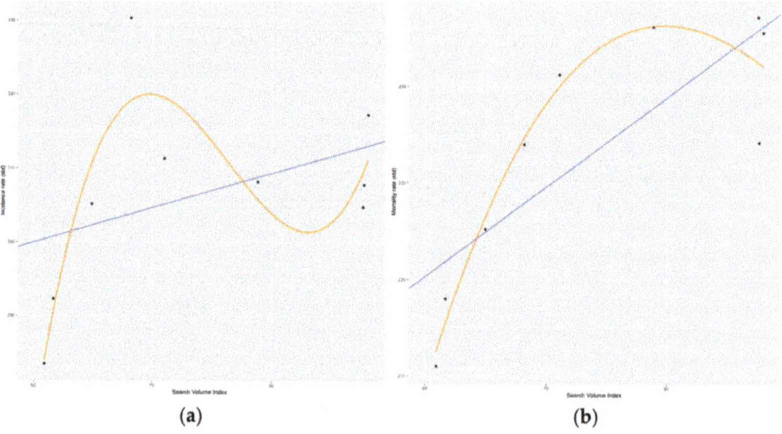

(a)   (b)

Figure 6. The relationship (linear—blue line, polynomial—orange line) between related web queries and the age-standardized cancer incidence rate in Romania (panel **a**). The relationship (linear—blue line, polynomial—orange line) between related web queries and the age-standardized cancer mortality rate in Romania (panel **b**). Source of data: Romanian Ministry of Health [32]. All estimations were performed in R software; plots were created in R software (i.e.,"ggplot" function).

3.2. Results from Modeling and Forecasting the Web-Query Index

Table 2 reports the estimated accuracy measures for the test-set data containing topic searches for "cancer" submitted in Romania that were issued through the statistical and machine-learning predictive models. Results indicate that NNAR has been able to accurately capture variations in data and thus provide the best forecast for the web query index over the testing window.

Table 2. Accuracy measures for the out-of-sample (test-set) forecasting performance.

Predictive Model	MAE	RMSE	MAPE	MASE
ARIMA	4.21	5.16	5.76	0.61
NNAR	3.96	4.71	5.32	0.57
TBATS	4.60	5.73	6.54	0.74

To assess the forecasting superiority of the feed-forward neural network autoregression model, we estimated the Diebold-Mariano (DM) test [56,57] to examine any significant differences between forecasts produced by NNAR and the second best-performing model (ARIMA). The DM test result (estimated with the "dm.test" function within the "forecast" package in R software) confirmed that there was a significant difference between the distribution of errors from ARIMA and NNAR, thus ensuring the forecasting superiority of the machine-learning method.

We next employed the best-performing model in terms of out-of-sample forecasting accuracy (i.e., NNAR) to produce the expected web-query volume in Romania for the next 48 months (4 years), corresponding to the period spanning April 2022 to March 2026. Figure 7 reflects the estimation results, showing (in blue color) the point estimates produced by NNAR for April 2022–March 2026. Of note, estimations corresponding to the 48-month forecasting horizon indicated a continuation of the increasing trend in web searches related to "cancer" issued from Romania.

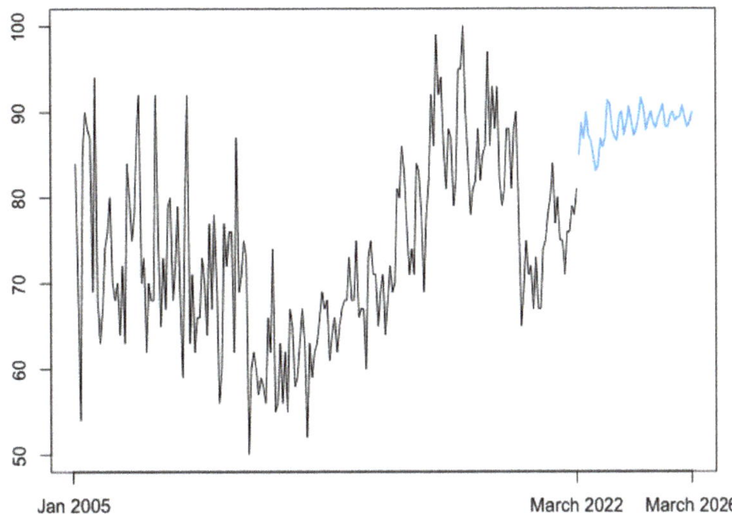

Figure 7. Forecasted trend over April 2022–March 2026 (48 months) for web queries for the term "cancer" in Romania issued with NNAR (12,6). Source: estimation results. Model information: average of 20 networks, each of which is a 12-6-1 network with 85 weight options.

3.3. Forecasts of Cancer Incidence and Cancer Mortality Rates in Romania over 2022–2026

Lastly, we put together the linear relationship model estimated before and the point estimates produced by the neural network autoregression model (NNAR) to estimate the cancer incidence and cancer mortality rate for the next four years. It is important to underline that all estimations are based on the status quo hypothesis, implying that these projections are expected if no changes in public health policy are implemented. Projected values for the two health indicators are centralized in Table 3.

Table 3. Forecasted values for cancer incidence and cancer mortality rates.

Year	Incidence Rate (Projected, Standardized)	Mortality Rate (Projected, Standardized)
2023	308.7	228.8
2024	313.0	233.0
2025	313.6	233.6
2026	313.8	233.8

We notice that estimations issued through the incidence and mortality models and based on NNAR projections of the related web-search index reflect a continuation of the increasing trends in cancer incidence and mortality in Romania, underlining the urgency to change the status quo in the Romanian public health system. Estimates thus indicate a standardized cancer incidence rate of 313.8 by 2026 and a standardized cancer mortality rate of 233.8 by the same horizon, increasing from levels of 307.7 and 227.1, respectively, registered in 2019.

4. Discussion

Approximately 10 million deaths have been attributed to cancer in 2020, or nearly one in six deaths. Concurrently, many cancer patients worldwide still lack access to timely, high-quality diagnosis and treatment. Within the EU, CEE countries consistently report higher mortality rates, mainly as a result of delayed diagnosis and suboptimal treatment. For Romania, statistics are particularly worrisome, with the country reporting significantly lower survival rates for all main types of cancer and also a divergent trend in mortality rates relative to its EU counterparts that have managed to reverse the increasing trend. Consequently, "oncotourism" is especially characteristic of Romania, as diagnosed patients move away from the inefficient public healthcare system toward the private system or the healthcare systems of more developed EU countries.

Consequently, accurate predictions of cancer incidence and mortality rates are keys to informing policymakers and assisting in the policymaking process. The main goal of this study is to develop a robust model capable of capturing the evolution of cancer incidence and mortality rates and forecasting their evolution. Concurrently, we build on [58] and acknowledge that over the past decade, the use of internet data has become an important aspect of health informatics, with online sources becoming more accessible and offering data that can be used to analyze and forecast human behavior. As a result, we also agree with [59] that data from tracking online information seekers' behavior is useful in public health surveillance and research.

Thus, in model development, the study made use of web-search data extracted from Google Trends, which was introduced as an independent variable in light of previous studies that acknowledge the population's internet search habits as a reliable proxy for health problem occurrence. Moreover, this approach overcomes the current issue of the unreliability of official statistics, caused on one hand by the numerous undiagnosed cases after the COVID-19 outbreak and, on the other hand, by the unavailability or tardiness of treatment for diagnosed patients, directly affecting the rate of incidence and mortality. Concurrently, point estimates for the web-query index were issued through the best performing predictive model over the test-set (i.e., out-of-sample), after an assessment of the in-sample fit and out-of-sample forecasting accuracy of various statistical and machine-learning models (ARIMA, TBATS, and NNAR) had been performed via several accuracy metrics (MAE, RMSE, MAPE, and MASE). Estimations indicated that NNAR was the most capable of capturing the time series characteristics and of producing the most accurate estimates. Consequently, estimations for the web-query index were automatically produced with the NNAR model and sourced into the incidence/mortality models that have been previously developed. Ultimately, forecasts of cancer incidence and mortality rates in Romania by 2026 were issued, indicating a continuation of the increasing trend for both variables. Our results are in line with projections of [32] the Romanian Ministry of Health (2021), confirming the ascendant trend, although our point estimates fall below the public ministry's predictions, indicating more conservative increasing rates.

Future predictions depend on multiple assumptions, most importantly on the status quo (i.e., no-change) hypothesis of the Romanian public health system. Similarly, forecasts do not consider the impact of relevant changes in impact factors, such as potential changes in smoking prevalence at the national level, changes in obesity, changes in alcohol consumption, changes in nutrition habits, increased funding of the public health system, increased screening, HPV vaccination, etc. As a result, we agree with [60] that it is critical to review predictions at regular intervals to incorporate the most recent trends in the data. However, similar to [42], we reason that current estimates do provide a useful baseline for the planning of cancer resources and for evaluating the impact of any changes produced in impact factors as a result of newly introduced public health policies and measures. Furthermore, as with most research, this study suffers from other limitations. Mostly, the use of Google Trends data does carry some vulnerabilities that should be acknowledged, including the construction of the search index itself [61]. As a consequence, the long-run stability of the time series is heavily dependent on the data's time frame and frequency. As per [61], this study used monthly data that was best able to accurately capture the long-term trend. Moreover, random sampling is an inherent bias in Google Trends data [44]. However, this issue is particularly troublesome in forecasting when dealing with topics that are less frequently searched for, which is not the case in the current study. Additionally, the performed resampling further mitigated the bias. Nonetheless, it should be acknowledged that this approach merely minimizes the bias and does not eliminate it. Finally, it should be mentioned that results should be interpreted with care, considering that data reflect the search habits of the population that has Internet access, which in turn depends on income level and other socioeconomic factors. Overall, I argue that the popularity of the search topic, together with the resampling strategy and previous studies that reinforce the usefulness of web search data as a powerful predictive instrument [44] does allow for confidence in the current research findings.

In addition, it should also be mentioned that the link between cancer variables and the web-query index has been assessed through the classical regression model. However, it has been increasingly acknowledged that the neutrosophic regression model [62–65], which issues the parameters in the indeterminacy interval rage, can be more efficient in the uncertainty environment than the classical regression model [66]. Thus, the assessment of the historical link between cancer incidence/cancer mortality rates and the web search index through the neutrosophic regression model constitutes a good avenue for future research. The implementation of the proposed method on data specific to different gender and age groups, as well as to geographic regions of the country, could also reveal particularly vulnerable groups and/or areas and offer relevant information to policymakers.

From a policy perspective, the findings highlight that cancer will continue to be a significant burden for Romania, which should be carefully planned for. Complementarily, results indicate the need for better policies aimed at mitigating main risk factors such as smoking, alcohol consumption, obesity and overweight, unhealthy nutrition, lack of physical exercise, etc., and at increasing the financing and efficiency of the public health system by allocating future resources for cancer research, treatment, and prevention.

5. Conclusions

In conclusion, this study developed two novel cancer incidence/cancer mortality models based on population web-search habits and historical links with official health variables. The models were empirically estimated using data from one of the most vulnerable European Union (EU) members, Romania, and further used to forecast cancer incidence and mortality rates in the country by employing estimates for the web-search query index issued through the best performing out-of-sample forecasting method (NNAR). Research findings have important policy implications, and the novel framework, owing to its generalizability, can be applied to the same task in other countries. It provides the important advantage of overcoming a current issue related to the quality of official statistics in the aftermath of the COVID-19 pandemic that disrupted public health systems and caused a significant number of cancers to remain undiagnosed. Overall, the results indicate a continuation of the increasing trends in cancer incidence and mortality in Romania and thus underline the urgency to change the status quo in the Romanian public health system.

Funding: This research received no external funding.

Institutional Review Board Statement: Not applicable.

Informed Consent Statement: Not applicable.

Data Availability Statement: Data employed in this study are publicly available from the Eurostat database, from Google Trends, and the Romanian Ministry of Health.

Conflicts of Interest: The authors declare no conflict of interest.

References

1. World Health Organization (WHO). Cancer. 2022. Available online: https://www.who.int/news-room/fact-sheets/detail/cancer (accessed on 4 April 2022).
2. World Cancer Research Fund. Worldwide Cancer Data. 2022. Available online: https://www.wcrf.org/dietandcancer/worldwide-cancer-data/ (accessed on 30 March 2022).

3. World Health Organization (WHO). Cancer. 2019. Available online: https://www.who.int/health-topics/cancer#tab=tab_1 (accessed on 4 April 2022).
4. Ma, X.; Yu, H. Cancer issue: Global burden of cancer. *Yale J. Biol. Med.* **2006**, *79*, 85.
5. Nagai, H.; Kim, Y.H. Cancer prevention from the perspective of global cancer burden patterns. *J. Thorac. Dis.* **2017**, *9*, 448. [CrossRef]
6. Zaorsky, N.G.; Churilla, T.M.; Egleston, B.L.; Fisher, S.G.; Ridge, J.A.; Horwitz, E.M.; Meyer, J.E. Causes of death among cancer patients. *Ann. Oncol.* **2017**, *28*, 400–407. [CrossRef]
7. Thun, M.J.; DeLancey, J.O.; Center, M.M.; Jemal, A.; Ward, E.M. The global burden of cancer: Priorities for prevention. *Carcinogenesis* **2010**, *31*, 100–110. [CrossRef]
8. World Cancer Day, Financial and Economic Impact of Cancer. 2022. Available online: https://www.worldcancerday.org/financial-and-economic-impact-0 (accessed on 30 March 2022).
9. Mariotto, A.B.; Enewold, L.; Zhao, J.; Zeruto, C.A.; Yabroff, K.R. Medical care costs associated with cancer survivorship in the United States. *Cancer Epidemiol. Prev. Biomark.* **2020**, *29*, 1304–1312. [CrossRef]
10. United Nations (UN). New WHO Platform Promotes Global Cancer Prevention. 2022. Available online: https://news.un.org/en/story/2022/02/1111312 (accessed on 7 April 2022).
11. White, M.C.; Peipins, L.A.; Watson, M.; Trivers, K.F.; Holman, D.M.; Rodriguez, J.L. Cancer prevention for the next generation. *J. Adolesc. Health* **2013**, *52*, S1–S7. [CrossRef]
12. Rapiti, E.; Guarnori, S.; Pastoors, B.; Miralbell, R.; Usel, M. Planning for the future: Cancer incidence projections in Switzerland up to 2019. *BMC Public Health* **2014**, *14*, 102. [CrossRef]
13. Petropoulos, F.; Spiliotis, E. The wisdom of the data: Getting the most out of univariate time series forecasting. *Forecasting* **2021**, *3*, 478–497. [CrossRef]
14. Hyndman, R.J.; Athanasopoulos, G. Forecasting: Principles and Practice; Otexts. 2018. Available online: https://otexts.com/fpp2/ (accessed on 30 March 2022).
15. Szilagyi, I.S.; Ullrich, T.; Lang-Illievich, K.; Klivinyi, C.; Schittek, G.A.; Simonis, H.; Bornemann-Cimenti, H. Google Trends for Pain Search Terms in the World's Most Populated Regions Before and After the First Recorded COVID-19 Case: Infodemiological Study. *J. Med. Internet Res.* **2021**, *23*, e27214. [CrossRef]
16. Polgreen, P.M.; Chen, Y.; Pennock, D.M.; Nelson, F.D.; Weinstein, R.A. Using internet searches for influenza surveillance. *Clin. Infect. Dis.* **2008**, *47*, 1443–1448. [CrossRef]
17. Pew Research Center, Health Online. 2013. Available online: http://www.pewinternet.org/2013/01/15/health-online-2013/ (accessed on 7 April 2022).
18. Nuti, S.V.; Wayda, B.; Ranasinghe, I.; Wang, S.; Dreyer, R.P.; Chen, S.I.; Murugiah, K. The use of google trends in health care research: A systematic review. *PLoS ONE* **2014**, *9*, e109583. [CrossRef] [PubMed]
19. Massicotte, P.; Eddelbuettel, D. Gtrendsr: Perform and Display Google Trends Queries. R Package Version 1.4.4. 2019. Available online: https://CRAN.R-project.org/package=gtrendsR (accessed on 30 March 2022).
20. Forbes. Understanding What You're Searching for in A Multilingual World. 2015. Available online: https://www.forbes.com/sites/kalevleetaru/2015/10/18/understanding-what-youre-searching-for-in-a-multilingual-world/?sh=9e2b3f23e0f4 (accessed on 25 May 2022).
21. Tennekes, M. tmap: Thematic Maps in R. *J. Stat. Softw.* **2018**, *84*, 1–39. [CrossRef]
22. Jacob, L.; Loosen, S.H.; Kalder, M.; Luedde, T.; Roderburg, C.; Kostev, K. Impact of the COVID-19 pandemic on cancer diagnoses in general and specialized practices in Germany. *Cancers* **2021**, *13*, 408. [CrossRef] [PubMed]
23. Marques, N.P.; Silveira, D.M.M.; Marques, N.C.T.; Martelli, D.R.B.; Oliveira, E.A.; Martelli-Júnior, H. Cancer diagnosis in Brazil in the COVID-19 era. *Semin. Oncol.* **2021**, *48*, 156–159. [CrossRef]
24. Becker's Hospital Review, As COVID-19 Dies Down, Undiagnosed Cancers Emerge. 2021. Available online: https://www.beckershospitalreview.com/oncology/as-covid-19-dies-down-undiagnosed-cancers-emerge.html (accessed on 6 April 2022).
25. Greiner, B.; Tipton, S.; Nelson, B.; Hartwell, M. Cancer screenings during the COVID-19 pandemic: An analysis of public interest trends. *Curr. Probl. Cancer* **2022**, *46*, 100766. [CrossRef]
26. Schootman, M.; Toor, A.; Cavazos-Rehg, P.; Jeffe, D.B.; McQueen, A.; Eberth, J.; Davidson, N.O. The utility of Google Trends data to examine interest in cancer screening. *BMJ Open* **2015**, *5*, e006678. [CrossRef]
27. Vrdoljak, E.; Wojtukiewicz, M.Z.; Pienkowski, T.; Bodoky, G.; Berzinec, P.; Finek, J.; Todorović, V.; Borojević, N.; Croitoru, A. Cancer epidemiology in Central and South Eastern European countries. *Croat. Med. J.* **2011**, *52*, 478–487. [CrossRef]
28. World Health Organization (WHO). Up to a quarter of Europeans Will Develop Cancer: From Prevention, Early Diagnosis, Screening and Treatment to Palliative Care, Countries Must Do More. 2020. Available online: https://www.euro.who.int/en/health-topics/noncommunicable-diseases/cancer/news/news/2020/2/up-to-a-quarter-of-europeans-will-develop-cancer-from-prevention,-early-diagnosis,-screening-and-treatment-to-palliative-care,-countries-must-do-more (accessed on 7 April 2022).
29. Tudor, C.; Sova, R. EU Net-Zero Policy Achievement Assessment in Selected Members through Automated Forecasting Algorithms. *ISPRS Int. J. Geo-Inf.* **2022**, *11*, 232. [CrossRef]
30. Furtunescu, F.; Bohiltea, R.E.; Voinea, S.; Georgescu, T.A.; Munteanu, O.; Neacsu, A.; Pop, C.S. Breast cancer mortality gaps in Romanian women compared to the EU after 10 years of accession: Is breast cancer screening a priority for action in Romania? (Review of the Statistics). *Exp. Ther. Med.* **2021**, *21*, 268. [CrossRef]

31. Azam, M.; Aslam, M.; Basharat, J.; Mughal, M.A.; Nadeem, M.S.; Anwar, F. An empirical study on quality of life and related factors of Pakistani breast cancer survivors. *Sci. Rep.* **2021**, *11*, 24391. [CrossRef]

32. Romanian Ministry of Health. Analysis of the Cancer Situation in 2021 (in Romanian). 2021. Available online: https://www.ms.ro/2021/06/30/cancerul-este-un-risc-pentru-o-forma-severa-de-covid-19-nu-lasa-boala-sa-te-afecteze-si-tu-poti-lua-masuri-si-tu-poti-preveni-si-tu-poti-proteja/ (accessed on 7 April 2022).

33. Gillis, D.; Edwards, B.P. The utility of joinpoint regression for estimating population parameters given changes in population structure. *Heliyon* **2019**, *5*, e02515. [CrossRef]

34. Qiu, D.; Katanoda, K.; Marugame, T.; Sobue, T. A Joinpoint regression analysis of long-term trends in cancer mortality in Japan (1958–2004). *Int. J. Cancer* **2009**, *124*, 443–448. [CrossRef]

35. Crispo, A.; Barba, M.; Malvezzi, M.; Arpino, G.; Grimaldi, M.; Rosso, T.; Esposito, E.; Sergi, D.; Ciliberto, G.; Giordano, A.; et al. Cancer mortality trends between 1988 and 2009 in the metropolitan area of Naples and Caserta, Southern Italy: Results from a joinpoint regression analysis. *Cancer Biol. Ther.* **2013**, *14*, 1113–1122. [CrossRef]

36. Zahmatkesh, B.; Keramat, A.; Alavi, N.; Khosravi, A.; Kousha, A.; Motlagh, A.G.; Darman, M.; Partovipour, E.; Chaman, R. Breast cancer trend in Iran from 2000 to 2009 and prediction till 2020 using a trend analysis method. *Asian Pac. J. Cancer Prev.* **2016**, *17*, 1493–1498. [CrossRef]

37. Sarakarn, P.; Suwanrungruang, K.; Vatanasapt, P.; Wiangnon, S.; Promthet, S.; Jenwitheesuk, K.; Koonmee, S.; Tipsunthonsak, N.; Chen, S.L.S.; Yen, A.M.F.; et al. Joinpoint analysis trends in the incidence of colorectal cancer in Khon Kaen, Thailand (1989–2012). *Asian Pac. J. Cancer Prev. APJCP* **2017**, *18*, 1039.

38. Wilson, L.; Bhatnagar, P.; Townsend, N. Comparing trends in mortality from cardiovascular disease and cancer in the United Kingdom, 1983–2013: Joinpoint regression analysis. *Popul. Health Metr.* **2017**, *15*, 23. [CrossRef]

39. Dragomirescu, I.; Llorca, J.; Gómez-Acebo, I.; Dierssen-Sotos, T. A join point regression analysis of trends in mortality due to osteoporosis in Spain. *Sci. Rep.* **2019**, *9*, 4264. [CrossRef]

40. Atlatszo. 2019. Available online: https://english.atlatszo.hu/2019/01/28/pay-or-die-onco-tourism-and-corruption-in-romania-and-hungary/ (accessed on 7 April 2022).

41. Investigative Journalism for Europe (IJ4EU). "Cancer Tourism" in Central and Eastern Europe. 2018. Available online: http://www.investigativejournalismforeu.net/projects/cancer-tourism-in-central-and-eastern-europe/ (accessed on 7 April 2022).

42. Mistry, M.; Parkin, D.M.; Ahmad, A.S.; Sasieni, P. Cancer incidence in the United Kingdom: Projections to the year 2030. *Br. J. Cancer* **2011**, *105*, 1795–1803. [CrossRef]

43. Narita, M.F.; Yin, R. In Search of Information: Use of Google Trends' Data to Narrow Information Gaps for Low-Income Developing Countries; International Monetary Fund: 2018. Available online: https://www.elibrary.imf.org/view/journals/001/2018/286/article-A001-en.xml (accessed on 10 May 2022).

44. Medeiros, M.C.; Pires, H.F. The Proper Use of Google Trends in Forecasting Models. *arXiv* **2021**, arXiv:2104.03065.

45. Stephens-Davidowitz, S.; Varian, H. A Hands-On Guide to Google Data. 2014. Available online: https://people.ischool.berkeley.edu/~{}hal/Papers/2015/primer.pdf (accessed on 20 May 2022).

46. Breiman, L. Statistical modeling: The two cultures (with comments and a rejoinder by the author). *Stat. Sci.* **2001**, *16*, 199–231.

47. Charpentier, A.; Flachaire, E.; Ly, A. Econometrics and machine learning. *Econ. Stat.* **2018**, *505*, 147–169. [CrossRef]

48. Tudor, C.; Sova, R. Flexible decision support system for algorithmic trading: Empirical application on crude oil markets. *IEEE Access* **2022**, *10*, 9628–9644. [CrossRef]

49. Box, G.; Jenkins, G. *Time Series Analysis: Forecasting and Control*; Holden-Day: San Francisco, CA, USA, 1970.

50. De Livera, A.M.; Hyndman, R.J.; Snyder, R.D. Forecasting time series with complex seasonal patterns using exponential smoothing. *J. Am. Stat. Assoc.* **2011**, *106*, 1513–1527. [CrossRef]

51. Yu, C.; Xu, C.; Li, Y.; Yao, S.; Bai, Y.; Li, J.; Wang, L.; Wu, W.; Wang, Y. Time series analysis and forecasting of the hand-foot-mouth disease morbidity in China using an advanced exponential smoothing state space TBATS model. *Infect. Drug Resist.* **2021**, *14*, 2809. [CrossRef]

52. Pasini, A. Artificial neural networks for small dataset analysis. *J. Thorac. Dis.* **2015**, *7*, 953.

53. Munim, Z.H.; Shakil, M.H.; Alon, I. Next-day bitcoin price forecast. *J. Risk Financ. Manag.* **2019**, *12*, 103. [CrossRef]

54. Tudor, C.; Sova, R. Benchmarking GHG Emissions Forecasting Models for Global Climate Policy. *Electronics* **2021**, *10*, 3149. [CrossRef]

55. Hyndman, R.J.; Khandakar, Y. Automatic time series forecasting: The forecast package for R. *J. Stat. Softw.* **2008**, *27*, 1–22. [CrossRef]

56. Diebold, F.X.; Mariano, R.S. Comparing predictive accuracy. *J. Bus. Econ. Stat.* **1995**, *13*, 253–263.

57. Harvey, D.; Leybourne, S.; Newbold, P. Testing the equality of prediction mean squared errors. *Int. J. Forecast.* **1997**, *13*, 281–291. [CrossRef]

58. Mavragani, A.; Ochoa, G. Google Trends in infodemiology and infoveillance: Methodology framework. *JMIR Public Health Surveill.* **2019**, *5*, e13439. [CrossRef] [PubMed]

59. Dehkordy, S.F.; Carlos, R.C.; Hall, K.S.; Dalton, V.K. Novel data sources for women's health research: Mapping breast screening online information seeking through Google trends. *Acad. Radiol.* **2014**, *21*, 1172–1176.

60. Smittenaar, C.R.; Petersen, K.A.; Stewart, K.; Moitt, N. Cancer incidence and mortality projections in the UK until 2035. *Br. J. Cancer* **2016**, *115*, 1147–1155. [CrossRef] [PubMed]

61. Eichenauer, V.Z.; Indergand, R.; Martínez, I.Z.; Sax, C. Obtaining consistent time series from Google Trends. *Econ. Inq.* **2022**, *60*, 694–705. [CrossRef]
62. Smarandache, F. Introduction to Neutrosophic Statistics. Infinite Study. 2014. Available online: https://arxiv.org/pdf/1406.2000 (accessed on 24 May 2022).
63. Chen, J.; Ye, J.; Du, S.; Yong, R. Expressions of rock joint roughness coefficient using neutrosophic interval statistical numbers. *Symmetry* **2017**, *9*, 123. [CrossRef]
64. Aslam, M. A new sampling plan using neutrosophic process loss consideration. *Symmetry* **2018**, *10*, 132. [CrossRef]
65. Aslam, M. Design of sampling plan for exponential distribution under neutrosophic statistical interval method. *IEEE Access* **2018**, *6*, 64153–64158. [CrossRef]
66. Aslam, M.; Albassam, M. Application of neutrosophic logic to evaluate correlation between prostate cancer mortality and dietary fat assumption. *Symmetry* **2019**, *11*, 330. [CrossRef]

 *biology*

Article

Prediction of Major Histocompatibility Complex Binding with Bilateral and Variable Long Short Term Memory Networks

Limin Jiang [1], Jijun Tang [2], Fei Guo [3,*] and Yan Guo [1,*]

1 Comprehensive Cancer Center, Department of Internal Medicine, University of New Mexico, Albuquerque, NM 87131, USA; jianglm@tju.edu.cn
2 Shenzhen Institute of Advanced Technology, Chinese Academy of Sciences, Shenzhen 518055, China; jtang@cse.sc.edu
3 School of Computer Science and Technology, College of Intelligence and Computing, Tianjin University, Tianjin 300350, China
* Correspondence: fguo@tju.edu.cn (F.G.); yaguo@salud.unm.edu (Y.G.)

Simple Summary: Major histocompatibility complex molecules are of significant biological and clinical importance due to their utility in immunotherapy. The prediction of potential MHC binding peptides can estimate a T-cell immune response. The variable length of existing MHC binding peptides creates difficulty for MHC binding prediction algorithms. Thus, we utilized a bilateral and variable long-short term memory neural network to address this specific problem and developed a novel MHC binding prediction tool.

Abstract: As an important part of immune surveillance, major histocompatibility complex (MHC) is a set of proteins that recognize foreign molecules. Computational prediction methods for MHC binding peptides have been developed. However, existing methods share the limitation of fixed peptide sequence length, which necessitates the training of models by peptide length or prediction with a length reduction technique. Using a bidirectional long short-term memory neural network, we constructed BVMHC, an MHC class I and II binding prediction tool that is independent of peptide length. The performance of BVMHC was compared to seven MHC class I prediction tools and three MHC class II prediction tools using eight performance criteria independently. BVMHC attained the best performance in three of the eight criteria for MHC class I, and the best performance in four of the eight criteria for MHC class II, including accuracy and AUC. Furthermore, models for non-human species were also trained using the same strategy and made available for applications in mice, chimpanzees, macaques, and rats. BVMHC is composed of a series of peptide length independent MHC class I and II binding predictors. Models from this study have been implemented in an online web portal for easy access and use.

Keywords: major histocompatibility complex; bidirectional long short-term memory neural network; deep learning

Citation: Jiang, L.; Tang, J.; Guo, F.; Guo, Y. Prediction of Major Histocompatibility Complex Binding with Bilateral and Variable Long Short Term Memory Networks. *Biology* **2022**, *11*, 848. https://doi.org/10.3390/biology11060848

Academic Editors: Shibiao Wan, Yiping Fan, Chunjie Jiang and Shengli Li

Received: 12 April 2022
Accepted: 27 May 2022
Published: 1 June 2022

1. Introduction

Major Histocompatibility Complex (MHC) genes code for proteins that recognize foreign molecules and play an important part in immune surveillance. Due to variation in molecular structure, function, and distribution, MHC molecules are divided into three subsets: MHC class I, II, and III. A MHC class I molecule may constitute the MHC heavy chain (alpha chain), which encompasses three alpha domains (alpha1, alpha2, and alpha3) [1]. Alpha1 and alpha2 form the recognition region, with an interval deep groove capturing the peptide antigen [2]. Alpha3 is adjacent to the transmembrane domain in the heavy chain and it interacts with antigen transporters to load and express antigens. A specific type of MHC class I molecules are encoded by the β2-microglobulin gene, and in MHC they constitute the MHC light chain (beta chain). MHC class I molecules are located at the surface of cells to present antigens, which trigger immune responses by attracting cytotoxic lymphocytes immune cells (TC cells) including CD8+, the cytotoxic T cells which express CD8+ receptors. These receptors recognize related MHC complexes at the cell surface: when an antigen peptide of foreign origin is bound, CD8+ immune cells are activated to trigger programmed apoptosis [1]. An MHC class II molecule encodes two membrane-spanning chains that are of similar size. While MHC I molecules are located on the surface of nearly all nucleated cells, MHC II glycoproteins are expressed on the surface of specialized immune cells (such as B cells, macrophages, and dendritic cells), where they present processed antigenic peptides to TH cells. MHC class III genes encode various secreted proteins that have immune functions, including components of the complement system and molecules involved in inflammation [3].

Of the three MHC classes, class I has attracted great attention in medical research. For example, reduced abundance in MCH class I is associated with poor prognosis in Hodgkin lymphoma [4]. Another study [5] demonstrated that cancer cells escape T-cell responses via losing MHC class I molecules. MHC molecules are highly polymorphic proteins. As one MHC protein can have many variants, and such variants are commonly referred to as "MHC alleles" [6], MHC alleles are organized into multiple categories for each MHC class. For instance, MHC class I proteins in humans are encoded as human leukocyte antigen (HLA) groups A, B, C, etc. by the gene name, and each HLA group is composed of many alleles by the variants. From the view of molecular structure, MHC molecules have pockets, and the antigenic peptides have anchors of which some are determined residues, and anchors have special properties to lead peptides to enter the pockets [7]. An antigenic peptide's MHC binding affinity can be measured experimentally by a variety of assays, including a competitive binding assay [8].

The accumulated experimentally verified MHC binding peptides have been curated into various databases during the last three decades. Around 13 MHC binding databases are currently available [9]. With more than 900,000 entries, the Epitope Database (IEDB) [10] contains the largest collection of MHC binding peptides, followed by MHCBN [11] curating 25,860 peptides. In addition to the experimental methods, a peptide's binding potential with regard to a particular MHC molecule can be estimated through computational algorithms. Computational methods can systematically prioritize credible candidates for a more favorable study design, thus helping reduce both financial cost and human labor of the wet-lab assay-based validation experiments. The experimentally verified MHC binding peptide sequences offer an understructure for the development of computational approaches to predict the binding affinity between an MHC allele and a novel peptide. More than 30 MHC binding prediction tools were developed based on the accumulated MHC binding databases over the years. The majority of these tools [12–24] were developed for MHC class I and II binding prediction.

A common limitation of the existing MHC binding prediction tools is the necessity to align all peptides to one fixed length. Specifically, to meet the requirement, developers must either train different models to tackle peptides of different lengths, or they must arbitrarily adjust the original peptide. There are two sequence selection strategies in the model training/predicting process, one of which is to select peptides with a fixed length, such as selecting 9-mer peptides to train a model for class I [25–27]. The other is to adjust the peptides sequence to a specific length, such as adjusting the peptide length of class I to 9-mer/15 mer by inserting "X" symbols (elongating) or deleting amino acids (shortening) [28–31]. For the first strategy, there are two disadvantages: (1) It is tedious to train multiple models out of the initial single allele set; (2) When dividing the whole training set into multiple length-specific training sets, some models of certain lengths may have insufficient training data and therefore result in undertraining and suboptimal performance. For the second strategy, one obvious disadvantage is that inserting or deleting amino acids inevitably leads to a loss of information; specifically, the neighbor amino acids at a perturbed position will not be the same post the elongating/shortening operation. To overcome this constraint, we developed BVMHC, a novel MHC binding prediction tool based on Bidirectional Long Short-Term Memory (biLSTM) neural network [32,33], a type of recurrent neural network (RNN), which has the major novelty of offering variable length MHC binding prediction. BVMHC is designed to make predictions for both MHC class I and class II alleles in humans, and models for non-human species were also trained using the same strategy. The performance of BVMHC has been thoroughly compared with popular MHC class I/II binding prediction tools.

2. Materials and Methods

2.1. Training and Validation Datasets

To establish a sizeable training dataset, we obtained from the IEDB database 122,129 and 45,440 human binding peptide sequences for 48 MHC class I alleles and 27 MHC class II alleles, respectively. Additionally, 15,740 MHC class I peptide sequences of four non-human species (mouse, rat, macaque, and chimpanzee) and 1041 MHC class II peptide sequences of mouse were also extracted from the IEDB database. Each peptide was associated with a binding affinity measured as IC_{50} in nM. A dichotomization of these binding affinity values was conducted as follows: peptides with $IC_{50} \geq 500$ nM were considered as negative binding and peptides with $IC_{50} < 500$ nM were considered as positive binding. All binding affinity values (*aff*) were standardized to the interval [0,1] through a function, i.e., $1 - \log (aff) / \log(50,000)$. The initial sequences underwent the following three aspects of filtration: (1) For sequences that are repeated and have the same IC50 value, we kept only one instance of the sequences and removed all duplicate instances. (2) For sequences that are repeated and have different IC50 values, we deleted all items. (3) For sequences that are repeated and have different allele information, we kept all items because we would train different predictors for different allele sequences. Five-fold cross-validation procedures were used on the training datasets to train models. An independent validation dataset consisting of 320 class I and 131 class II human peptide sequences was constructed from the databases MHCBN [34] and SYFPEITHI [35], where we made sure that items co-existing in the IEDB were removed.

2.2. Feature Representation at Evolutionary Level

The BVMHC model involves two major components: feature representation and the computation model (Figure 1). For a numerical representation of training/testing data, each peptide sequence was first encoded as a $20 \times L$ matrix through one-hot encoding [36], where L is the length of the peptide. The dynamic convolutional neural network with twenty 1×20 convolution kernels was used to process one-hot coding matrices. BLOSUM is a 20×20 matrix that represents evolutionary conservation information between amino acids [37], and we used it to initialize the twenty convolution kernels. The overall method can be represented with Equation (1), where X denotes the One-hot encoding matrix, i the index of amino acid in peptide, k the index of kernel, $M = 1$ the window size, and $n = 20$ the number of kernels. Of note, the two indices, i and j, start from an initial value of 0.

$$\text{Evo}(X)_{i,k} = \sum\nolimits_{m=0}^{M-1} \sum\nolimits_{m=0}^{m=0} W_{m,n}^k X_{i+m,n} \tag{1}$$

As the kernels were updated in the training process, an updated presentment matrix in the evolutionary level was obtained and was input into a biLSTM model. After training, a novel BLOSUM matrix can be obtained by using the twenty trained convolution kernels.

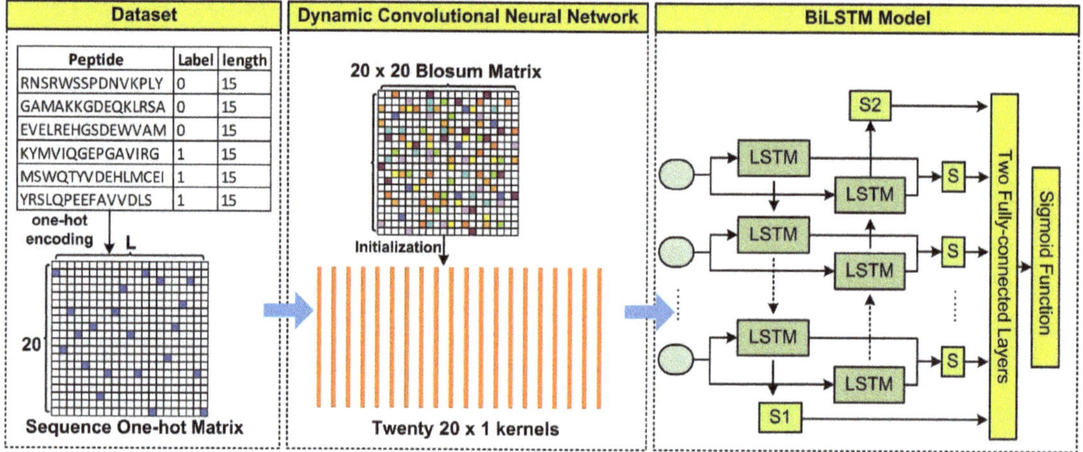

Figure 1. Overview of BVMHC. One-hot encoding was used to convert a peptide sequence to a matrix. BLOSUM was applied to initialize kernels in the convolutional neural network that was used to extract the peptide sequence feature at the evolutionary level. The biLSTMmodel was then applied to process the merged matrix at the sequential level.

2.3. Feature Representation at Sequential Level

The advantage of biLSTM (Figure 1) is the ability to handle peptides with variable lengths. Long short-term memory (LSTM) [33] is a type of recurrent neural network and all connections between units in LSTM form a directed cycle. This cycle is conducive to modeling dynamic temporal or spatial behavior. LSTM block is dynamically changed with the sequence length. An LSTM unit includes input, forget, and output gates. The calculation process is defined as Equations (2)–(6), where, x_t denotes the input vector, f_t the forget gate's activation vector, o_t the output gate's activation vector, h_t a 128-dimenstion hidden state vector, and C_t the cell state vector. In these equations, the common notations W and U refer to parameter matrices and b designates a bias vector.

$$f_t = \sigma\left(W_f x_t + U_f h_{t-1} + b_f\right) \tag{2}$$

$$i_t = \sigma(W_i x_t + U_i h_{t-1} + b_i) \tag{3}$$

$$o_t = \sigma(W_o x_t + U_o h_{t-1} + b_o) \tag{4}$$

$$C_t = i_t{}^\circ\tanh(W_c x_t + U_c h_{t-1} + b_c) + f_t{}^\circ C_{t-1} \tag{5}$$

$$h_t = o_t{}^\circ\tanh(C_t) \tag{6}$$

In our biLSTM model, one set of LSTMs merged the feature matrix from left to right, and another set of LSTMs merged the feature matrix from right to left. A dropout layer was applied to avoid over-fitting. A vector with 128 dimensions from biLSTM was obtained first. Afterward, a regression output value was obtained from two fully-connected layers and converted into a probability through the sigmoid function. In the process of training, we chose binary cross-entropy as the loss function and set the learning rate at 0.0001, and the dropout rate at 0.8.

2.4. Evaluation Criteria

Eight evaluation criteria, including Accuracy, Sensitivity, Specificity, F1, Matthew's correlation coefficient (MCC), Precision, Area Under the receiver-operating-characteristic Curve (AUC), and Area Under the Precision-Recall curve (AUPR), were used to evaluate the performance of the models. The calculation of the first six criteria is illustrated in

Equations (7)–(12), where *TP* represents the number of true positive MHC binders, false negative represents the number of true negative MHC binders, *FP* represents the number of false positive binders, and false negative represents the number of false negative MHC binders.

$$\text{Accuracy} = \frac{TP + TN}{TP + TN + FP + FN} \tag{7}$$

$$\text{Sensitivity} = \frac{TP}{TP + FN} \tag{8}$$

$$\text{Specificity} = \frac{TN}{TN + FP} \tag{9}$$

$$\text{F1} = \frac{2 \times (\text{Precision} \times \text{Sensitivity})}{(\text{Precision} + \text{Sensitivity})} \tag{10}$$

$$\text{MCC} = \frac{TP \times TN - FP \times FN}{\sqrt{(TP + FP) \times (TN + FN) \times (TP + FN) \times (TN + FP)}} \tag{11}$$

$$\text{Precision} = \frac{TP}{TP + FP} \tag{12}$$

3. Results

3.1. Human Dataset Description

The numbers of binding (positive examples) and non-binding (negative examples) peptides for MHC class I and II alleles making up the training and independent validation datasets are available in Supplementary Table S1. Overall, the human training dataset consisted of 75 alleles and entailed multiple (n) distinct peptide sequence lengths. For each of the 75 alleles, traditional approaches would have trained n length-dependent models to tackle different peptide lengths, or trained one fixed-length model which would necessitate a pre-procedure of length adjustment. Using the length-independent approach biLSTM, we trained 75 length-independent models and validated them with five-fold cross-validation. All 48 models for MHC class I binding and 12 of 27 models for MHC class II binding achieved over 0.8 accuracy and AUC values (Figure 2A,B). Overall, there exists a considerable difference in the performance levels between MHC Class I and Class II models, with the latter exceeding the former. Performances of MHC Class I models are generally acceptable except for a few outliers, such as HLA-B*15:02.

We identified a few models of extremity performances and went on to characterize the sequence motifs. Specifically, the performance values of HLA-DQB1*05:01 in Figure 2B and HLA-A*02:50 in Figure 2A are nearly one. By contrast, the MCC and Specificity associated with HLA-B*15:02 in Figure 2A are merely 0.25. We analyzed the difference between motifs of Binders and Non-Binders for HLA-A*02:50, HLA-B*15:02, and HLA-DQB1*05:01 (Figure 2C–E), respectively. Figure 2C,E describe the motifs for the well-performing models HLA-A*02:50 and HLA-DQB1*05:01, and we can see that the amino acid motifs are distinct between Binders and Non-Binders. Figure 2D describes the bad-performing model HLA-B*15:02, which shows non-differential motifs between Binders and Non-Binders. Therefore, the unsatisfactory prediction performance might be due to the weak distinction in motif patterns between positive and negative examples, which may hint at the contamination of binders by many false positives (non-binders). The good performance of BVMHC is attributed to the exploitation of the positional conservation and the preservation of intact peptide sequences. The detailed performance evaluation results by peptide length can be found in Table 1.

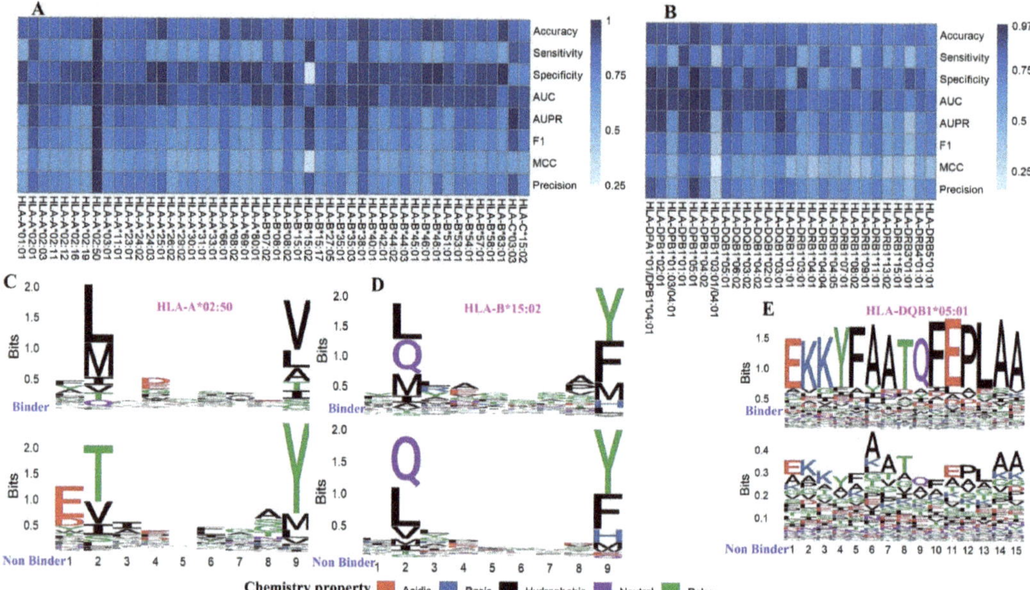

Figure 2. BVMHC performance on human datasets and binding motifs of a few extremity models. (**A,B**) The performance of BVMHC on the training dataset for predicting human MHC Class I (**A**) and II binders (**B**) in five-fold cross-validation. (**C**) The motifs of binders and non-binders for MHC Class I allele HLA-A*02:50. (**D**) The motifs of binders and non-binders for MHC Class I allele HLA-B*15:02. (**E**) The motifs of binders and non-binders for MHC Class II allele HLA-DQB1*05:01.

Table 1. Five-fold cross-validation results stratified by peptide length.

	Length	Accuracy	AUC	F1	MCC	Specificity	Sensitivity	Precision	AUPR	Positive [1]	Negative [2]
	8 mer	0.891	0.924	0.783	0.531	0.887	0.677	0.525	0.785	229	1879
	9 mer	0.883	0.915	0.745	0.650	0.902	0.735	0.760	0.800	23,000	72,963
Class I	10 mer	0.813	0.850	0.693	0.527	0.842	0.690	0.661	0.725	7263	14,024
	11 mer	0.879	0.905	0.768	0.608	0.881	0.756	0.651	0.755	310	1604
	Others	0.986	1.000	0.992	0.564	0.750	1.000	0.985	1.000	54	803
	13 mer	0.857	0.879	0.883	0.700	0.833	0.872	0.895	0.923	232	205
	14 mer	0.898	0.907	0.880	0.792	0.912	0.880	0.880	0.873	131	239
	15 mer	0.868	0.906	0.781	0.687	0.912	0.769	0.794	0.840	16,743	25,683
	16 mer	0.776	0.846	0.802	0.545	0.718	0.823	0.782	0.878	563	569
Class II	17 mer	0.680	0.673	0.429	0.312	0.933	0.300	0.750	0.643	106	257
	18 mer	0.643	0.939	0.706	0.452	1.000	0.545	1.000	0.986	71	40
	19 mer	0.875	0.938	0.857	0.775	1.000	0.750	1.000	0.950	55	75
	20 mer	0.750	0.900	0.500	0.488	1.000	0.333	1.000	0.886	65	66
	Other	0.690	0.640	0.381	0.183	0.758	0.444	0.333	0.566	81	259

[1] Number of positives; [2] Number of negatives

3.2. Independent Validation and Comparison with Other MHC Binding Predictors

An independent dataset extracted from MHCBN and SYFPEITHIwas used for validation and comparison with other MHC binding predictors. Seven popular MHC class I binding predictors (comblib_sidney2008 [21], ANN [19], SMM [17], NetMHCcons [16], NetMHCpan [18], PickPocket [20] and NetMHCpan EL [24]) for class I and three well-accepted MHC class II binding predictors (NETMHCIIPan [23], NN-align [15] and SMM-align [22]) were selected for the comparison. A common limitation of these existing tools is that the established model is bounded by a fixed peptide sequence length, which means that investigators have to distort the sequence structure when they take special actions (insertion or deletion) to ensure that the peptide length meets the model requirement. Moreover, in the above section, we have demonstrated that a model's prediction performance benefits from the positional conservation, the phenomenon of which is generally neglected in existing methods. The performance of BVMHC and the other MHC class I/II tools was measured with the eight aforementioned criteria (Table 2), and the complete results are displayed in Table S2. Models of HLA-DRB1*03:01 trained to predict MHC class II binding peptide achieved accuracy and AUC over 0.8 on the five-fold cross-validation; we downloaded the HLA-DRB1*03:01 peptide data from MHCBN. Of all eight evaluation indices, BVMHC achieved the best performance in three of the eight criteria for MHC class I prediction and the best performance in four criteria for MHC class II prediction. For example, BVMHC obtained the best overall AUC of 0.887 (Figure 3A), and the best average AUC for 9-mer models in MHC class I prediction (Figure 3B).

Table 2. The comparison results of the BVMHC model against seven other prediction tools on the independent validation dataset. The best performance value in each comparison track is highlighted in bold text.

	Methods	Accuracy	Sensitivity	Specificity	AUC	AUPR	F1	MCC	Precision	Positive [1]	Negative [2]
	BVMHC	0.597	0.371	**0.959**	**0.887**	0.866	0.531	**0.374**	**0.936**	197	123
	NetMHCcons [16]	**0.600**	**0.386**	0.943	0.865	0.890	**0.543**	0.365	0.916	197	123
	SMM [17]	0.584	0.350	**0.959**	0.859	**0.891**	0.509	0.357	0.932	197	123
Class I	NetMHCpan [18]	0.566	0.330	0.943	0.867	0.886	0.483	0.318	0.903	197	123
	ANN [19]	0.563	0.325	0.943	0.867	0.880	0.478	0.314	0.901	197	123
	PickPocket [20]	0.563	0.345	0.911	0.813	0.833	0.493	0.289	0.861	197	123
	NetMHCpan EL [24]	0.553	0.335	0.902	0.816	0.856	0.480	0.269	0.846	197	123
	comblib_sidney2008 [21]	NAN [§]	NAN [§]	NAN [§]	0.744	NAN [§]	NAN [§]	NAN [§]	NAN [§]	68	46
	BVMHC	**0.878**	**0.333**	0.965	0.718	0.417	**0.429**	**0.386**	0.600	18	113
Class II	NN-align [15]	0.863	0.278	0.956	**0.866**	**0.484**	0.357	0.303	0.500	18	113
	NETMHCIIPan [23]	0.870	0.111	**0.991**	0.795	0.423	0.190	0.235	**0.667**	18	113
	SMM-align [22]	0.840	0.000	0.973	0.787	0.319	NA [§]	−0.061	0.000	18	113

[1] Number of positives [2] Number of negatives [§] NA: the sum of Sensitivity and Precision is zero, thus F1 is NA. [§] NAN: the evaluation indices cannot be obtained because the original score threshold is not available. The value in bold are the best for each column.

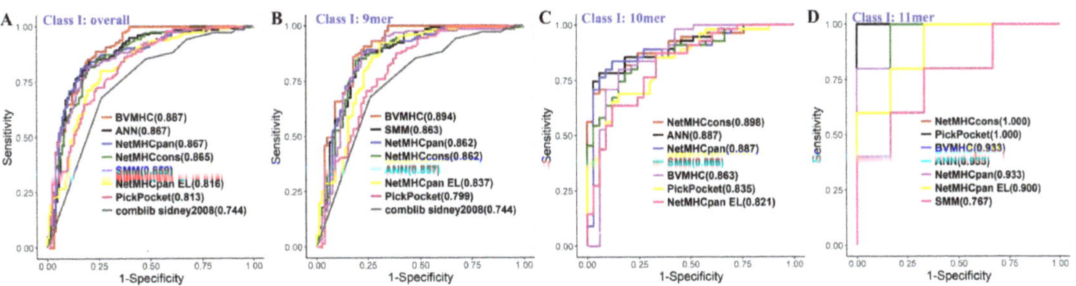

Figure 3. Receiver-Operating-Characteristic (ROC) curves of the eight tools for predicting MHC class I binders on the independent validation dataset. (**A–D**). BVMHC and seven existing prediction tools for overall (**A**), 9-mer (**B**), 10-mer (**C**), and 11-mer (**D**) MHC class I binders, respectively.

3.3. Performance of Non-Human Species

Using the same strategy as in humans, BVMHC models were also trained for MHC class I prediction for three mouse alleles, eight macaque alleles, five chimpanzee alleles, and one rat allele; in addition, two mouse MHC class II alleles were also covered. Results of five-fold cross-validation of these non-human MHC prediction models are available in Table 3. All 17 MHC class I models achieved greater than 0.8 accuracy and AUC. Both MHC class II models obtained greater than 0.80 accuracy. Due to the limitations of non-human data availability, independent validation was not performed.

Table 3. Performance evaluation results of BVMHC model on non-human species.

	Alleles	Accuracy	AUC	F1	MCC	Specificity	Sensitivity	Precision	AUPR
Class I	H-2-Db	0.829	0.855	0.573	0.466	0.897	0.564	0.583	0.602
	H-2-Dd	0.924	0.870	0.696	0.660	0.975	0.615	0.800	0.751
	H-2-Ld	0.814	0.852	0.698	0.564	0.875	0.682	0.714	0.779
	Mamu-A07	0.905	0.949	0.854	0.783	0.929	0.854	0.854	0.902
	Mamu-A11	0.822	0.899	0.726	0.595	0.880	0.707	0.747	0.805
	Mamu-A2201	0.908	0.957	0.854	0.789	0.955	0.814	0.897	0.943
	Mamu-B01	0.942	0.865	0.667	0.654	0.988	0.550	0.846	0.767
	Mamu-B03	0.857	0.921	0.769	0.666	0.903	0.758	0.781	0.843
	Mamu-B08	0.852	0.911	0.690	0.600	0.875	0.769	0.625	0.776
	Mamu-B17	0.822	0.882	0.717	0.592	0.838	0.782	0.662	0.710
	Mamu-B52	0.827	0.870	0.870	0.617	0.677	0.912	0.832	0.884
	Patr-A0101	0.816	0.838	0.619	0.520	0.935	0.520	0.765	0.688
	Patr-A0401	0.881	0.904	0.636	0.565	0.929	0.636	0.636	0.616
	Patr-A0701	0.825	0.820	0.545	0.438	0.901	0.522	0.571	0.682
	Patr-B0101	0.911	0.947	0.794	0.759	0.991	0.675	0.964	0.894
	Patr-B1301	0.875	0.917	0.903	0.727	0.824	0.903	0.903	0.951
	RT1A	0.893	0.923	0.400	0.352	0.923	0.500	0.333	0.667
Class II	H-2-IAb	0.826	0.797	0.489	0.394	0.925	0.423	0.579	0.627
	H-2-IAd	0.810	0.810	0.571	0.452	0.896	0.533	0.615	0.632

3.4. Web Server Implementation

A web server for the BVMHC models was developed using the combination of R, PHP, and Python, which is freely accessible at http://www.innovebioinfo.com/Proteomics/MHC/home.php. The website can conduct predictions for MHC class I and II binding peptides of multiple species. For MHC class I prediction, BVMHC covers 48 human alleles, three mouse alleles, eight macaque alleles, five chimpanzee alleles, and one rat allele; for MHC class II prediction, BVMHC covers 12 human alleles and two mouse alleles.

4. Discussion

MHC binding prediction is a crucial step toward identifying potential novel therapeutic strategies. For example, MHC class I molecules were found to be tumor suppressor genes [38] and can served as targets for immunotherapy [39]. Similar to MHC class I, the class II antigens can also serve as targets in cancer immunotherapy [40]. The prediction of MHC binding peptides is biologically and clinically important because it predicts the binding affinity of a T-cell immune response. Factors such as the polymorphic nature of MHC molecules, the variable length of peptides, etc. make it difficult to accurately predict MHC binding. However, advances in machine learning, especially those based on neural networks, have propelled substantial advancement in MHC binding prediction research. In this study, we proposed an approach using the Bilateral and Variable Long-Short Term Memory Networks to tackle the variable length issue in MHC binding prediction. By thoroughly comparing to other fixed-length-constrained MHC binding prediction tools, we show that BVMHC has the advantage in several performance measurements. However, In this paper, we just use the peptide sequences information to construct predictors. Inspired by NetMHCpan [18] and NetMHCIIpan [23], in the future we will incorporate the MHC protein sequence information to augment the feature representation of binders. As AlphaFold [41] becomes the focus of research about protein structure, we look to discern the differences between different MHC allele proteins at the protein structure level, which may hold promises for an even improved prediction of MHC protein binders. Additionally, a BVMHC predictor can be used to quickly screen potential binders—an effective strategy is to dissect a complete protein sequence into equal-sized segments and run the predictor over these segments across the whole span of the protein sequence. Considering the computational time complexity, such screening workflows must be optimized to reduce the running time to the minimum.

5. Conclusions

BVMHC is an MHC binding prediction tool that supports five species (human, chimpanzee, macaque, mouse, and rat). Compared to existing MHC prediction tools, BVMHC can use peptides of variable lengths to train a predictor, which allows for the reservation of the innate primary structure of the sequence. The combination of analyses at the conservatory level and the sequential level is vital for the superior performance of the resultant BVMHC model. In independent validation and comparison, BVMHC showed the best overall performance compared to seven other popular MHC class I predictors and three well-accepted MHC class II predictors. BVMHC was developed into a web server and can be accessed freely online.

Supplementary Materials: The following supporting information can be downloaded at: https://www.mdpi.com/article/10.3390/biology11060848/s1, Table S1: Summary of datasets; Table S2: The detailed results based on independent dataset.

Author Contributions: Conceptualization: Y.G., J.T. and F.G.; Analysis: L.J.; Manuscript writing: L.J. and Y.G.; Funding acquisition: Y.G., J.T. and F.G. All authors have read and agreed to the published version of the manuscript.

Funding: YG was supported by Cancer Center Support Grants from the National Cancer Institute (P30CA118100). JT was supported by The National Natural Science Foundation of China (NSFC 61772362, 61972280) and Shenzhen KQTD Project [KQTD20200820113106007].

Institutional Review Board Statement: Not applicable.

Informed Consent Statement: Not applicable.

Data Availability Statement: All data used in this study were obtained from public repositories.

Conflicts of Interest: The authors declare that they have no conflict of interest.

References

1. Bjorkman, P.J.; Saper, M.A.; Samraoui, B.; Bennett, W.S.; Strominger, J.L.; Wiley, D.C. Structure of the human class I histocompatibility antigen, HLA-A2. *Nature* **1987**, *329*, 506–512. [CrossRef] [PubMed]
2. Tanaka, K.; Kasahara, M. The MHC class I ligand-generating system: Roles of immunoproteasomes and the interferon-gamma-inducible proteasome activator PA28. *Immunol. Rev.* **1998**, *163*, 161–176. [CrossRef] [PubMed]
3. Schott, G.; Garcia-Blanco, M.A. MHC Class III RNA Binding Proteins and Immunity. *RNA Biol.* **2021**, *18*, 640–646. [CrossRef] [PubMed]
4. Roemer, M.G.M.; Advani, R.H.; Redd, R.A.; Pinkus, G.S.; Natkunam, Y.; Ligon, A.H.; Connelly, C.F.; Pak, C.J.; Carey, C.D.; Daadi, S.E.; et al. Classical Hodgkin Lymphoma with Reduced beta M-2/MHC Class I Expression Is Associated with Inferior Outcome Independent of 9p24.1 Status. *Cancer Immunol. Res.* **2016**, *4*, 910–916. [CrossRef] [PubMed]
5. Garrido, F.; Aptsiauri, N. Cancer immune escape: MHC expression in primary tumours versus metastases. *Immunology* **2019**, *158*, 255–266. [CrossRef] [PubMed]
6. Falk, K.; Rotzschke, O.; Stevanovie, S.; Jung, G.; Rammensee, H.-G. Allele-specific motifs revealed by sequencing of self-peptides eluted from MHC molecules. *Nature* **1991**, *351*, 290–296. [CrossRef]
7. Hobohm, U.; Meyerhans, A. A pattern search method for putative anchor residues in T cell epitopes. *Eur. J. Immunol.* **1993**, *23*, 1271–1276. [CrossRef]
8. Kessler, J.H.; Benckhuijsen, W.E.; Mutis, T.; Melief, C.J.; van der Burg, S.H.; Drijfhout, J.W. Competition-based cellular peptide binding assay for HLA class I. *Curr. Protoc. Immunol.* **2004**, *18*. [CrossRef]
9. Jiang, L.; Yu, H.; Li, J.; Tang, J.; Guo, Y.; Guo, F. Predicting MHC class I binder: Existing approaches and a novel recurrent neural network solution. *Brief Bioinform.* **2021**, *22*, bbab216. [CrossRef]
10. Vita, R.; Mahajan, S.; Overton, J.A.; Dhanda, S.K.; Martini, S.; Cantrell, J.R.; Wheeler, D.K.; Sette, A.; Peters, B. The Immune Epitope Database (IEDB): 2018 update. *Nucleic Acids Res.* **2019**, *47*, D339–D343. [CrossRef]
11. Bhasin, M.; Singh, H.; Raghava, G.P. MHCBN: A comprehensive database of MHC binding and non-binding peptides. *Bioinformatics* **2003**, *19*, 665–666. [CrossRef] [PubMed]
12. Hu, Y.; Wang, Z.; Hu, H.; Wan, F.; Chen, L.; Xiong, Y.; Wang, X.; Zhao, D.; Huang, W.; Zeng, J. ACME: Pan-specific peptide-MHC class I binding prediction through attention-based deep neural networks. *Bioinformatics* **2019**, *35*, 4946–4954. [CrossRef] [PubMed]
13. Zeng, H.; Gifford, D.K. DeepLigand: Accurate prediction of MHC class I ligands using peptide embedding. *Bioinformatics* **2019**, *35*, i278–i283. [CrossRef] [PubMed]

14. Wilson, E.A.; Krishna, S.; Anderson, K.S. A Random Forest based approach to MHC class I epitope prediction and analysis. *J. Immunol.* **2018**, *200*.

15. Jensen, K.K.; Andreatta, M. Improved methods for predicting peptide binding affinity to MHC class II molecules. *Immunology* **2018**, *154*, 394–406. [CrossRef] [PubMed]

16. Karosiene, E.; Lundegaard, C.; Lund, O.; Nielsen, M. NetMHCcons: A consensus method for the major histocompatibility complex class I predictions. *Immunogenetics* **2012**, *64*, 177–186. [CrossRef]

17. Peters, B.; Sette, A. Generating quantitative models describing the sequence specificity of biological processes with the stabilized matrix method. *BMC Bioinform.* **2005**, *6*, 132. [CrossRef]

18. Hoof, I.; Peters, B.; Sidney, J.; Pedersen, L.E.; Sette, A.; Lund, O.; Buus, S.; Nielsen, M. NetMHCpan, a method for MHC class I binding prediction beyond humans. *Immunogenetics* **2009**, *61*, 1. [CrossRef]

19. Lundegaard, C.; Lund, O.; Nielsen, M. Accurate approximation method for prediction of class I MHC affinities for peptides of length 8, 10 and 11 using prediction tools trained on 9mers. *Bioinformatics* **2008**, *24*, 1397–1398. [CrossRef]

20. Zhang, H.; Lund, O.; Nielsen, M. The PickPocket method for predicting binding specificities for receptors based on receptor pocket similarities: Application to MHC-peptide binding. *Bioinformatics* **2009**, *25*, 1293–1299. [CrossRef]

21. Sidney, J.; Assarsson, E.; Moore, C.; Ngo, S.; Pinilla, C.; Sette, A.; Peters, B. Quantitative peptide binding motifs for 19 human and mouse MHC class I molecules derived using positional scanning combinatorial peptide libraries. *Immunome Res.* **2008**, *4*, 2. [CrossRef] [PubMed]

22. Nielsen, M.; Lundegaard, C.; Lund, O. Prediction of MHC class II binding affinity using SMM-align, a novel stabilization matrix alignment method. *BMC Bioinform.* **2007**, *8*, 238. [CrossRef] [PubMed]

23. Andreatta, M.; Karosiene, E.; Rasmussen, M.; Stryhn, A.; Buus, S.; Nielsen, M. Accurate pan-specific prediction of peptide-MHC class II binding affinity with improved binding core identification. *Immunogenetics* **2015**, *67*, 641–650. [CrossRef] [PubMed]

24. Vanessa, J.; Sinu, P.; Massimo, A.; Paolo, M.; Bjoern, P.; Morten, N. NetMHCpan-4.0: Improved Peptide-MHC Class I Interaction Predictions Integrating Eluted Ligand and Peptide Binding Affinity Data. *J. Immunol.* **2017**, *199*, 3360–3368. [CrossRef]

25. Singh, H.; Raghava, G.P. ProPred1: Prediction of promiscuous MHC Class-I binding sites. *Bioinformatics* **2003**, *19*, 1009–1014. [CrossRef]

26. Dönnes, P.; Elofsson, A. Prediction of MHC class I binding peptides, using SVMHC. *BMC Bioinform.* **2002**, *3*, 25. [CrossRef]

27. Rasmussen, M.; Fenoy, E. Pan-Specific Prediction of Peptide-MHC Class I Complex Stability, a Correlate of T Cell Immunogenicity. *J. Immunol.* **2016**, *197*, 1517–1524. [CrossRef]

28. Liu, G.; Li, D.; Li, Z.; Qiu, S.; Li, W.; Chao, C.C.; Yang, N.; Li, H.; Cheng, Z.; Song, X.; et al. PSSMHCpan: A novel PSSM-based software for predicting class I peptide-HLA binding affinity. *GigaScience* **2017**, *6*, gix017. [CrossRef]

29. Nielsen, M.; Andreatta, M. NetMHCpan-3.0; improved prediction of binding to MHC class I molecules integrating information from multiple receptor and peptide length datasets. *Genome Med.* **2016**, *8*, 33. [CrossRef]

30. O'Donnell, T.J.; Rubinsteyn, A.; Bonsack, M.; Riemer, A.B.; Laserson, U.; Hammerbacher, J. MHCflurry: Open-Source Class I MHC Binding Affinity Prediction. *Cell Systems* **2018**, *7*, 129–132.e4. [CrossRef]

31. Zhao, T.; Cheng, L.; Zang, T.; Hu, Y. Peptide-Major Histocompatibility Complex Class I Binding Prediction Based on Deep Learning With Novel Feature. *Front. Genet.* **2019**, *10*, 1191. [CrossRef] [PubMed]

32. Graves, A.; Schmidhuber, J. Framewise phoneme classification with bidirectional LSTM and other neural network architectures. *Neural Netw.* **2005**, *18*, 602–610. [CrossRef] [PubMed]

33. Hochreiter, S.; Schmidhuber, J. Long Short-Term Memory. *Neural Comput.* **1997**, *9*, 1735–1780. [CrossRef]

34. Lata, S.; Bhasin, M.; Raghava, G.P.S. MHCBN 4.0: A database of MHC/TAP binding peptides and T-cell epitopes. *BMC Res. Notes* **2009**, *2*, 61. [CrossRef] [PubMed]

35. Rammensee, H.G.; Bachmann, J.; Emmerich, N.P.N.; Bachor, O.A.; Stevanović, S. SYFPEITHI: Database for MHC ligands and peptide motifs. *Immunogenetics* **1999**, *50*, 213–219. [CrossRef] [PubMed]

36. Davis, M.J. Contrast Coding in Multiple Regression Analysis: Strengths, Weaknesses, and Utility of Popular Coding Structures. *J. Data Sci.* **2010**, *8*, 61–73. [CrossRef]

37. Henikoff, S.; Henikoff, J.G. Amino acid substitution matrices from protein blocks. *Proc. Natl. Acad. Sci. USA* **1992**, *89*, 10915–10919. [CrossRef]

38. Garrido, C.; Paco, L.; Romero, I.; Berruguilla, E.; Stefansky, J.; Collado, A.; Algarra, I.; Garrido, F.; Garcia-Lora, A.M. MHC class I molecules act as tumor suppressor genes regulating the cell cycle gene expression, invasion and intrinsic tumorigenicity of melanoma cells. *Carcinogenesis* **2012**, *33*, 687–693. [CrossRef]

39. Cornel, A.M.; Mimpen, I.L.; Nierkens, S. MHC Class I Downregulation in Cancer: Underlying Mechanisms and Potential Targets for Cancer Immunotherapy. *Cancers* **2020**, *12*, 1760. [CrossRef]

40. Sun, Z.; Chen, F.; Meng, F.; Wei, J.; Liu, B. MHC class II restricted neoantigen: A promising target in tumor immunotherapy. *Cancer Lett.* **2017**, *392*, 17–25. [CrossRef]

41. Jumper, J.; Evans, R.; Pritzel, A.; Green, T.; Figurnov, M.; Ronneberger, O.; Tunyasuvunakool, K.; Bates, R.; Žídek, A.; Potapenko, A.; et al. Highly accurate protein structure prediction with AlphaFold. *Nature* **2021**, *596*, 583–589. [CrossRef] [PubMed]

biology

Article

Identification of DPP4/CTNNB1/MET as a Theranostic Signature of Thyroid Cancer and Evaluation of the Therapeutic Potential of Sitagliptin

Sheng-Yao Cheng [1,†], Alexander T. H. Wu [2,3,4,5,†], Gaber El-Saber Batiha [6], Ching-Liang Ho [7], Jih-Chin Lee [1], Halimat Yusuf Lukman [8], Mohammed Alorabi [9], Abdullah N. AlRasheedi [10] and Jia-Hong Chen [7,*]

[1] Department of Otolaryngology-Head and Neck Surgery, Tri-Service General Hospital, National Defense Medical Center, 325, Section 2, Chenggong Road, Taipei 114, Taiwan; gjcheng5032@gmail.com (S.-Y.C.); doc30450@gmail.com (J.-C.L.)
[2] TMU Research Center of Cancer Translational Medicine, Taipei Medical University, Taipei 110, Taiwan; chaw1211@tmu.edu.tw
[3] The PhD Program of Translational Medicine, College of Science and Technology, Taipei Medical University, Taipei 110, Taiwan
[4] Clinical Research Center, Taipei Medical University Hospital, Taipei Medical University, Taipei 110, Taiwan
[5] Graduate Institute of Medical Sciences, National Defense Medical Center, Taipei 110, Taiwan
[6] Department of Pharmacology and Therapeutics, Faculty of Veterinary Medicine, Damanhour University, Damanhour 22511, Egypt; gaberbatiha@gmail.com
[7] Division of Hematology/Oncology, Department of Medicine, Tri-Service General Hospital, National Defense Medical Center, Taipei 11490, Taiwan; charileho22623@gmail.com
[8] Department of Chemical Sciences, Biochemistry Unit, College of Natural and Applied Sciences, Summit University Offa, Offa PMB 4412, Nigeria; halimatyusuf40@summituniversity.edu.ng
[9] Department of Biotechnology, College of Sciences, Taif University, Taif P.O. Box 11099, Saudi Arabia; maorabi@tu.edu.sa
[10] Otolaryngology-Head & Neck Surgery Department, College of Medicine, Jouf University, Sakaka P.O. Box 2014, Saudi Arabia; analrashedi@ju.edu.sa
* Correspondence: ndmc_tw.tw@yahoo.com.tw
† These authors contributed equally to this work.

Citation: Cheng, S.-Y.; Wu, A.T.H.; Batiha, G.E.-S.; Ho, C.-L.; Lee, J.-C.; Lukman, H.Y.; Alorabi, M.; AlRasheedi, A.N.; Chen, J.-H. Identification of DPP4/CTNNB1/MET as a Theranostic Signature of Thyroid Cancer and Evaluation of the Therapeutic Potential of Sitagliptin. *Biology* **2022**, *11*, 324. https://doi.org/10.3390/biology11020324

Academic Editors: Shibiao Wan, Yiping Fan, Chunjie Jiang and Shengli Li

Received: 11 December 2021
Accepted: 14 February 2022
Published: 17 February 2022

Publisher's Note: MDPI stays neutral with regard to jurisdictional claims in published maps and institutional affiliations.

Simple Summary: In recent years, the incidence of thyroid cancer has been increasing globally, with papillary thyroid cancer (PTCa) being the most prevalent pathological type. Although PTCa has been regarded to be slow growing and has a good prognosis, in some cases, PTCa can be aggressive and progress despite surgery and radioactive iodine treatment. Therefore, searching for new targets and therapies is required. We utilized bioinformatics analyses to identify critical theranostic markers for PTCa. We found that DPP4/CTNNB1/MET is an oncogenic signature that is overexpressed in PTCa and associated with disease progression, distant metastasis, treatment resistance, immuno-evasive phenotypes, and poor clinical outcomes. Interestingly, our in silico molecular docking results revealed that sitagliptin, an antidiabetic drug, has strong affinities and potential for targeting DPP4/CTNNB1/MET signatures, even higher than standard inhibitors of these genes. Collectively, our findings suggest that sitagliptin could be repurposed for treating PTCa.

Abstract: In recent years, the incidence of thyroid cancer has been increasing globally, with papillary thyroid cancer (PTCa) being the most prevalent pathological type, accounting for approximately 80% of all cases. Although PTCa has been regarded to be slow growing and has a good prognosis, in some cases, PTCa can be aggressive and progress despite surgery and radioactive iodine treatment. In addition, most cancer treatment drugs have been shown to be cytotoxic and nonspecific to cancer cells, as they also affect normal cells and consequently cause harm to the body. Therefore, searching for new targets and therapies is required. Herein, we explored a bioinformatics analysis to identify important theranostic markers for THCA. Interestingly, we identified that the *DPP4/CTNNB1/MET* gene signature was overexpressed in PTCa, which, according to our analysis, is associated with immuno-invasive phenotypes, cancer progression, metastasis, resistance, and unfavorable clinical outcomes of thyroid cancer cohorts. Since most cancer drugs were shown to exhibit cytotoxicity and to be nonspecific, herein, we evaluated the anticancer effects of the antidiabetic drug sitagliptin, which was recently shown to possess anticancer activities, and is well tolerated and effective. Interestingly, our in silico molecular docking results exhibited putative binding affinities of sitagliptin with *DPP4/CTNNB1/MET* signatures, even higher than standard inhibitors of these genes. This suggests that sitagliptin is a potential THCA therapeutic, worthy of further investigation both in vitro and in vivo and in clinical settings.

Keywords: sitagliptin; thyroid cancer (THCA); papillary thyroid cancer (PTCa); thyroidectomy; metastasis; drug resistance

---

1. Introduction

Thyroid cancer (THCA) is the most prevalent malignancy of the endocrine system, and the 9th most common cancer in the world [1,2], accounting for approximately 600,000 newly diagnosed cases annually on a global scale [3], with high rates of morbidity reported in recent years [4]. THCA is divided into various subtypes, including anaplastic thyroid cancer (ATC), papillary thyroid carcinoma (PTCa), and follicular thyroid carcinoma (FTC), with PTCa being the most prevalent, as it accounts for approximately 85% of THCA [5,6]. PTC and FTC are well-differentiated thyroid cancers with an optimal prognosis of about 10 years disease-specific survival [7]. However, the ATC is poorly differentiated with proliferative stem-cell-like properties, resistance to therapies, and accounts for the majority of thyroid-cancer-related deaths [8,9]. The rapid increase in thyroid cancer, particularly PTCa, has been accredited to the availability and sensitive use of ultrasonography and other diagnostic imaging modalities [10,11], which have likely led to a massive detection and diagnosis of a large reservoir of subclinical, indolent lesions of the thyroid [12,13]. Studies have also implicated obesity, hormonal imbalance, metabolic syndromes, and environmental pollutants in the development of PTCa [14].

Patients with PTCa usually show good clinical outcomes compared with other cancers; however, there is also a very high rate of relapse post-treatment, leading to distant metastasis [15,16]. About 11% of patients with PTC present with distant metastases outside the neck and mediastinum [17]. Moreover, long-term survival outcomes for aggressive PTC subgroups exhibit heterogeneous clinical behavior and a wide range of mortality risks, suggesting that treatment should be tailored to specific histologic subtypes [18]. The diagnostic criteria for PTC allow it to demonstrate various histological features and growth patterns; different variants of PTCa are recognized, including classic, microcarcinoma, encapsulated, follicular, diffuse sclerosing, tall cell, columnar cell, cribriform-morular, hobnail, solid, oncocytic, spindle cell, clear cell, and Warthin-like variants [19]. However, among these variants, tall cell, columnar cells, and hobnail variants are of undoubted clinical significance, since they are aggressive variants associated with aggressive clinicopathological features and worse prognosis than for classic and encapsulated PTC [20–22].

Surgery, endocrine therapy, and radioiodine therapy are well-known therapy regimens for PTCa, offering a good prognosis; however, the aggressive variants of PTCa progress despite surgery and radioactive iodine treatment [23]. In addition, tumor recurrence in PTCa is associated with therapeutic resistance which increases the death toll in patients [24–26]. Unfortunately, an upsurge in the incidence of aggressive PTCs was observed at a rate higher than that seen in well-differentiated PTCs or anaplastic thyroid carcinomas (ATCs) in the past two decades in a study of a large cohort of thyroid cancers [22]; therefore, there is an urgent need to identify novel diagnostic and prognostic molecular biomarkers that could also be used as molecular targets for the development of new drugs or in repurposing existing drugs for the treatment of PTCa.

Increasing evidence shows that dipeptidyl aminopeptidase IV (*DPP IV*) is associated with cancer development and progression [27,28]; DPP4 is an adenosine deaminase complex protein, and was demonstrated to be upregulated in THCA, particularly in PTCa, and is associated with tumor aggression and poor prognoses [29–31]. Moreover, high expression of *DPP4* was shown to promote distance metastasis and stemness in esophageal adenocarcinoma and colorectal cancer [32,33]. However, the prognostic role of *DPP4* expression and its role in THCA metastasis remains elusive [7,29,31]. Studies have shown that DPP4 and b-catenin crosstalk to regulate critical cellular processes, including motility and invasion [34]. A study involving lung cancer patients has revealed that the expression levels of β-catenin correlate with DPP4 expression [35] and contributed to tumor metastasis [34,36]. An experimental study has also reported that activating mutation of *Ctnnb1* induced DPP4 overexpression in epidermal keratinocytes of LRIG1⁺ stem cells [37]. Research has illuminated that inhibitors of DPP4 exert their therapeutic effect via modulation of the Wnt/β-catenin signaling pathway [38]. Sitagliptin, an inhibitor of DPP4, has also been reported to provide renal protection via inhibition of the tubulointerstitial Wnt/β-catenin signaling pathway in diabetic nephropathy [39].

Accumulating studies demonstrated a pivotal correlation between distant metastasis in PTCa and *MET* (MET proto-oncogenic receptor tyrosine kinase) [40]. Approximately 70% of PTCas were reported to overexpress the *MET* gene, and it is associated with poor prognoses [41]. In addition, Rossana et al. also demonstrated that higher expression levels of *MET* in PTCa promoted cancer growth and distance metastasis [42,43]. *MET* is a transmembrane tyrosine kinase identified as a high-affinity receptor for hepatocyte growth factor (HGF), and both *MET* and *HGF* were demonstrated to be expressed in PTCa [42], and consequently promote progression and secondary metastasis [44]. Additionally, *MET* was shown to activate β-catenin (*CTNNB1*), an important component of the canonical Wnt pathway [45,46]. *CTNNB1* was recently reported to be mutated in PTCa, and to ultimately promote cancer development and stemness [47,48]. Moreover, upregulated *MET* was also demonstrated to regulate the expression of mitogen-activated protein kinase (*MAPK*), phosphatidylinositol 3-kinase (*PI3K*)/*AKT*, signal transducer and activator of transcription 3 (*STAT3*), and nuclear factor (*NF*)-κB pathways in THCA [40,49]. This suggests that *MET* is a crucial target gene in THCA, and worthy of further investigation. To date, most drugs used for cancer treatment are cytotoxic and usually not specific to cancer cells, but also affect normal cells; therefore, there is still a huge gap in finding more sensitive and specific drugs for cancer. Recent studies suggested an association between cancer occurrence and antidiabetic medicaments. Sitagliptin is a standard inhibitor of *DPP4*, widely used for treating diabetes, and was shown to possess anticancer activities, as well as being efficacious and well tolerated [50]. In the present study, we predicted the potential anticancer activities of sitagliptin as a target for *DPP4/CTNNB1/MET* oncogenic signatures, which are overexpressed in THCA.

2. Materials and Methods

2.1. Microarray Data Acquisition and Identification of Differentially Expressed Genes (DEGs)

Gene expressions of four THCA datasets (GEO3467, GEO3678, GEO6004, and GEO33630) were extracted from the NCBI gene expression omnibus. The acquired datasets were further analyzed using GEO2R (https://www.ncbi.nlm.nih.gov/geo/geo2r/ accessed on 5 September 2021), and results contained DEG profiles from THCA patients compared to normal samples. To control the false discovery rate (FDR), the Benjamini–Hochberg adjustment was applied to *p* values (adjusted (adj.) *p* values), to moderate the balance between detection of significant genes and possible false-positive values. The fold-change (FC) threshold was set to 1.5, and adj. $p < 0.05$ was considered statistically significant. Venn diagrams were constructed using the Bioinformatics and Evolutionary Genomics (BEG) online tool (http://bioinformatics.psb.ugent.be/webtools/Venn/ accessed on 6 September 2021).

2.2. Differential Expression of the THCA Gene Hub

Differential expressions of THCA gene profiles between tumor tissues and normal adjacent tissues of the Cancer Genome Atlas (TCGA) database were analyzed using UALCAN (http://ualcan.path.uab.edu accessed on 12 September 2021), an online web portal used to identify gene expression levels between primary tumors compared to normal tissue samples [51]. Moreover, we explored the cBioPortal online web tool (https://www.cbioportal.org accessed on 19 September 2021), which categorizes gene alterations based on percentages of individual genes due to amplification [52]. For further analysis, we used the cBioPortal correlation sub-tool to determine gene expression correlations with positive Spearman and Pearson correlation coefficients with $p < 0.05$ as statistically significant.

2.3. Comparisons of DPP4/CTNNB1/MET Expressions in Normal, Primary, and Metastatic Tumor of Thyroid Cancer Cohorts

To compare expression levels of the *DPP4/CTNNB1/MET* oncogenes among normal, tumor, and metastatic tissues, we explored the tumor, normal, and metastatic plot (TNMplot), (https://tnmplot.com/analysis/ accessed on 21 September 2021), an RNA-sequence-based rapid analysis, which is used to compare data of selected genes [53]. Data were compared using the Kruskal–Wallis test, which is a method used to test samples originally from the same distribution of specimens, followed by Dunn's test, which assesses the significance of gene expressions in promoting THCA tumor metastasis, with $p < 0.05$ considered statistically significant.

2.4. Interaction Network and Gene Enrichment Analysis

An interaction network analysis was constructed using the Search Tool for the Retrieval of Interacting Genes/Proteins (STRING, https://string-db.org/ accessed on 25 September 2021) database [54], and GeneMANIA [55] (http://genemania.org/data accessed on 28 September 2021), which are online web tools developed to analyze interaction networks. The STRING database was used under a high confidence of 0.700, and protein enrichment of $p < 6.0 \times 10^{-03}$ was obtained. Interactions among genes were analyzed according to correlations based on experimental data (pink), gene neighborhoods (green), gene fusion (red), gene co-occurrences (blue), and gene co-expression (black). Moreover, we explored the Network Analyst user-friendly online tool (https://www.networkanalyst.ca/ accessed on 5 October 2021) to analyze co-expressed gene enrichment from the biological processes databases; herein we applied the Igraph R package visualization tool for analysis [56]. Furthermore, gene ontology (GO), biological processes (BPs), and Kyoto Encyclopedia of Genes and Genomes (KEGG) enrichment analyses were analyzed using FunRich software (http://www.funrich.org accessed on 9 October 2021), an open access, stand-alone functional enrichment and network analytical tool [57].

2.5. Analysis of Genomic Alterations and Mutations of the DPP4/CTNNB1/MET Oncogenes in THCA

Mutations of *DPP4/CTNNB1/MET* oncogenic expressions in THCA were analyzed using cBioPortal software. Herein, we analyzed altered frequencies of these oncogenes in THCA. Furthermore, we explored the muTarget platform (https://www.mutarget.com/ accessed on 11 October 2021), a platform linking changes in gene expressions and the mutation status of solid tumors, based on a genotype analysis, to determine associations between *DPP4/CTNNB1/MET* and alterations in gene expressions in THCA. Differences in expressions between the mutant group and wild-type (WT) group were considered statistically significant at $p < 0.05$.

2.6. Correlations of DPP4/CTNNB1/MET Expressions and Tumor Infiltration Levels of Immune and Immunosuppressive Cells in THCA

The Tumor Immune Estimation Resource (TIMER) (https://cistrome.shinyapps.io/timer/ accessed on 18 October 2021) is an online computational tool used to analyze the nature of tumor immune interactions across different cancer types [58]. Herein, we determined correlations of *DPP4/CTNNB1/MET* expressions and tumor infiltration levels of tumor associated macrophages (M2 TAM), regulatory T cell (Treg), cancer-associated fibroblast (CAF), and cluster of differentiation 8-positive (CD8$^+$ T cell), using a set of gene markers of immune infiltration model, as described previously [59,60]. The strength of correlations between the genes and immune cells is reflected by the purity-adjusted partial Spearman's rho value, where a value of r $\geq$ 1 means a perfect positive correlation and a value of r $\leq$ -1 means a perfect negative correlation, with $p < 0.05$ considered statistically significant.

2.7. In Silico Molecular Docking of the DPP4/CTNNB1/MET Oncogenes with Sitagliptin

The potential inhibitory effects of sitagliptin on THCA hub genes of *DPP4*, *CTNNB1*, and *MET* were analyzed by molecular docking simulations, compared to the standard inhibitors of CTNNB1 and MET of PNU-74654 and crizotinib, respectively. The 3D structures of sitagliptin (CID: 4369359), PNU-74654 (CID:9836739), and crizotinib (CID:116250) were retrieved from the pubchem database (https://pubchem.ncbi.nlm.nih.gov/ accessed on 22 October 2021), in the spatial data file (SDF) format, and consequently converted to PDB file format using the PyMOL visualization tool [61] (https://pymol.org/2/ accessed on 22 October 2021), while the crystal structures of DPP4 (PDB:2ONC), CTNNB1 (PDB:1JDH), and MET (PDB:3DKF) were downloaded from the protein database (PDB), (https://www.rcsb.org/ accessed on 22 October 2021), in PDF file format. File preparation for molecular docking was as described in previous studies [62–64]. Using autodock software, an in silico molecular docking tool [65], all PDB files were converted to PDBQT file formats, and docking was accordingly performed using autodock, as described previously [66,67]. For further analysis, we used PyMol to analyze ligand–receptor interactions in 3D view, and finally used the discovery studio web tool [68] for data interpretation.

3. Results

3.1. Identification of Common Oncogenes in THCA

Microarray datasets were downloaded from the NCBI-GEO database to identify DEGs in THCA. Commonly expressed oncogenes were identified from THCA tissues compared to adjacent normal tissues obtained from different studies. Volcano plots were used to show all DEGs from all selected datasets, and accordingly, the GSE3467, GSE3678, GSE6004, and GSE33630 datasets, respectively, displayed 691, 449, 1455, and 789 upregulated genes and 1088, 1232, 2890, and 1568 downregulated genes (Figure 1A–D). The relatedness of all samples in each dataset to each other was analyzed by uniform manifold approximation and projection (UMAP), in which the number of nearest neighbors was used for calculations as indicated in each plot (Figure 1E–H). In total, 123 overlapping genes were obtained using Venn diagrams, as observed from THCA tissues compared with normal tissues (Figure 1I,J). We further used these genes for further analysis of THCA in this study.

3.2. DPP4/CTNNB1/MET Expressions Are Associated with THCA Progression, Metastasis, and Worse Prognosis of THCA Cohorts

Our differential expression analysis revealed that the (m)RNA expression levels of *DPP4/CTNNB1/MET* were higher in THCA tumor tissues compared with adjacent normal tissues (Figure 2A). We further analyzed the role of *DPP4*, *CTNNB1*, and *MET* expressions in promoting THCA progression and tumor metastasis. Interestingly, our analysis revealed that the mRNA expressions levels of *DPP4/CTNNB1/MET* were more elevated in stage IV of THCA cancer (Figure 2B), and were significantly elevated in metastasis tumor compared with the primary tumor (Figure 2C). In addition, we found expression correlation among the *DPP4/CTNNB1/MET* signature in THCA cohorts (Figure 2D). Furthermore, we constructed a Kaplan–Meier (KM) plot of patients' survival and found that higher expression levels of the *DPP4/CTNNB1/MET* genes were associated with shorter survival duration of the cohorts (Figure 2E). Although the KM plot revealed no significant ($p > 0.05$) difference in the overall survival between cohorts with high and cohorts with low expression levels of DPP4, our analysis revealed that the disease-free survival of the cohorts was significantly ($p < 0.048$) higher in the low-DPP4-expression group when compared with the high-expression group. Collectively, our findings strongly suggested that the expression levels of DPP4/CTNNB1/MET signature are associated with THCA progression, metastasis, and worse prognosis of THCA cohorts, hence serving as important biomarker for diagnosis, prognosis, and therapeutic exploration in THCA.

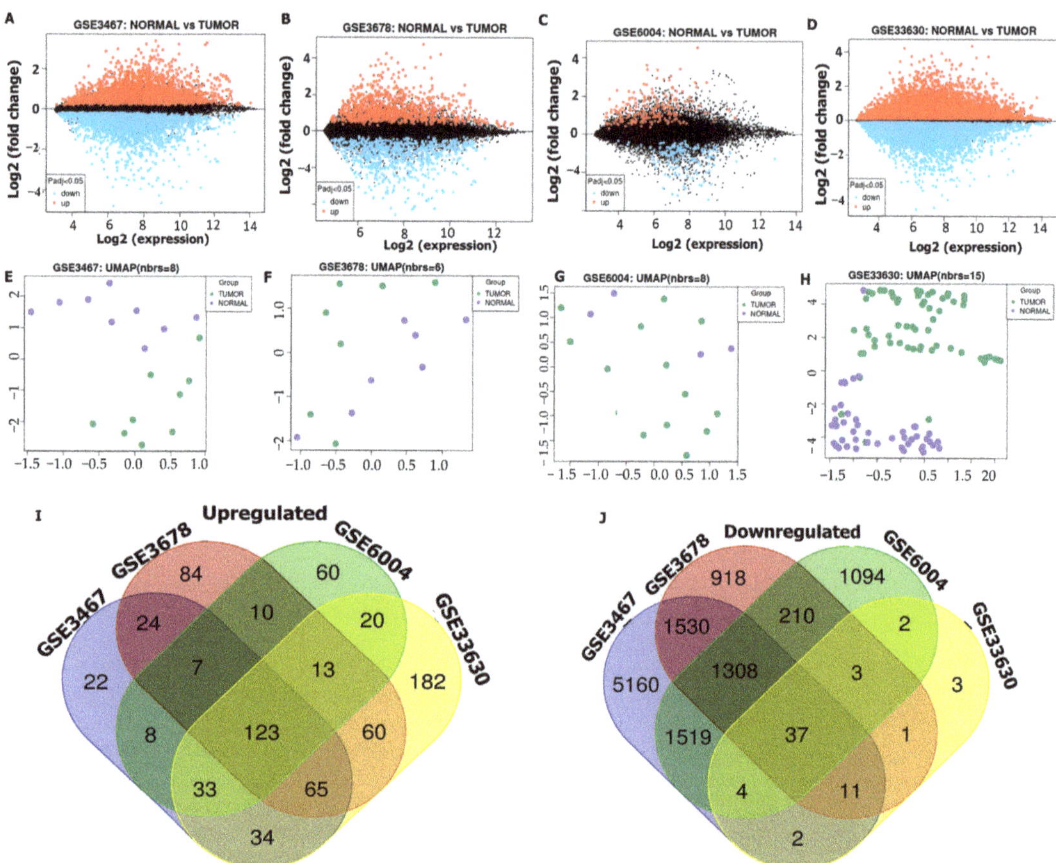

Figure 1. Differentially expressed genes (DEGs) in thyroid cancer (THCA). (**A–D**) Volcano plots showing DEGs extracted from the GSE3467, GSE3678, GSE6004, and GSE33630 microarray datasets, between cancer tissues compared with normal adjacent tissues, with upregulated genes (red), downregulated genes (blue), and non-significant genes (black). (**E–H**) Two-dimensional (2D) visualization of UMAP dimensionality reduction in THCA tumor tissues (green) compared with normal tissues (purple). (**I,J**) Venn diagram of 123 overlapping DEGs between normal colon tissues and tumor tissues.

3.3. DPP4/CTNNB1/MET Genes Are Frequently Altered and Their Mutations Are Linked to Genetic Expressions in THCA

Mutations of *DPP4/CTNNB1/MET* oncogenes in THCA were analyzed using the cBioPortal tool, and altered frequencies were based on percentages of individual genes due to amplification. Analytical results showed respective amplification of *DPP4*, *CTNNB1*, and *MET* occur in 3%, 6%, and 6% of THCA cohorts respectively. These included deep deletions (blue), mRNAs (red), proteins (red), mutations (green), and structural variants (purple) (Figure 3A–D). For further analysis, we compared associations between alterations in *DPP4* and *MET* oncogenic expressions with mutations of the top genes expressed in THCA at the target level, and according to our findings, *BRAF* mutations promoted increased expression levels of *DPP4* and MET compared with the WT. Patients with high expression levels of

DPP4 and *MET* signatures exhibited worse clinical outcomes compared with patients with low expression levels (Figure 3E,F).

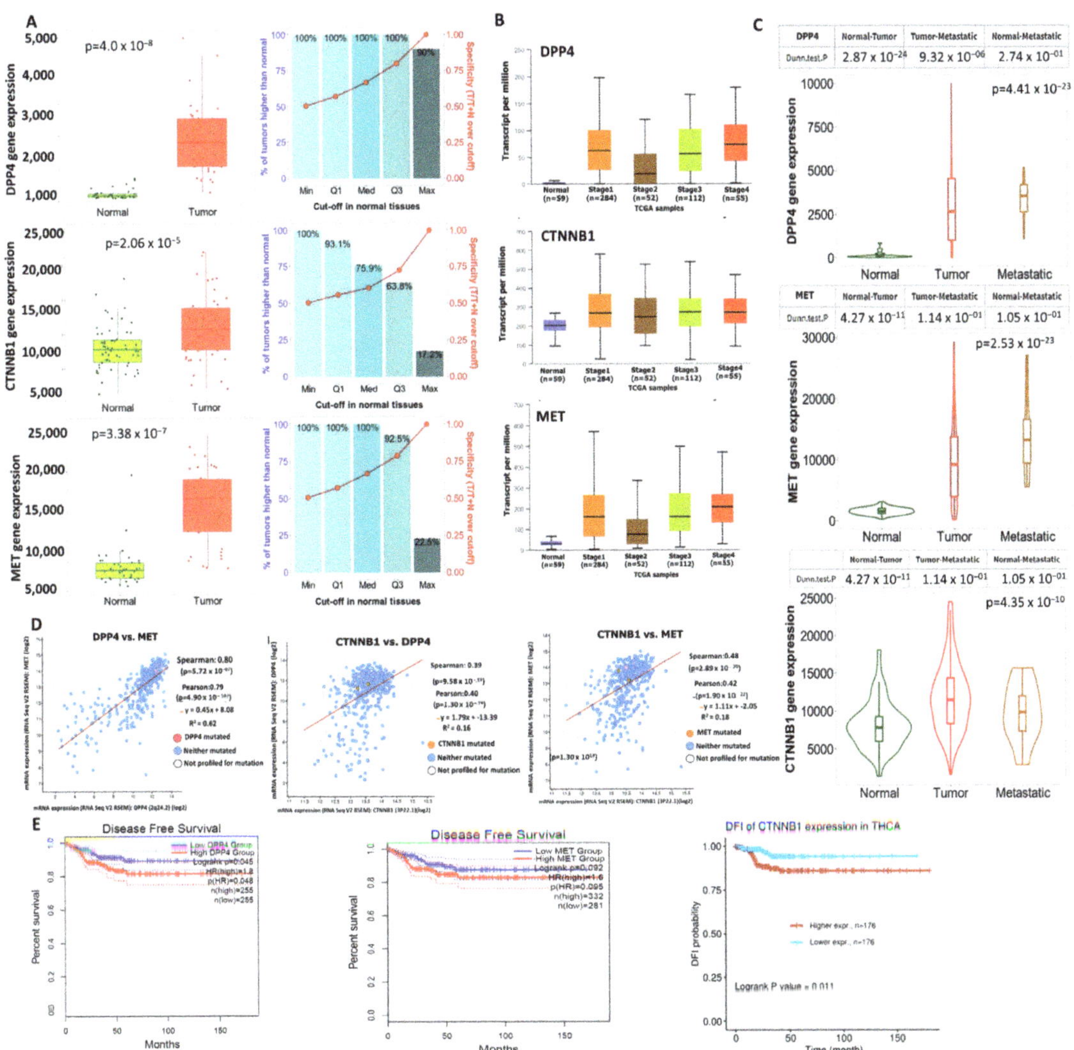

Figure 2. Overexpression of *DPP4/CTNNB1/MET* mRNAs, associated with thyroid cancer (THCA) progression. Differential expression levels of *DPP4/CTNNB1/MET* between (**A**) THCA tumor and adjacent normal tissue, (**B**) tumor stages, and (**C**) between primary and metastatic tumor of TCGA cohort. (**D**) Correlations of *DPP4* with *MET, CTNNB1* with *DPP4*, and *MET* with *CTNNB1* oncogenic expressions in THCA. (**E**) KPM plots of survival ratio between THCA cohorts with high and those with low expression levels of *DPP4/CTNNB1/MET*.

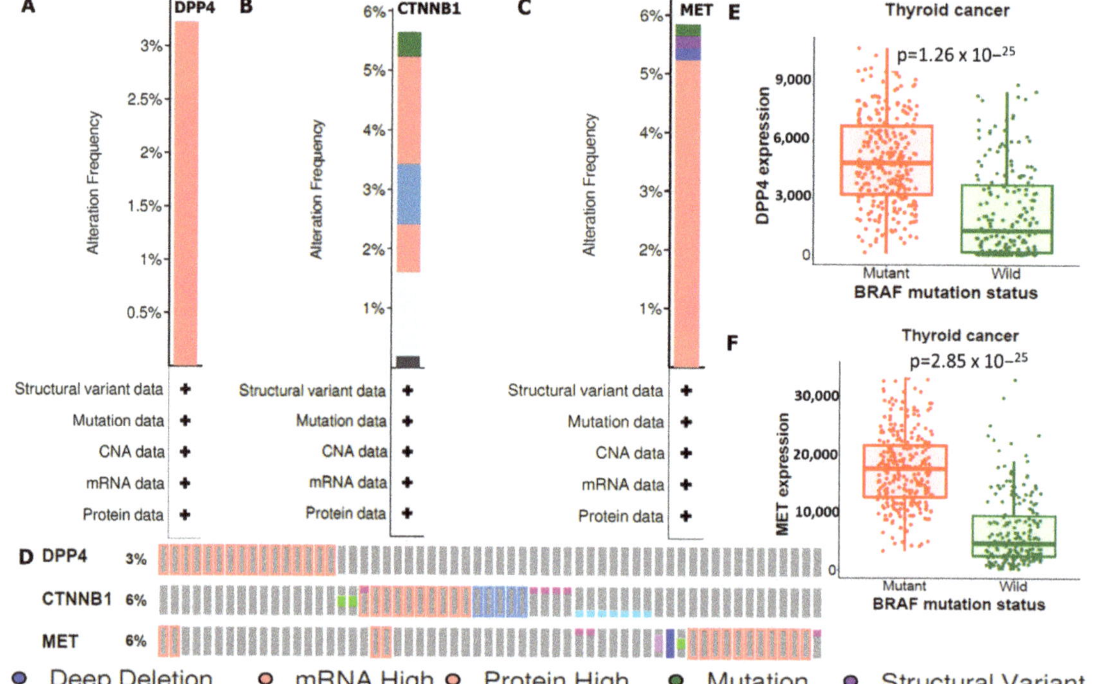

Figure 3. Genetic mutations based on percentages due to amplification of (**A**) *DPP4* (3%), (**B**) *CTNNB1* (6%), and (**C**) *MET* (6%), including deep deletions (blue), mRNAs (red), proteins (red), mutations (green), and structural variants (purple). (**D**) Individual genetic alteration profile of DPP4/CTNNB1/MET in THCA. (**E,F**) *BRAF* mutations promoted overexpression of *DPP4* and *MET* compared with the wild type, with $p < 0.05$ considered statistically significant.

3.4. DPP4/CTNNB1/MET Genes Potentially Promote Tumor Growth by Interacting with Different Oncogenic Targets/Pathways

We applied the STRING database and GeneMANIA online web tools developed to analyze interaction networks among four selected oncogenes. Herein, we considered experimental data (pink), gene neighborhoods (green), gene fusion (red), gene co-occurrences (blue), and gene co-expressions (black) when analyzing interactions. As expected, interaction networks were identified between *DPP4* and *CTNNB1*, *MET* and *DPP4*, *CTNNB1* and *MET*, *HFG* and *MET*, *DPP4* and *CTNND1*, and *GSK3B* and *CTNND1* within the network clustering. An average local clustering coefficient of 0.787 was obtained, with an expected number of edges of 21 and interaction *p* value of 0.006 (Figure 4A,B). For further analysis, we conducted a gene enrichment analysis and predicted GO processes using network analytical software, which showed co-expressions of *CTNNB1*, *GSK3B*, *AXIN1*, and *MET* to be enriched in the BP databases. Herein, we applied the Igraph R package visualization tool for analysis (Figure 4C). For more analysis, we used FunRich software to validate GO including BPs and KEGG enrichment analyses. The top five enriched BPs included chromosomal segregation, signaling transduction, cell communication, regulation of the cell cycle, and protein metabolism, while pathways involved in interactions included E-cadherin signaling in the nascent cadherin junction, stabilization and expression of adherens junctions, E-cadherin signaling events, posttranscriptional regulation of adherens junction stability, and N-cadherin signaling events (Figure 4D,E), with $p < 0.05$ considered significant.

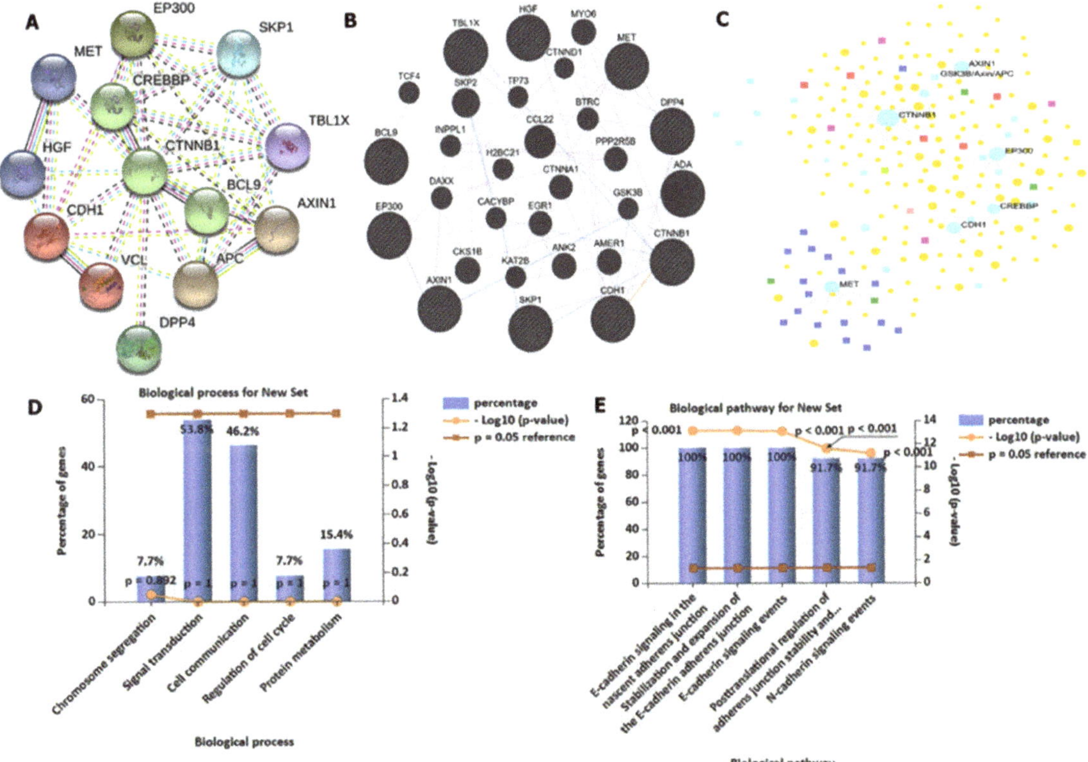

Figure 4. *DPP4/CTNNB1/MET* gene interactions co-expressed in the same clustering network. (**A**,**B**) Interaction networks showing co-expression between *DPP4* and *CTNNB1*, *MET* and *DPP4*, *CTNNB1* and *MET*, *HFG* and *MET*, *DPP4* and *CTNND1*, and *GSK3B* and *CTNND1* within the network clustering. An average local clustering coefficient of 0.787 was obtained, with an expected number of edges of 21 and an interaction *p* value of 0.006. (**C**) Gene enrichment analysis gene ontology (GO) showed enrichment in co-expressions of *CTNNB1*, *GSK3B*, *AXIN1*, and *MET* in biological processes. (**D**,**E**) Validation of GO, involving enrichment of the top five pathways involved, with *p* < 0.05 considered significant.

3.5. High Expression Levels of DPP4/CTNNB1/MET Are Associated with Immunosuppressive Phenotypes of THCA Tissues

We queried the association between the mRNA expression levels of DPP4/CTNNB1/MET and tumor infiltrations of immunosuppressive cells using the TCGA cohorts. Interestingly, we found that the mRNA expression levels of DPP4/CTNNB1/MET are inversely associated with tumor purity (Figure 5A). In addition, the high expression levels of the DPP4/CTNNB1/MET correlate positively (all *p* < 0.001, cor > 0.3) with the infiltration levels of tumor-associated macrophages (M2 TAM Figure 5B), regulatory T cell (Treg, Figure 5C), and cancer-associated fibroblast (CAF, Figure 5D) in thyroid cancer cohorts (Figure 5). In contrast, a strong negative association (all *p* < 0.001, cor < 0) was observed between the mRNA expression levels of *DPP4/CTNNB1/MET* and the immune infiltration level of CD8[+] T cell (Figure 5E), an anti-tumor T cell subtype. Collectively, these findings strongly suggested that high expression levels of DPP4/CTNNB1/MET are associated with immunosuppressive phenotypes via a mechanism involving T cell exclusion in THCA tissues.

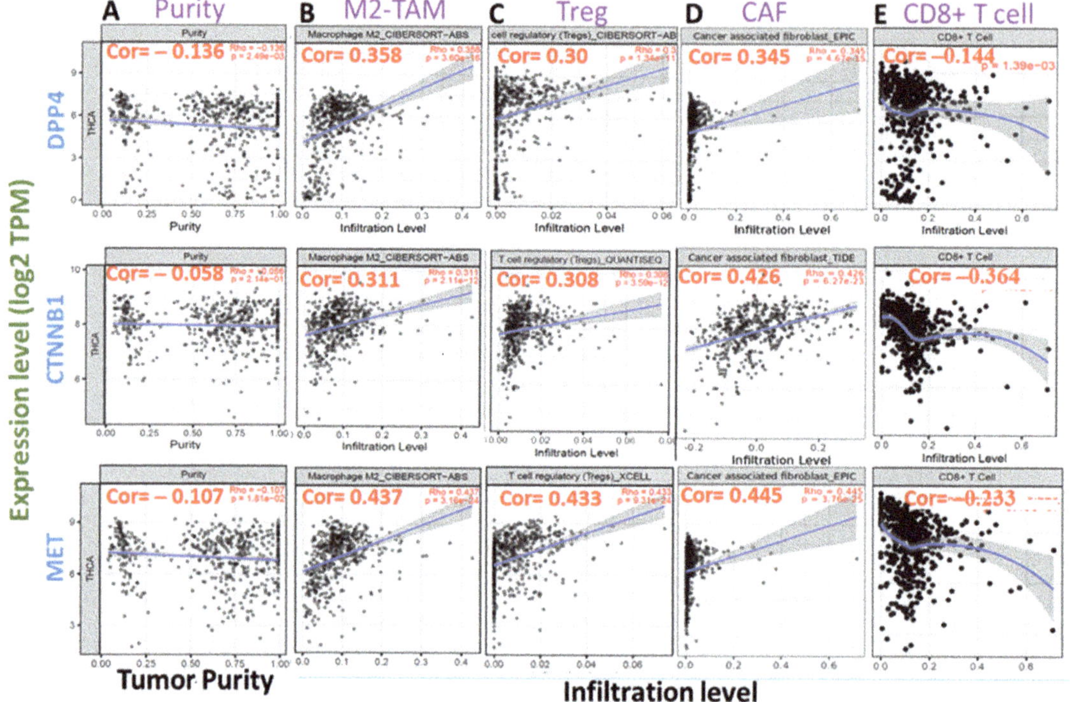

Figure 5. High expression levels of DPP4/CTNNB1/MET are associated with immunosuppressive phenotypes of THCA tissues. Scatterplots of *DPP4/CTNNB1/MET* expression correlations with t(**A**) tumor purity, and infiltration levels of (**B**) tumor-associated macrophages (M2 TAM), (**C**) regulatory T cell (Treg), (**D**) cancer-associated fibroblast (CAF), and (**E**) CD8[+] T cell. The strength of correlations between the genes and immune cells is reflected by the purity-adjusted partial Spearman's rho value, where a value of $r \geq 1$ means a perfect positive correlation and a value of $r \leq -1$ means a perfect negative correlation, with $p < 0.05$ considered statistically significant.

3.6. Molecular Docking Reveals Higher Inhibitory Effects of Sitagliptin on the DPP4 Oncogene

Our in silico molecular docking analysis revealed that sitagliptin exhibited higher binding energy of −8.6 kcal/mol with the *DPP4* oncogene. Further analysis of the docking results showed that sitagliptin bound to the binding pocket of the *DPP4* gene by hydrogen bonds with shorter binding distances at TRY631 (2.07 Å) and ARG125 (2.71 Å), and was further stabilized by a salt bridge, van der Waals forces, carbon–hydrogen bonds, Pi-Pi stacked, Pi-Pi T-shaped, amide Pi-stacked, and Pi-alkyl around the sitagliptin backbone (Figure 6).

3.7. Molecular Docking Revealed Potential Inhibitory Effects of Sitagliptin on the CTNNB1 Oncogene

Our docking analysis revealed that sitagliptin exhibited high binding energy of −7.3 kcal/mol with the *CTNNB1* oncogene, compared with its Food and Drug Administration (FDA)-approved inhibitor, PNU-74654, which showed a lower binding affinity of −6.7 kcal/mol. Further analysis of the docking results showed that sitagliptin bound to the binding pocket of the *CTNNB1* oncogene by 4 conventional hydrogen bonds and shorter binding distances with CYS466 (2.03 Å), LYS508 (2.51 Å), SER20 (1.87 Å), and ARG469 (1.03 Å). The interactions were further stabilized by van der Waals forces with ALA463, PRO463, PHE21, ASP459, and LEU18, halogen (fluorine) with GLU17, PRO505, GLU462, and GLU24, and Pi-alkyl with VAL564 and ILE17 around the sitagliptin backbone. The results were further compared with the PNU-74654/*CTNNB1* complex, which is bound to the binding pocket of the *CTNNB1* oncogene by only two conventional hydrogen bonds and longer binding distances compared with the sitagliptin/*CTNNB1* complex. The interactions were further stabilized by van der Waals forces with SER32, TYR306, and SER335, amide Pi-stacked with GLU375, and Pi-cation with GLU28, LYS345, and ARG342 around the PNU-74654 backbone. This suggests that sitagliptin has a high potential to target β-catenin (*CTNNB1*), compared with its standard inhibitor, PNU-74654 (Figure 7).

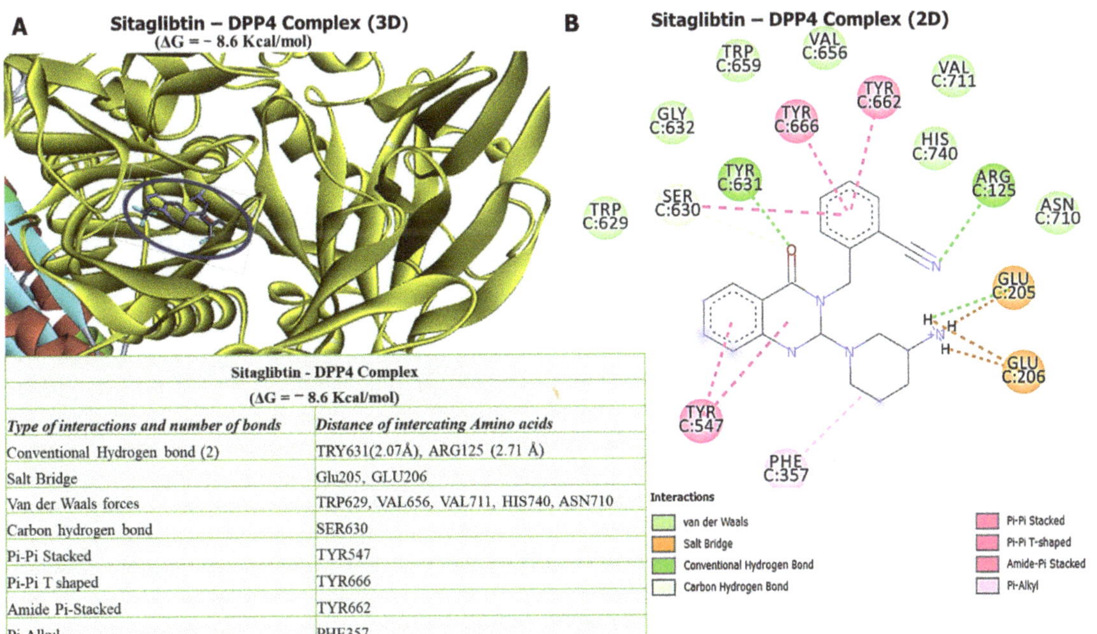

Sitaglibtin - DPP4 Complex	
(ΔG = – 8.6 Kcal/mol)	
Type of interactions and number of bonds	*Distance of intercating Amino acids*
Conventional Hydrogen bond (2)	TRY631(2.07Å), ARG125 (2.71 Å)
Salt Bridge	Glu205, GLU206
Van der Waals forces	TRP629, VAL656, VAL711, HIS740, ASN710
Carbon hydrogen bond	SER630
Pi-Pi Stacked	TYR547
Pi-Pi T shaped	TYR666
Amide Pi-Stacked	TYR662
Pi-Alkyl	PHE357

Figure 6. Ligand–receptor interaction results of sitagliptin with *DPP4*. (**A**) Three-dimensional (3D) representation of sitagliptin in complex with *DPP4* with the highest binding energy of −8.6 kcal/mol. (**B**) Two-dimensional (2D) representation of sitagliptin in complex with *DPP4*, showing interactions with two conventional H-bonds, with interactions further stabilized by different amino acids around the sitagliptin backbone. The accompanying table shows summary results of the analysis.

3.8. Molecular Docking Revealed Potential Inhibitory Effects of Sitagliptin on the MET Oncogene

Our docking analysis revealed that sitagliptin exhibited a high binding energy of −7.6 kcal/mol with the *MET* oncogene, the same as its FDA-approved inhibitor, crizotinib, which showed a binding affinity of −7.6 kcal/mol. Further analysis of the docking results showed that sitagliptin bound to the binding pocket of the *MET* oncogene by 4 conventional hydrogen bonds with shorter binding distances with TRY631 (2.07 Å) and ARG125 (2.71 Å). Interactions were further stabilized by a salt bridge (GLU205 and GLU206), van der Waals forces (TRP629, VAL656, VAL711, HIS740, and ASN710), carbon–hydrogen bond (SER630), Pi-Pi stacked (TYR547), Pi-Pi T-shaped (TYR666), amide Pi-stacked (TYR662), and Pi-alkyl (PHE357) around the sitagliptin backbone. However, results displayed of the crizotinib/*MET* complex did not exhibit conventional hydrogen bonds in the binding pocket of the *MET* oncogene. This suggests that sitagliptin has high potential to target *MET*, compared with its standard inhibitor, crizotinib (Figure 8).

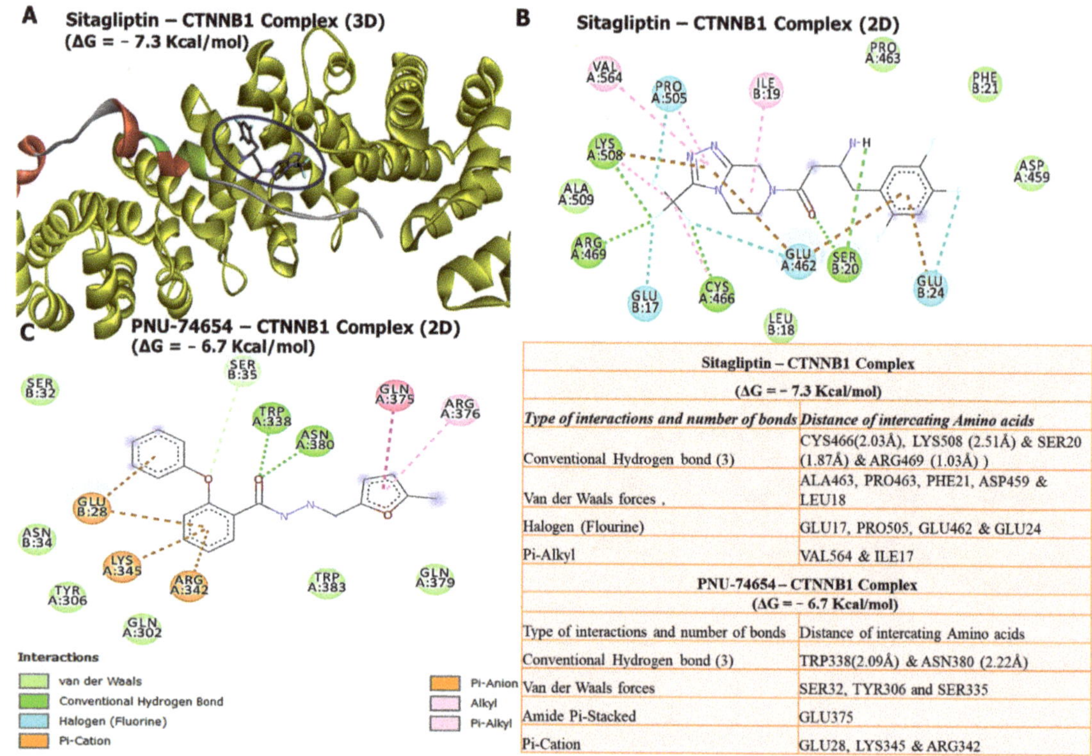

Sitagliptin – CTNNB1 Complex	
(ΔG = − 7.3 Kcal/mol)	
Type of interactions and number of bonds	*Distance of intercating Amino acids*
Conventional Hydrogen bond (3)	CYS466(2.03Å), LYS508 (2.51Å) & SER20 (1.87Å) & ARG469 (1.03Å))
Van der Waals forces ,	ALA463, PRO463, PHE21, ASP459 & LEU18
Halogen (Flourine)	GLU17, PRO505, GLU462 & GLU24
Pi-Alkyl	VAL564 & ILE17
PNU-74654 – CTNNB1 Complex	
(ΔG = − 6.7 Kcal/mol)	
Type of interactions and number of bonds	Distance of intercating Amino acids
Conventional Hydrogen bond (3)	TRP338(2.09Å) & ASN380 (2.22Å)
Van der Waals forces	SER32, TYR306 and SER335
Amide Pi-Stacked	GLU375
Pi-Cation	GLU28, LYS345 & ARG342

Figure 7. In silico molecular docking analysis of ligand–protein interactions. (**A**) Three-dimensional (3D) representation of sitagliptin in complex with *CTNNB1* with a binding energy of −7.3 kcal/mol. (**B**) Two-dimensional (2D) representation of sitagliptin in complex with *CTNNB1*, showing interactions with four conventional H-bonds and shorter binding distances, with interactions further stabilized by different amino acids around the sitagliptin backbone. (**C**) Two-dimensional (2D) representation of PNU-74654 in complex with *CTNNB1*, displaying lower binding energy of −6.7 kcal/mol, and interactions with (2) conventional hydrogen bonds with longer binding distances compared with that of sitagliptin, in the binding pockets of *CTNNB1*. The accompanying table shows a summary of the results.

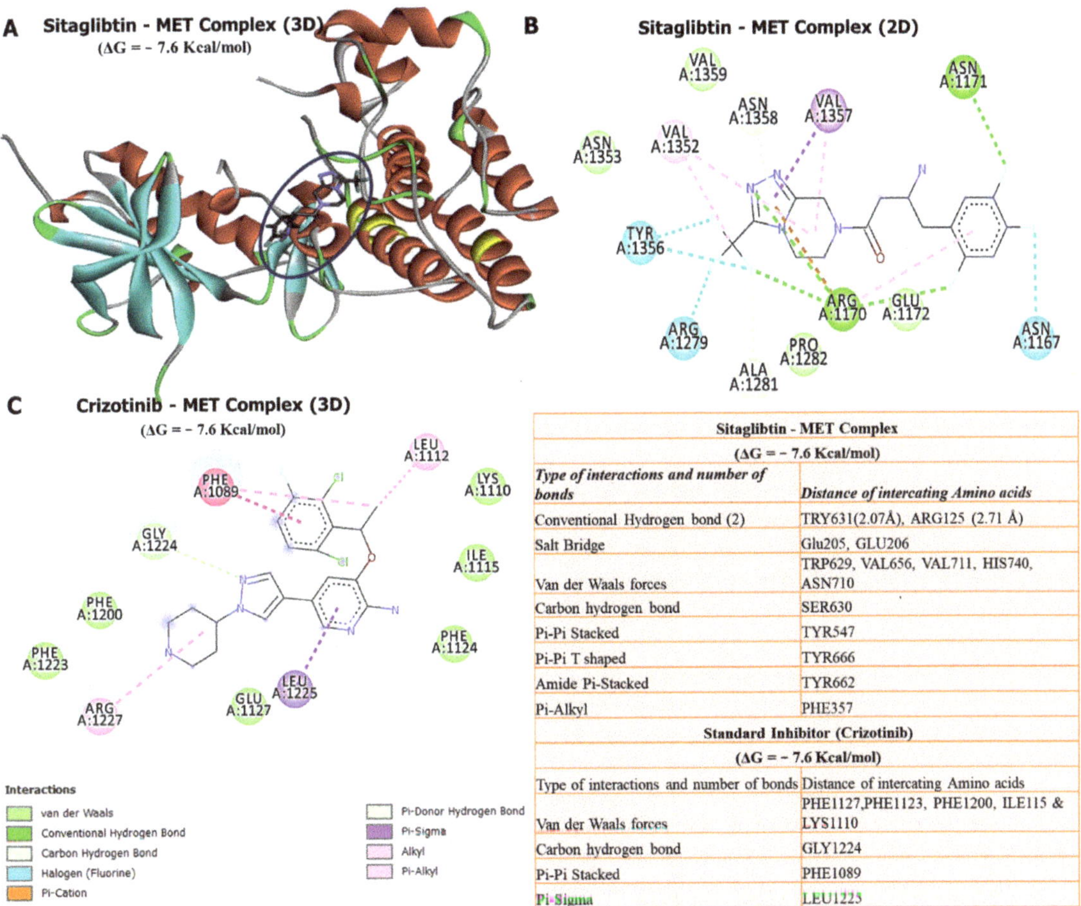

Sitaglibtin - MET Complex	
($\Delta G = -7.6$ Kcal/mol)	
Type of interactions and number of bonds	*Distance of intercating Amino acids*
Conventional Hydrogen bond (2)	TRY631(2.07Å), ARG125 (2.71 Å)
Salt Bridge	Glu205, GLU206
Van der Waals forces	TRP629, VAL656, VAL711, HIS740, ASN710
Carbon hydrogen bond	SER630
Pi-Pi Stacked	TYR547
Pi-Pi T shaped	TYR666
Amide Pi-Stacked	TYR662
Pi-Alkyl	PHE357
Standard Inhibitor (Crizotinib)	
($\Delta G = -7.6$ Kcal/mol)	
Type of interactions and number of bonds	Distance of intercating Amino acids
Van der Waals forces	PHE1127,PHE1123, PHE1200, ILE115 & LYS1110
Carbon hydrogen bond	GLY1224
Pi-Pi Stacked	PHE1089
Pi-Sigma	LEU1225

Figure 8. In silico molecular docking analysis of ligand–protein interactions. (**A**) Three-dimensional (3D) representation of sitagliptin in complex with *MET* with a binding energy of −7.6 kcal/mol. (**B**) Two-dimensional (2D) representation of sitagliptin in complex with *MET*, showing interactions with conventional H-bonds and different amino acids. (**C**) Two-dimensional (2D) representation of crizotinib in complex with *MET*, exhibiting the same binding energy as sitagliptin, but no interaction with conventional hydrogen bonds. The accompanying table shows a summary of the results.

4. Discussion

PTCa is the most prevalent type of THCA, which accounts for approximately 80% of all THCAs, consequently promoting cancer invasion, metastasis, and mortality in patients [69,70]. PTCa has recently been managed with a thyroidectomy; however, due to distant metastasis, THCA tends to be extremely aggressive, and resistant to treatment leading to poor prognoses [71–73]. Treatment modalities for THCA include the use of doxorubicin, but this has proven not to be very effective due to the development of resistance [1,74–76]. As a result, there is an urgent need to understand the molecular mechanisms associated with THCA metastasis, which will help in developing more effective treatments [15,77]. Identification of reliable biomarkers which can be used as diagnostic measures is urgently needed in PTCa. Most cancer therapeutic drugs have been shown to be cytotoxic and nonspecific to cancer cells, as they also affect normal cells and consequently cause harm to the body.

In the present study, we evaluated the anticancer effects of the antidiabetic drug sitagliptin, which was recently shown to possess anticancer activities, and is well tolerated and effective. Sitagliptin is an FDA-approved *DPP4* oncogene [78]. To further analyze sitagliptin, we explored computer-based simulations to identify and predict target genes, which are commonly overexpressed and associated with THCA invasion, progression, metastasis, poor prognosis, and resistance to therapeutics. We utilized microarray datasets from the NCBI-GEO, and identified DEGs in THCA compared to normal tissues. Among the top upregulated genes, were the *DPP4*, *CTNNB1*, and *MET* oncogenes. To validate their expressions, we used the UALCAN online bioinformatics tool with default settings, which showed that mRNA levels of *DPP4/CTNNB1/MET* were higher in THCA tumor tissues compared with adjacent normal tissues. Moreover, after exploring the TNMplot software, for further analysis, we identified that overexpression of *DPP4/CTNNB1/MET* gene signatures promoted THCA metastasis, and were associated with poor disease-free survival and poor prognoses.

The complex and dynamic interactions of immune cells, stoma, and cancer cells within the tumor microenvironment (TME) play a pivotal role in tumor invasion, cancer progression, and host immune response [62,79]. Consequently, our analysis of tumor immune infiltrating cells within the TME of THCA tumor revealed that the high expression levels of the DPP4/CTNNB1/MET signature correlate positively with the infiltration levels of tumor-associated macrophages, regulatory T cell, and cancer-associated fibroblast. These immunosuppressive cells are known to exert an inhibitory role on cytotoxic lymphocytes' function leading to T cell exclusion and tumor invasive phenotype [59,80]. In contrast, we found a strong negative association was observed between the mRNA expression levels of DPP4/CTNNB1/MET and immune infiltration level of CD8$^+$ T cell, suggesting that high expression levels of DPP4/CTNNB1/MET are associated with immunosuppressive phenotypes via a mechanism involving T cell exclusion in THCA tissues

Molecular docking has become an increasingly important tool commonly used to understand drug bimolecular interactions with the target proteins for rational drug design and development [62,81,82]. It is useful in estimating binding affinities of the ligand to the proteins and in providing preliminary mechanistic insight into the behavior of a small molecule drug in the binding cavity of target proteins [83,84], as well as elucidating the potential drug-regulated biochemical processes [79,85]. Consequently, we conducted a molecular docking analysis of interactions of *DPP4/CTNNB1/MET* gene signatures with sitagliptin. As expected, sitagliptin exhibited a higher binding energy of −8.6 kcal/mol with the *DPP4* oncogene. Furthermore, our docking analysis revealed that sitagliptin exhibited a higher binding energy of −7.3 kcal/mol with the *CTNNB1* oncogene compared with its FDA-approved inhibitor, PNU-74654, which showed a lower binding affinity of −6.7 kcal/mol. Our analysis showed that sitagliptin bound to the binding pocket of the *CTNNB1* oncogene by 4 conventional hydrogen bonds and had shorter binding distances with CYS466 (2.03 Å), LYS508 (2.51 Å), SER20 (1.87 Å), and ARG469 (1.03 Å) compared with PNU-74654, which bound to the binding pocket of the *CTNNB1* oncogene by only 2 conventional hydrogen bonds, and had longer binding distances compared with sitagliptin. In addition, analytical results of sitagliptin in complex with *MET* exhibited the same binding energy of −7.6 kcal/mol as the *MET* FDA-approved inhibitor, crizotinib. Sitagliptin bound to the binding pocket of the *MET* oncogene by 4 conventional hydrogen bonds and shorter binding distances with TRY631 (2.07 Å) and ARG125 (2.71 Å). However, results displayed from the crizotinib/*MET* complex did not exhibit conventional hydrogen bonds in the binding pocket of the *MET* oncogene.

In summary, these docking results suggest that sitagliptin has high potential to target *DPP4/CTNNB1/MET* signaling pathways in THCA compared with their standard inhibitors. Since recent studies have shown the efficacy and tolerance of sitagliptin as cancer therapeutic, it would be interesting to further investigate its activities as a target for *DPP4/CTNNB1/MET* signaling pathways in THCA, both in vitro and in vitro in tumor-bearing mice.

5. Conclusions

In summary, we revealed that *DPP4*, *CTNNB1*, and *MET* oncogenic signatures are overexpressed in THCA, and are associated with cancer progression, metastasis, resistance, poor disease-free survival, and unfavorable clinical outcomes. Moreover, an in silico molecular docking study exhibited putative binding affinities of sitagliptin with the abovementioned oncogenes, which were higher than the standard inhibitors of these genes. This suggests that sitagliptin could be a potential THCA therapeutic, since it has been shown to be more tolerable and effective in different cancers.

Author Contributions: S.-Y.C., A.T.H.W., G.E.-S.B., C.-L.H., J.-C.L., H.Y.L., M.A., A.N.A. and J.-H.C. contributed to data analysis, and drafting or revising the article. All authors have read and agreed to the published version of the manuscript.

Funding: The current work was partly supported by Taif University Researchers Supporting Project number (TURSP-2020/310), Taif University, Taif, Saudi Arabia. Jia-Hong Chen was supported by TSGH-C03-111030; Ching-Liang Ho was funded by TSGH-C03-110022 and TSGH-C02-111024; Jih-Chin Lee was supported by TSGH-D-110079 and TSGH-D-111060; Alexander TH Wu was financially supported by the "TMU Research Center of Cancer Translational Medicine" from The Featured Areas Research Center Program within the framework of the Higher Education Sprout Project by the Ministry of Education (MOE) in Taiwan. (Grant number: DP2-110-21121-03-C-09).

Institutional Review Board Statement: Not applicable.

Informed Consent Statement: Not applicable.

Data Availability Statement: The datasets generated and analyzed in this study can be made available upon reasonable request.

Acknowledgments: The authors express their thanks to Taif University for supporting the project—TURSP 2020/310.

Conflicts of Interest: The authors declare no conflict of interest.

References

1. Nguyen, Q.T.; Lee, E.J.; Huang, M.G.; Park, Y.I.; Khullar, A.; Plodkowski, R.A. Diagnosis and treatment of patients with thyroid cancer. *Am. Health Drug Benefits* **2015**, *8*, 30–40. [PubMed]
2. Siegel, R.L.; Miller, K.D.; Jemal, A. Cancer statistics, 2020. *CA Cancer J. Clin.* **2020**, *70*, 7–30. [CrossRef] [PubMed]
3. Bray, F.; Ferlay, J.; Soerjomataram, I.; Siegel, R.L.; Torre, L.A.; Jemal, A. Global cancer statistics 2018: GLOBOCAN estimates of incidence and mortality worldwide for 36 cancers in 185 countries. *CA Cancer J. Clin.* **2018**, *68*, 394–424. [CrossRef] [PubMed]
4. La Vecchia, C.; Malvezzi, M.; Bosetti, C.; Garavello, W.; Bertuccio, P.; Levi, F.; Negri, E. Thyroid cancer mortality and incidence: A global overview. *Int. J. Cancer* **2015**, *136*, 2187–2195. [CrossRef] [PubMed]
5. Roman, B.R.; Morris, L.G.; Davies, L. The thyroid cancer epidemic, 2017 perspective. *Curr. Opin. Endocrinol. Diabetes Obes.* **2017**, *24*, 332–336. [CrossRef]
6. Siegel, R.L.; Miller, K.D.; Fuchs, H.E.; Jemal, A. Cancer Statistics, 2021. *CA Cancer J. Clin.* **2021**, *71*, 7–33. [CrossRef]
7. Fagin, J.A.; Wells, S.A., Jr. Biologic and clinical perspectives on thyroid cancer. *N. Engl. J. Med.* **2016**, *375*, 1054–1067. [CrossRef]
8. Molinaro, E.; Romei, C.; Biagini, A.; Sabini, E.; Agate, L.; Mazzeo, S.; Materazzi, G.; Sellari-Franceschini, S.; Ribechini, A.; Torregrossa, L. Anaplastic thyroid carcinoma: From clinicopathology to genetics and advanced therapies. *Nat. Rev. Endocrinol.* **2017**, *13*, 644–660. [CrossRef]
9. Knauf, J.A.; Ma, X.; Smith, E.P.; Zhang, L.; Mitsutake, N.; Liao, X.-H.; Refetoff, S.; Nikiforov, Y.E.; Fagin, J.A. Targeted expression of BRAFV600E in thyroid cells of transgenic mice results in papillary thyroid cancers that undergo dedifferentiation. *Cancer Res.* **2005**, *65*, 4238–4245. [CrossRef]
10. Brito, J.P.; Morris, J.C.; Montori, V.M. Thyroid cancer: Zealous imaging has increased detection and treatment of low risk tumours. *BMJ* **2013**, *347*, f4706. [CrossRef]
11. Udelsman, R.; Zhang, Y. The epidemic of thyroid cancer in the United States: The role of endocrinologists and ultrasounds. *Thyroid* **2014**, *24*, 472–479. [CrossRef]
12. Furuya-Kanamori, L.; Bell, K.J.; Clark, J.; Glasziou, P. Prevalence of differentiated thyroid cancer in autopsy studies over six decades: A meta-analysis. *J. Clin. Oncol.* **2016**, *34*, 3672–3679. [CrossRef] [PubMed]
13. Ahn, H.S.; Welch, H.G. South Korea's thyroid-cancer "epidemic"—Turning the tide. *N. Engl. J. Med.* **2015**, *373*, 2389–2390. [CrossRef] [PubMed]
14. Kitahara, C.M.; Schneider, A.; Brenner, A.V. Chapter 44: Thyroid cancer. In *Cancer Epidemiology and Prevention*, 4th ed.; Schottenfeld, D., Fraumeni, J., Eds.; Oxford University Press: New York, NY, USA, 2016; pp. 839–860.

15. Wei, W.J.; Zhang, G.Q.; Luo, Q.Y. Postsurgical Management of Differentiated Thyroid Cancer in China. *Trends Endocrinol. Metab.* **2018**, *29*, 71–73. [CrossRef] [PubMed]
16. Nikiforov, Y.E.; Nikiforova, M.N. Molecular genetics and diagnosis of thyroid cancer. *Nat. Rev. Endocrinol.* **2011**, *7*, 569–580. [CrossRef] [PubMed]
17. Long, K.L.; Grubbs, E.G. Carcinoma of the thyroid gland and neoplasms of the parathyroid glands. In *The MD Anderson Surgical Oncology Handbook*, 6th ed.; Wolters Kluwer Health Adis (ESP): London, UK, 2018; pp. 463–491.
18. Ho, A.S.; Luu, M.; Barrios, L.; Chen, I.; Melany, M.; Ali, N.; Patio, C.; Chen, Y.; Bose, S.; Fan, X.; et al. Incidence and Mortality Risk Spectrum Across Aggressive Variants of Papillary Thyroid Carcinoma. *JAMA Oncol.* **2020**, *6*, 706–713. [CrossRef] [PubMed]
19. Bai, Y.; Kakudo, K.; Jung, C.K. Updates in the Pathologic Classification of Thyroid Neoplasms: A Review of the World Health Organization Classification. *Endocrinol. Metab.* **2020**, *35*, 696–715. [CrossRef]
20. WHO. *WHO Classification of Tumours of Endocrine Organs*; IARC: Lion, France, 2017.
21. Nath, M.C.; Erickson, L.A. Aggressive Variants of Papillary Thyroid Carcinoma: Hobnail, Tall Cell, Columnar, and Solid. *Adv. Anat. Pathol.* **2018**, *25*, 172–179. [CrossRef]
22. Cavaco, D.; Martins, A.F.; Cabrera, R.; Vilar, H.; Leite, V. Diffuse sclerosing variant of papillary thyroid carcinoma: Outcomes of 33 cases. *Eur. Thyroid. J.* **2022**, *11*, e210020. [CrossRef]
23. Roman, S.; Sosa, J.A. Aggressive variants of papillary thyroid cancer. *Curr. Opin. Oncol.* **2013**, *25*, 33–38. [CrossRef]
24. Besic, N.; Auersperg, M.; Dremelj, M.; Vidergar-Kralj, B.; Gazic, B. Neoadjuvant chemotherapy in 16 patients with locally advanced papillary thyroid carcinoma. *Thyroid* **2013**, *23*, 178–184. [CrossRef] [PubMed]
25. Giuffrida, R.; Adamo, L.; Iannolo, G.; Vicari, L.; Giuffrida, D.; Eramo, A.; Gulisano, M.; Memeo, L.; Conticello, C. Resistance of papillary thyroid cancer stem cells to chemotherapy. *Oncol. Lett.* **2016**, *12*, 687–691. [CrossRef] [PubMed]
26. Kim, K.B.; Cabanillas, M.E.; Lazar, A.J.; Williams, M.D.; Sanders, D.L.; Ilagan, J.L.; Nolop, K.; Lee, R.J.; Sherman, S.I. Clinical responses to vemurafenib in patients with metastatic papillary thyroid cancer harboring BRAF(V600E) mutation. *Thyroid* **2013**, *23*, 1277–1283. [CrossRef] [PubMed]
27. Noels, H.; Theelen, W.; Sternkopf, M.; Jankowski, V.; Moellmann, J.; Kraemer, S.; Lehrke, M.; Marx, N.; Martin, L.; Marx, G.; et al. Reduced post-operative DPP4 activity associated with worse patient outcome after cardiac surgery. *Sci. Rep.* **2018**, *8*, 11820. [CrossRef]
28. Javidroozi, M.; Zucker, S.; Chen, W.T. Plasma seprase and DPP4 levels as markers of disease and prognosis in cancer. *Dis. Markers* **2012**, *32*, 309–320. [CrossRef] [PubMed]
29. Lee, J.J.; Wang, T.Y.; Liu, C.L.; Chien, M.N.; Chen, M.J.; Hsu, Y.C.; Leung, C.H.; Cheng, S.P. Dipeptidyl Peptidase IV as a Prognostic Marker and Therapeutic Target in Papillary Thyroid Carcinoma. *J. Clin. Endocrinol. Metab.* **2017**, *102*, 2930–2940. [CrossRef]
30. Kotani, T.; Aratake, Y.; Ogata, Y.; Umeki, K.; Araki, Y.; Hirai, K.; Kuma, K.; Ohtaki, S. Expression of dipeptidyl aminopeptidase IV activity in thyroid carcinoma. *Cancer Lett.* **1991**, *57*, 203–208. [CrossRef]
31. Aratake, Y.; Kotani, T.; Tamura, K.; Araki, Y.; Kuribayashi, T.; Konoe, K.; Ohtaki, S. Dipeptidyl aminopeptidase IV staining of cytologic preparations to distinguish benign from malignant thyroid diseases. *Am. J. Clin. Pathol.* **1991**, *96*, 306–310. [CrossRef]
32. Nouraee, N.; Van Roosbroeck, K.; Vasei, M.; Semnani, S.; Samaei, N.M.; Naghshvar, F.; Omidi, A.A.; Calin, G.A.; Mowla, S.J. Expression, tissue distribution and function of miR-21 in esophageal squamous cell carcinoma. *PLoS ONE* **2013**, *8*, e73009. [CrossRef]
33. Pang, R.; Law, W.L.; Chu, A.C.; Poon, J.T.; Lam, C.S.; Chow, A.K.; Ng, L.; Cheung, L.W.; Lan, X.R.; Lan, H.Y.; et al. A subpopulation of CD26+ cancer stem cells with metastatic capacity in human colorectal cancer. *Cell Stem Cell* **2010**, *6*, 603–615. [CrossRef]
34. Liu, L.; Yan, M.; Zhao, F.; Li, J.; Ge, C.; Geng, Q.; Zhu, M.; Sun, L.; He, X.; Li, J. CD26/dipeptidyl peptidase IV contributes to tumor metastasis in human lung adenocarcinoma. *Bangladesh J. Pharmacol.* **2013**, *8*, 198–206. [CrossRef]
35. Jang, J.; Haberecker, M.; Curioni, A.; Janker, F.; Soltermann, A.; Gil-Bazo, I.; Hwang, I.; Kwon, K.; Weder, W.; Jungraithmayr, W. EP1.03–33 CD26/DPP4 as a Novel Prognostic Marker for Lung Adenocarcinoma. *J. Thorac. Oncol.* **2019**, *14*, S965. [CrossRef]
36. Lu, Z.; Qi, L.; Bo, X.J.; Liu, G.D.; Wang, J.M.; Li, G. Expression of CD26 and CXCR4 in prostate carcinoma and its relationship with clinical parameters. *J. Res. Med.* **2013**, *18*, 647–652.
37. Kretzschmar, K.; Weber, C.; Driskell, R.R.; Calonje, E.; Watt, F.M. Compartmentalized Epidermal Activation of β-Catenin Differentially Affects Lineage Reprogramming and Underlies Tumor Heterogeneity. *Cell Rep.* **2016**, *14*, 269–281. [CrossRef]
38. Dong, C.; Yang, H.; Wang, Y.; Yan, X.; Li, D.; Cao, Z.; Ning, Y.; Zhang, C. Anagliptin stimulates osteoblastic cell differentiation and mineralization. *Biomed. Pharmacother.* **2020**, *129*, 109796. [CrossRef]
39. Ren, X.; Zhu, R.; Liu, G.; Xue, F.; Wang, Y.; Xu, J.; Zhang, W.; Yu, W.; Li, R. Effect of sitagliptin on tubulointerstitial Wnt/β-catenin signalling in diabetic nephropathy. *Nephrology* **2019**, *24*, 1189–1197. [CrossRef]
40. Garcia, C.; Buffet, C.; El Khattabi, L.; Rizk-Rabin, M.; Perlemoine, K.; Ragazzon, B.; Bertherat, J.; Cormier, F.; Groussin, L. MET overexpression and activation favors invasiveness in a model of anaplastic thyroid cancer. *Oncotarget* **2019**, *10*, 2320–2334. [CrossRef]
41. Trovato, M.; Campennì, A.; Giovinazzo, S.; Siracusa, M.; Ruggeri, R.M. Hepatocyte Growth Factor/C-Met Axis in Thyroid Cancer: From Diagnostic Biomarker to Therapeutic Target. *Biomark. Insights* **2017**, *12*, 1177271917701126. [CrossRef]
42. Mineo, R.; Costantino, A.; Frasca, F.; Sciacca, L.; Russo, S.; Vigneri, R.; Belfiore, A. Activation of the hepatocyte growth factor (HGF)-Met system in papillary thyroid cancer: Biological effects of HGF in thyroid cancer cells depend on Met expression levels. *Endocrinology* **2004**, *145*, 4355–4365. [CrossRef]

43. Di Renzo, M.F.; Olivero, M.; Ferro, S.; Prat, M.; Bongarzone, I.; Pilotti, S.; Belfiore, A.; Costantino, A.; Vigneri, R.; Pierotti, M.A.; et al. Overexpression of the c-MET/HGF receptor gene in human thyroid carcinomas. *Oncogene* **1992**, *7*, 2549–2553. [PubMed]
44. Lesko, E.; Majka, M. The biological role of HGF-MET axis in tumor growth and development of metastasis. *Front. Biosci.* **2008**, *13*, 1271–1280. [CrossRef] [PubMed]
45. Monga, S.P.; Mars, W.M.; Pediaditakis, P.; Bell, A.; Mulé, K.; Bowen, W.C.; Wang, X.; Zarnegar, R.; Michalopoulos, G.K. Hepatocyte growth factor induces Wnt-independent nuclear translocation of beta-catenin after Met-beta-catenin dissociation in hepatocytes. *Cancer Res.* **2002**, *62*, 2064–2071. [PubMed]
46. Tward, A.D.; Jones, K.D.; Yant, S.; Cheung, S.T.; Fan, S.T.; Chen, X.; Kay, M.A.; Wang, R.; Bishop, J.M. Distinct pathways of genomic progression to benign and malignant tumors of the liver. *Proc. Natl. Acad. Sci. USA* **2007**, *104*, 14771–14776. [CrossRef]
47. Sastre-Perona, A.; Santisteban, P. Role of the wnt pathway in thyroid cancer. *Front. Endocrinol.* **2012**, *3*, 31. [CrossRef]
48. Rezk, S.; Brynes, R.K.; Nelson, V.; Thein, M.; Patwardhan, N.; Fischer, A.; Khan, A. beta-Catenin expression in thyroid follicular lesions: Potential role in nuclear envelope changes in papillary carcinomas. *Endocr. Pathol.* **2004**, *15*, 329–337. [CrossRef]
49. Buchanan, S.G.; Hendle, J.; Lee, P.S.; Smith, C.R.; Bounaud, P.Y.; Jessen, K.A.; Tang, C.M.; Huser, N.H.; Felce, J.D.; Froning, K.J.; et al. SGX523 is an exquisitely selective, ATP-competitive inhibitor of the MET receptor tyrosine kinase with antitumor activity in vivo. *Mol. Cancer* **2009**, *8*, 3181–3190. [CrossRef]
50. Amritha, C.A.; Kumaravelu, P.; Chellathai, D.D. Evaluation of Anti Cancer Effects of DPP-4 Inhibitors in Colon Cancer- An Invitro Study. *J. Clin. Diagn. Res.* **2015**, *9*, Fc14–Fc16. [CrossRef]
51. Chandrashekar, D.S.; Bashel, B.; Balasubramanya, S.A.H.; Creighton, C.J.; Ponce-Rodriguez, I.; Chakravarthi, B.; Varambally, S. UALCAN: A Portal for Facilitating Tumor Subgroup Gene Expression and Survival Analyses. *Neoplasia* **2017**, *19*, 649–658. [CrossRef]
52. Gao, J.; Aksoy, B.A.; Dogrusoz, U.; Dresdner, G.; Gross, B.; Sumer, S.O.; Sun, Y.; Jacobsen, A.; Sinha, R.; Larsson, E.; et al. Integrative analysis of complex cancer genomics and clinical profiles using the cBioPortal. *Sci. Signal.* **2013**, *6*, pl1. [CrossRef]
53. Bartha, Á.; Győrffy, B. TNMplot.com: A Web Tool for the Comparison of Gene Expression in Normal, Tumor and Metastatic Tissues. *Int. J. Mol. Sci.* **2021**, *22*, 2622. [CrossRef] [PubMed]
54. Szklarczyk, D.; Gable, A.L.; Lyon, D.; Junge, A.; Wyder, S.; Huerta-Cepas, J.; Simonovic, M.; Doncheva, N.T.; Morris, J.H.; Bork, P.; et al. STRING v11: Protein-protein association networks with increased coverage, supporting functional discovery in genome-wide experimental datasets. *Nucleic Acids Res.* **2019**, *47*, D607–D613. [CrossRef] [PubMed]
55. Franz, M.; Rodriguez, H.; Lopes, C.; Zuberi, K.; Montojo, J.; Bader, G.D.; Morris, Q. GeneMANIA update 2018. *Nucleic Acids Res.* **2018**, *46*, W60–W64. [CrossRef] [PubMed]
56. Zhou, G.; Soufan, O.; Ewald, J.; Hancock, R.E.W.; Basu, N.; Xia, J. NetworkAnalyst 3.0: A visual analytics platform for comprehensive gene expression profiling and meta-analysis. *Nucleic Acids Res.* **2019**, *47*, W234–W241. [CrossRef] [PubMed]
57. Pathan, M.; Keerthikumar, S.; Ang, C.S.; Gangoda, L.; Quek, C.Y.; Williamson, N.A.; Mouradov, D.; Sieber, O.M.; Simpson, R.J.; Salim, A.; et al. FunRich: An open access standalone functional enrichment and interaction network analysis tool. *Proteomics* **2015**, *15*, 2597–2601. [CrossRef] [PubMed]
58. Juneja, M.; Kobelt, D.; Walther, W.; Voss, C.; Smith, J.; Specker, E.; Neuenschwander, M.; Gohlke, B.-O.; Dahlmann, M.; Radetzki, S.; et al. Statin and rottlerin small-molecule inhibitors restrict colon cancer progression and metastasis via MACC1. *PLoS Biol.* **2017**, *15*, e2000784. [CrossRef]
59. Lawal, B.; Lin, L.-C.; Lee, J.-C.; Chen, J.-H.; Bekaii-Saab, T.S.; Wu, A.T.H.; Ho, C.-L. Multi-Omics Data Analysis of Gene Expressions and Alterations, Cancer-Associated Fibroblast and Immune Infiltrations, Reveals the Onco-Immune Prognostic Relevance of STAT3/CDK2/4/6 in Human Malignancies. *Cancers* **2021**, *13*, 954. [CrossRef]
60. Lawal, B.; Tseng, S.-H.; Olugbodi, J.O.; Iamsaard, S.; Ilesanmi, O.B.; Mahmoud, M.H.; Ahmed, S.H.; Batiha, G.E.-S.; Wu, A.T.H. Pan-Cancer Analysis of Immune Complement Signature C3/C5/C3AR1/C5AR1 in Association with Tumor Immune Evasion and Therapy Resistance. *Cancers* **2021**, *13*, 4124. [CrossRef]
61. Seeliger, D.; de Groot, B.L. Ligand docking and binding site analysis with PyMOL and Autodock/Vina. *J. Comput. Aided Mol. Des.* **2010**, *24*, 417–422. [CrossRef]
62. Wu, S.-Y.; Lin, K.-C.; Lawal, B.; Wu, A.T.H.; Wu, C.-Z. MXD3 as an onco-immunological biomarker encompassing the tumor microenvironment, disease staging, prognoses, and therapeutic responses in multiple cancer types. *Comput. Struct. Biotechnol. J.* **2021**, *19*, 4970–4983. [CrossRef]
63. Wu, A.T.H.; Lawal, B.; Tzeng, Y.-M.; Shih, C.-C.; Shih, C.-M. Identification of a Novel Theranostic Signature of Metabolic and Immune-Inflammatory Dysregulation in Myocardial Infarction, and the Potential Therapeutic Properties of Ovatodiolide, a Diterpenoid Derivative. *Int. J. Mol. Sci.* **2022**, *23*, 1281. [CrossRef] [PubMed]
64. Wu, A.T.H.; Lawal, B.; Wei, L.; Wen, Y.-T.; Tzeng, D.T.W.; Lo, W.-C. Multiomics Identification of Potential Targets for Alzheimer Disease and Antrocin as a Therapeutic Candidate. *Pharmaceutics* **2021**, *13*, 1555. [CrossRef] [PubMed]
65. Goodsell, D.S.; Sanner, M.F.; Olson, A.J.; Forli, S. The AutoDock suite at 30. *Protein Sci.* **2021**, *30*, 31–43. [CrossRef]
66. Mokgautsi, N.; Wen, Y.-T.; Lawal, B.; Khedkar, H.; Sumitra, M.R.; Wu, A.T.; Huang, H.-S. An integrated bioinformatics study of a novel niclosamide derivative, nsc765689, a potential gsk3β/β-catenin/stat3/cd44 suppressor with anti-glioblastoma properties. *Int. J. Mol. Sci.* **2021**, *22*, 2464. [CrossRef] [PubMed]

67. Lawal, B.; Liu, Y.-L.; Mokgautsi, N.; Khedkar, H.; Sumitra, M.R.; Wu, A.T.H.; Huang, H.-S. Pharmacoinformatics and Preclinical Studies of NSC765690 and NSC765599, Potential STAT3/CDK2/4/6 Inhibitors with Antitumor Activities against NCI60 Human Tumor Cell Lines. *Biomedicines* **2021**, *9*, 92. [CrossRef] [PubMed]

68. Raman, E.P.; Paul, T.J.; Hayes, R.L.; Brooks, C.L., 3rd. Automated, Accurate, and Scalable Relative Protein-Ligand Binding Free-Energy Calculations Using Lambda Dynamics. *J. Chem. Theory Comput.* **2020**, *16*, 7895–7914. [CrossRef]

69. Huang, Y.; Prasad, M.; Lemon, W.J.; Hampel, H.; Wright, F.A.; Kornacker, K.; LiVolsi, V.; Frankel, W.; Kloos, R.T.; Eng, C.; et al. Gene expression in papillary thyroid carcinoma reveals highly consistent profiles. *Proc. Natl. Acad. Sci. USA* **2001**, *98*, 15044–15049. [CrossRef]

70. Petrakis, D.; Vassilopoulou, L.; Mamoulakis, C.; Psycharakis, C.; Anifantaki, A.; Sifakis, S.; Docea, A.O.; Tsiaoussis, J.; Makrigiannakis, A.; Tsatsakis, A.M. Endocrine Disruptors Leading to Obesity and Related Diseases. *Int. J. Environ. Res. Public Health* **2017**, *14*, 1282. [CrossRef]

71. Takano, T. Natural history of thyroid cancer [Review]. *Endocr. J.* **2017**, *64*, 237–244. [CrossRef]

72. Schlumberger, M.; Parmentier, C.; Delisle, M.J.; Couette, J.E.; Droz, J.P.; Sarrazin, D. Combination therapy for anaplastic giant cell thyroid carcinoma. *Cancer* **1991**, *67*, 564–566. [CrossRef]

73. Tennvall, J.; Lundell, G.; Wahlberg, P.; Bergenfelz, A.; Grimelius, L.; Akerman, M.; Hjelm Skog, A.L.; Wallin, G. Anaplastic thyroid carcinoma: Three protocols combining doxorubicin, hyperfractionated radiotherapy and surgery. *Br. J. Cancer* **2002**, *86*, 1848–1853. [CrossRef] [PubMed]

74. Ancker, O.V.; Krüger, M.; Wehland, M.; Infanger, M.; Grimm, D. Multikinase Inhibitor Treatment in Thyroid Cancer. *Int. J. Mol. Sci.* **2019**, *21*, 10. [CrossRef] [PubMed]

75. O'Neill, C.J.; Oucharek, J.; Learoyd, D.; Sidhu, S.B. Standard and emerging therapies for metastatic differentiated thyroid cancer. *Oncologist* **2010**, *15*, 146–156. [CrossRef] [PubMed]

76. Kapiteijn, E.; Schneider, T.C.; Morreau, H.; Gelderblom, H.; Nortier, J.W.R.; Smit, J.W.A. New treatment modalities in advanced thyroid cancer. *Ann. Oncol.* **2012**, *23*, 10–18. [CrossRef] [PubMed]

77. Tang, M.; Wang, Q.; Wang, K.; Wang, F. Mesenchymal stem cells-originated exosomal microRNA-152 impairs proliferation, invasion and migration of thyroid carcinoma cells by interacting with DPP4. *J. Endocrinol. Investig.* **2020**, *43*, 1787–1796. [CrossRef] [PubMed]

78. Tseng, C.H. Sitagliptin use and thyroid cancer risk in patients with type 2 diabetes. *Oncotarget* **2016**, *7*, 24871–24879. [CrossRef]

79. Chen, J.-H.; Wu, A.T.H.; Lawal, B.; Tzeng, D.T.W.; Lee, J.-C.; Ho, C.-L.; Chao, T.-Y. Identification of Cancer Hub Gene Signatures Associated with Immune-Suppressive Tumor Microenvironment and Ovatodiolide as a Potential Cancer Immunotherapeutic Agent. *Cancers* **2021**, *13*, 3847. [CrossRef]

80. Lawal, B.; Lee, C.-Y.; Mokgautsi, N.; Sumitra, M.R.; Khedkar, H.; Wu, A.T.H.; Huang, H.-S. mTOR/EGFR/iNOS/MAP2K1/FGFR/ TGFB1 Are Druggable Candidates for N-(2,4-Difluorophenyl)-2′,4′-Difluoro-4-Hydroxybiphenyl-3-Carboxamide (NSC765598), With Consequent Anticancer Implications. *Front. Oncol.* **2021**, *11*, 656738. [CrossRef]

81. Yeh, Y.-C.; Lawal, B.; Hsiao, M.; Huang, T.-H.; Huang, C.-Y.F. Identification of NSP3 (SH2D3C) as a Prognostic Biomarker of Tumor Progression and Immune Evasion for Lung Cancer and Evaluation of Organosulfur Compounds from Allium sativum L. as Therapeutic Candidates. *Biomedicines* **2021**, *9*, 1582. [CrossRef]

82. Lawal, B.; Wang, Y.-C.; Wu, A.T.H.; Huang, H.-S. Pro-Oncogenic c-Met/EGFR, Biomarker Signatures of the Tumor Microenvironment are Clinical and Therapy Response Prognosticators in Colorectal Cancer, and Therapeutic Targets of 3-Phenyl-2H-benzo[e][1,3]-Oxazine-2,4(3H)-Dione Derivatives. *Front. Pharmacol.* **2021**, *12*, 691234. [CrossRef]

83. Lawal, B.; Kuo, Y.-C.; Tang, S.-L.; Liu, F.-C.; Wu, A.T.H.; Lin, H.-Y.; Huang, H.-S. Transcriptomic-Based Identification of the Immuno-Oncogenic Signature of Cholangiocarcinoma for HLC-018 Multi-Target Therapy Exploration. *Cells* **2021**, *10*, 2873. [CrossRef] [PubMed]

84. Lawal, B.; Kuo, Y.-C.; Sumitra, M.R.; Wu, A.T.; Huang, H.-S. In vivo Pharmacokinetic and Anticancer Studies of HH-N25, a Selective Inhibitor of Topoisomerase I, and Hormonal Signaling for Treating Breast Cancer. *J. Inflamm. Res.* **2021**, *14*, 1–13. [CrossRef] [PubMed]

85. Kitchen, D.B.; Decornez, H.; Furr, J.R.; Bajorath, J. Docking and scoring in virtual screening for drug discovery: Methods and applications. *Nat. Rev. Drug Discov.* **2004**, *3*, 935–949. [CrossRef] [PubMed]

Technical Note

DRPPM-EASY: A Web-Based Framework for Integrative Analysis of Multi-Omics Cancer Datasets

Alyssa Obermayer [1], Li Dong [2], Qianqian Hu [3], Michael Golden [4], Jerald D. Noble [5], Paulo Rodriguez [6], Timothy J. Robinson [5], Mingxiang Teng [1], Aik-Choon Tan [1] and Timothy I. Shaw [1,*]

[1] Department of Biostatistics and Bioinformatics, H. Lee Moffitt Cancer Center, Tampa, FL 33612, USA; alyssa.obermayer@moffitt.org (A.O.); Mingxiang.Teng@moffitt.org (M.T.); AikChoon.Tan@moffitt.org (A.-C.T.)
[2] Computational Biology Department, St Jude Children's Research Hospital, Memphis, TN 38105, USA; li.dong@stjude.org
[3] Department of Drug Discovery, Moffitt Cancer Center, Tampa, FL 33612, USA; Qianqian.Hu@moffitt.org
[4] University of Central Florida, Orlando, FL 32816, USA; michaelgolden00true@gmail.com
[5] Department of Radiation Oncology, Moffitt Cancer Center, Tampa, FL 33612, USA; Jerald.Noble@moffitt.org (J.D.N.); Timothy.Robinson@moffitt.org (T.J.R.)
[6] Department of Immunology, Moffitt Cancer Center, Tampa, FL 33612, USA; Paulo.Rodriguez@moffitt.org
* Correspondence: timothy.shaw@moffitt.org

Simple Summary: With the influx of multi-omics profiling, effective integration of these data remains the bottleneck for omics-driven discovery. Thus, we developed DRPPM-EASY, an R Shiny framework for integrative multi-omics analysis of cancer datasets. Our tool enables the exploration of multi-omics data by providing a simple user interface that minimizes the need for computational experience. Furthermore, the interface can be deployed locally or on a webserver to facilitate scientific collaboration and discovery.

Abstract: High-throughput transcriptomic and proteomic analyses are now routinely applied to study cancer biology. However, complex omics integration remains challenging and often time-consuming. Here, we developed DRPPM-EASY, an R Shiny framework for integrative multi-omics analysis. We applied our application to analyze RNA-seq data generated from a USP7 knockdown in T-cell acute lymphoblastic leukemia (T-ALL) cell line, which identified upregulated expression of a TAL1-associated proliferative signature in T-cell acute lymphoblastic leukemia cell lines. Next, we performed proteomic profiling of the USP7 knockdown samples. Through DRPPM-EASY-Integration, we performed a concurrent analysis of the transcriptome and proteome and identified consistent disruption of the protein degradation machinery and spliceosome in samples with USP7 silencing. To further illustrate the utility of the R Shiny framework, we developed DRPPM-EASY-CCLE, a Shiny extension preloaded with the Cancer Cell Line Encyclopedia (CCLE) data. The DRPPM-EASY-CCLE app facilitates the sample querying and phenotype assignment by incorporating meta information, such as genetic mutation, metastasis status, sex, and collection site. As proof of concept, we verified the expression of TP53 associated DNA damage signature in TP53 mutated ovary cancer cells. Altogether, our open-source application provides an easy-to-use framework for omics exploration and discovery.

Keywords: R Shiny application; RNA-seq; proteomics; multi-omics analysis; T-cell acute lymphoblastic leukemia; CCLE

Citation: Obermayer, A.; Dong, L.; Hu, Q.; Golden, M.; Noble, J.D.; Rodriguez, P.; Robinson, T.J.; Teng, M.; Tan, A.-C.; Shaw, T.I. DRPPM-EASY: A Web-Based Framework for Integrative Analysis of Multi-Omics Cancer Datasets. *Biology* **2022**, *11*, 260. https://doi.org/10.3390/biology11020260

Academic Editors: Shibiao Wan, Yiping Fan, Chunjie Jiang and Shengli Li

Received: 31 December 2021
Accepted: 4 February 2022
Published: 8 February 2022

Publisher's Note: MDPI stays neutral with regard to jurisdictional claims in published maps and institutional affiliations.

1. Introduction

Multi-omics profiling of cancer patient samples and cell lines is becoming a staple of cancer research [1]. These technologies have a high potential for advancing our understanding of tumor biology and, in turn, reveal novel targets for treatment and diagnosis [2,3]. To date, a brief survey of the existing database reveals more than 500K cancer samples from GEO [4,5] and 90K pre-computed cancer expression data from recount3 [6]. Additionally, there are close to 4K mass spectrometry profiling of cancer patient samples from the Clinical Proteomic Tumor Analysis Consortium (CPTAC) data [7]. Large consortium projects, such as the Cancer Cell Line Encyclopedia (CCLE), have also generated many high-throughput datasets, such as transcript expression, RNA splicing, proteome profiling, drug response, and genetic screening data [8].

With the influx of multi-omics profiling, effective integration of these data remains the bottleneck for omics-driven discovery. The development of a simple user interface that minimizes the need for computational experience is of high interest to the community [9]. Several web-based tools are now available to perform general expression analysis of proteomics (e.g., POMAShiny [10]) and transcriptome data (e.g., TCC-GUI [11], START App [12], and GENAVi [13]). Multi-omics approaches for network analysis (e.g., MiBiOmics [14] and JUMPn [15]) are also available as a Shiny app. Web tools also exist for analyzing large datasets from the Gene Expression Omnibus (GEO) data (e.g., shinyGEO [16], ImaGEO [17]) and the cancer dependency map (e.g., shinyDepMap [18]). However, these applications tend to have limited features for analyzing complex heterogeneous phenotypes in cell lines and patients, such as mutation of genomic drivers, cell line characteristics, sex, or metastasis status. Additionally, none of these tools provides a streamlined pipeline to assess similarities and differences between omics datasets, such as transcriptome and proteome comparisons, or comparisons between mouse and human cancer models.

To address these challenges, we have developed DRPPM-EASY, a Shiny app built with an open-source R programming language that can be run as a local instance or deployed online. Here, our app is divided into two major modules: (1) a one-stop expression analysis for gene expression analysis and (2) an integrative framework for comparing omics data. As a proof of concept, we further implemented an app for querying and automating extraction of sample groupings of CCLE data for downstream analysis. The source code of our application can be downloaded from https://github.com/shawlab-moffitt/DRPPM-EASY-ExprAnalysisShinY (accessed on 1 February 2022).

2. Materials and Methods

2.1. Module 1. DRPPM-EASY APP Implementation

The DRPPM-EASY app is a Shiny web app built with an open-source R programming language (V.4.1.0). The Shiny framework leverages existing RNA-seq analysis packages to put together a one-stop analysis framework (Figure 1A) for data exploration (Table 1), differential expression analysis (Table 2), and gene set enrichment analysis (Table 3). The data exploration section allows the user to perform unsupervised and supervised hierarchical clustering. Clustering can be further evaluated by different types of distance calculations (i.e., ward, average, complete, centroid) or variable gene ranking strategy (mean absolute deviation or variance). The relative gene expression can be examined across sample groups by a boxplot or scatter plot to examine the gene expression of the positive control associated with the experimental design. Differential gene expression is performed by LIMMA [19] and can be visualized as a volcano plot and MA-plot. The list of differentially expressed genes can be further examined by pathway enrichment analysis (Figure 1A). Finally, the user can perform gene set enrichment analysis (GSEA), which ranks the genes based on signal-to-noise between the user-selected phenotype to examine enriched genes associated with a gene set signature (Figure 1A). A complementary strategy to estimate enrichment scores for individual samples can be performed by single-sample GSEA (ssGSEA) implemented in the GSVA library [20]. Finally, these single-sample enrichment scores can be downloaded as a tab-delimited table or visualized as a boxplot.

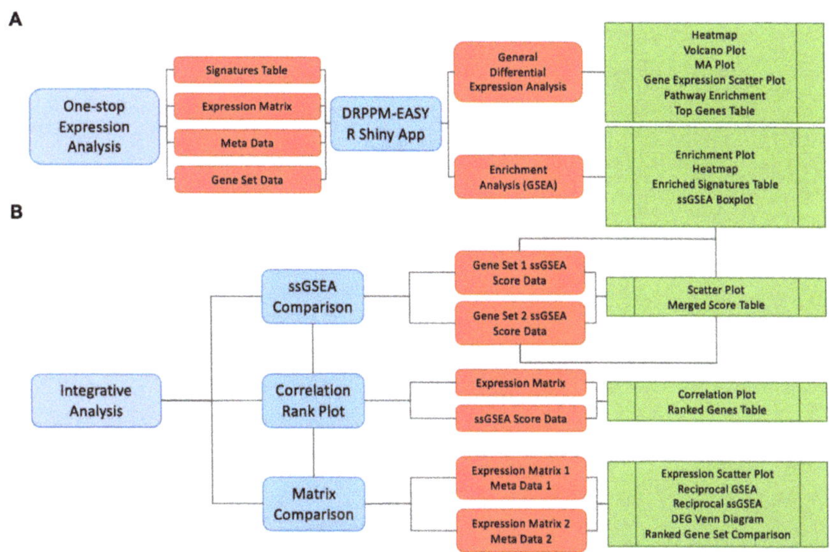

Figure 1. DRPPM-EASY expression analysis pipeline. (**A**) Schematic workflow of DRPPM-EASY. The pipeline takes in input files of an expression matrix, a sample meta-file specifying sample grouping, and a gene set database for GSEA. A GSEA enriched signature table is generated as a preprocessing step, which is used as input to the R Shiny app. The app generates two modes of exploring the data: (1) general differential gene expression analysis and (2) gene set enrichment analysis. The result from the analysis can be downloaded as output tables. (**B**) Schematic of the integrative analysis with three major features for pathway signature comparison. The app has three modes of integrative analysis: (1) scatter plot mode, (2) correlation plot mode, and (3) paired multi-omics analysis.

Table 1. Data Exploration Module.

	App Function	Description
E1	Unsupervised Heatmap	• Top variable gene selection • Expression data is log2 transformed then z-normalized • User-specified clustering method
E2	Scatter Plot	• User selects two genes of interest • Expression values compared via interactive scatter plot (log2 transformation is optional)
E3	Custom Heatmap	• Visualize user-selected genes and samples • Expression data is log2 transformed and z-normalized • User-specified clustering method
E4	Box Plot	• Gene expression in each group are shown • Expression values are log2 transformed • Comparing groups for statistical differences

Table 2. Differential Expression Analysis Module.

	App Function	Description
DEA1	Volcano Plot	• User selects comparison groups • Differential gene expression analysis with LIMMA • Up- and downregulated differentially expressed genes determined with user input
DEA2	MA Plot	• User selects comparison groups • Differential gene expression analysis with LIMMA • Up- and downregulated differentially expressed genes determined with user input
DEA4	Pathway Enrichment Analysis	• User selects comparison groups and gene set/pathway • Differential gene expression analysis with LIMMA • Pathway enrichment analysis using enrichR

Table 3. Gene Set Enrichment Analysis Module.

	App Function	Description
GA1	Enrichment Plot	• User selects comparison groups • Signal-to-noise ranking performed on expression data • GSEA function performed with chosen gene set
GA2	Gene Expression Heatmap	• User selects comparison groups • Signal-to-noise ranking performed on expression data • GSEA function performed with chosen gene set • Expression data log2 transformed and scaled • Genes from chosen gene set displayed in the heatmap
GA3	GSEA Summary Table	• Displays user pre-generated enriched signatures table
GA4	Generate Summary Table	• GSEA function performed on expression data with user input GMT file • Enriched signatures table produced is displayed
GA5	ssGSEA Boxplots	• User-selects gene set and single-sample GSEA method • Comparing groups for statistical differences

2.2. Module 2. The DRPPM-EASY-Integration App Implementation

The DRPPM-EASY-Integration provides an explorer for the user to upload normalized RNA expression, proteomic quantification, or ssGSEA scores to evaluate the potential relationship between these features (Figure 1B). These can be evaluated by either a 1:1 scatter plot or 1:n rank of Spearman correlation rho values (Table 4). The integrative app also allows the user to perform concurrent differential expression analysis and integration of two expression matrices, for example, to compare RNA and protein expression matrices. The fold change can be compared between the two datasets (Table 4), and differentially expressed genes can be compared by reciprocal GSEA or ssGSEA. Direct overlap between the differentially expressed genes is shown as a Venn diagram and further compared to existing gene set databases by Fisher's exact test, Cohen's kappa score, and the Jaccard index.

Table 4. Integrative Analysis.

	App Function	Description
IA1	Scatter Plot Comparison	• User input features are merged and plotted • Samples are colored based on metadata type
IA2	Correlation Rank Plot	• Assessing the relationship between ssGSEA score and gene expression performed • Correlation can be performed as Spearman, Pearson, or Kendall • Correlation values plotted by rank from lowest to highest
IA3	Matrix Comparison File Upload	• Upload two expression matrices and two metadata files
IA4	Log2FC Comparison Scatter Plot	• Differential gene expression analysis with LIMMA performed on both matrices • Log2 fold change values subset and difference between matrices calculated • Expression data displayed as scatter plot
IA5	Reciprocal GSEA	• Differential gene expression analysis with LIMMA • Four gene sets derived differentially expressed genes (two upregulated, and two downregulated gene set) • GSEA performed on the reciprocal data
IA6	Reciprocal ssGSEA	• Differential gene expression analysis with LIMMA • Four gene sets derived differentially expressed genes (two upregulated, and two downregulated gene set) • ssGSEA performed on the reciprocal data
IA7	Venn Diagram	• Differential gene expression analysis with LIMMA • Overlapping differentially expressed genes • Perform Fisher's exact test. Calculate Cohen's kappa, and Jaccard index to compare between the two matrix and across user selected pathways.

2.3. Installation and User Guide

The source code and user guide are available for download on the project's GitHub page. The GitHub page includes the list of individual R packages and their version along with an installation script for all package dependencies.

2.4. RNA Sequencing Analysis

USP7 samples were prepared as described in Shaw et al. [21]. Briefly, human T-ALL cell lines Jurkat (ATCC) cells were transduced with USP7 shRNA lentivirus and sorted for GFP positive cells or selected by puromycin. RNA samples were isolated using RNeasy Mini Kit (QIAGEN) and subjected to paired-end 2 × 151 base-pair RNA-seq sequencing (Illumina), 10 Jurkat samples—of which 6 were treated with shRNA and 4 were treated with a scramble RNA—were profiled by RNA-seq. RNA-seq data were processed by a custom pipeline (WRAP, https://github.com/gatechatl/DRPPM_Example_Input_Output/tree/master/WRAP:Wrapper-for-my-RNAseq-Analysis-Pipeline (accessed on 1 August 2021. RNA-seq reads were aligned using the STAR 2.7.1a aligner [22] in the two-pass mode to the human hg38 genome build using gene annotations provided by the Gencode v31 gene models. Read count for each gene was obtained with HT-seq [23]. Reads were normalized to fragments per kilobase million (FPKM) for each gene.

2.5. Whole Proteomics Mass Spectrometry and Data Analysis

The 10-plex TMT labeled mass spectrometry experiment was performed with a previously published protocol with slight modification [24,25] (See Supplementary Method,

mechanism of USP7 to drive the NOTCH association leukemia signature will need to be carefully examined in future studies.

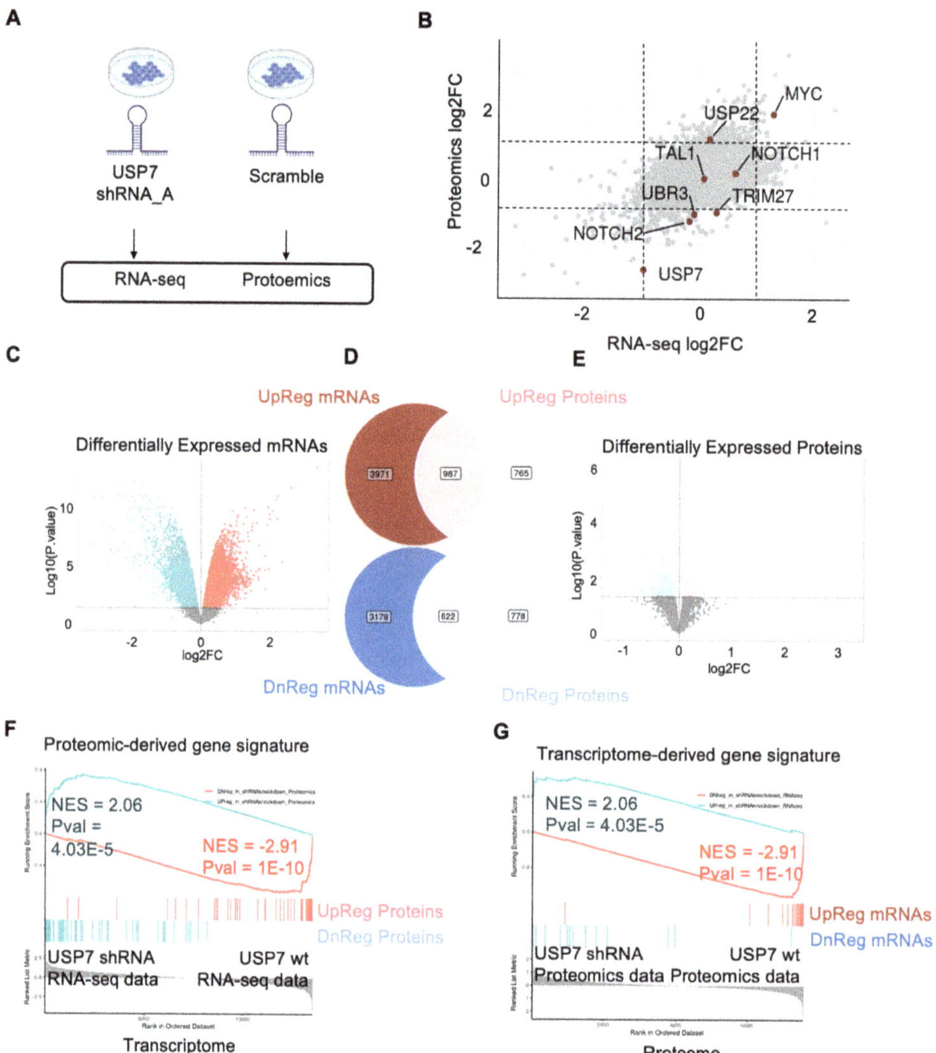

Figure 3. Integrated analysis example of proteomics and transcriptomics USP7 silenced Jurkat cells. (**A**) Jurkat samples treated with USP7 shRNA and scramble were profiled by RNA sequencing and TMT mass spectrometry. (**B**) The log2 fold change from the differential expression analyses is plotted. Positive log2FC indicates upregulated expression after USP7 silencing. Negative log2FC indicates downregulated expression after USP7 knockdown. Dotted line indicates the −1 and 1 log2FC cutoff. (**C**) Upregulated and downregulated gene signatures derived from differentially expressed mRNAs. (**D**) Venn diagram of genes differentially upregulated (top panel) and downregulated (bottom panel) in the transcriptome (left) and proteome (right). (**E**) Up-regulated and downregulated gene signatures derived from differentially expressed proteins. (**F,G**) Reciprocal GSEA of differentially expressed genes derived from the transcriptome and examined in the proteomics data (**F**). Similarly, differentially expressed proteins were first derived then examined in the transcriptome data by GSEA (**G**).

The DRPPM-EASY-Integration includes features assessing the consistency between two datasets. Using the RNA-seq and proteomic data as proof of concept, DRPPM-EASY-Integration found 987 genes consistently upregulated, and 622 genes consistently down-regulated in both datasets (Figure 3C–E). A connectivity map-inspired strategy [29,30] was applied to compare the consistency between the two datasets using reciprocal enrichment. Specifically, differential expressed genes in one dataset was used to derive a gene signature for GSEA to test in the other dataset. For example, differentially expressed proteins (Figure 3F) were applied as a GSEA gene set and tested for enrichment in the transcriptome data (Figure 3G). Similarly, gene sets derived from differentially expressed transcripts (Figure 3C) were tested for enrichment in the proteome data (Figure 3H). We then compared the significance of the overlapping differentially expressed genes against other pathway databases, such as Hallmark and KEGG. The overlap was evaluated by Fisher's exact test, Cohen's kappa, and Jaccard index. Consistently, the RNA and protein were most significantly overlapped compared to other gene sets. Moreover, the spliceosome and ubiquitin-mediated proteolysis pathways from KEGG and the unfolded protein response and MYC pathway from Hallmark were consistently enriched in both datasets (Supplementary Figure S3B,C; Supplementary Tables S1 and S2).

3.2. DRPPM-EASY-CCLE Use Case 2

To further illustrate the DRPPM-EASY functionality, we developed DRRPM-EASY-CCLE, an extended app with features to select samples from the Cancer Cell Line Encyclopedia (CCLE) data. The app is preloaded with 1379 CCLE samples spanning 37 lineages, 96 lineage sub-types, and 33 diseases. For the genetic characterization, 299 cancer drivers [31] were selected and further divided based on the damaging and non-damaging variant status from DepMap [32] (see Supplementary Table S3 for the complete phenotype categories). As an example, we extracted ovary cancer cell lines and performed expression analysis comparing *TP53* mutation status to its wild-type counterpart (Figure 4A). In *TP53* mutated ovary cancer cells, we found a decreased DNA damage response gene signature (Figure 4B), thereby solidifying the role of *TP53* loss-of-function for regulating DNA damage in these ovarian cancer cells.

Previously, KRAS was found to be frequently mutated in non-small cell lung cancer (NSCLC) and is associated with drug resistance [33]. Thus, we analyzed NSCLC cell lines and compared KRAS mutation status to its wild-type counterpart (Figure 4C). By pathway analysis, the MsigDB defined KRAS signature was consistently upregulated in our KRAS mutated samples (Supplementary Figure S4A). Interestingly, top pathways enriched in the KRAS mutated samples are associated with an anti-apoptosis signature (Supplementary Figure S4B). By ssGSEA, amplified expression in KRAS mutated NSCLC cells were enriched with genes that negatively regulate apoptosis (Figure 4D) and upregulating genes that associated with stress granule assembly and disassembly (Figure 4E), which is a dynamic process fundamental to surviving under stress [34]. Interestingly, oncogenic KRAS-driven stress granules were previously identified in pancreatic and colorectal adenocarcinoma [35]; thus, our result suggests a similar stress response in NSCLC cells.

To further expand our functionality for exploring these large project data, we have also implemented features that enable users to upload their own expression matrix to perform an integrative analysis in CCLE and lung squamous cell carcinoma CPTAC datasets https://github.com/shawlab-moffitt/DRPPM-EASY-LargeProject-Integration (accessed on 1 February 2022) (Supplementary Figures S5A–C). Altogether, our framework provides a user-friendly environment to categorize the samples for downstream analysis with a high potential for novel discovery.

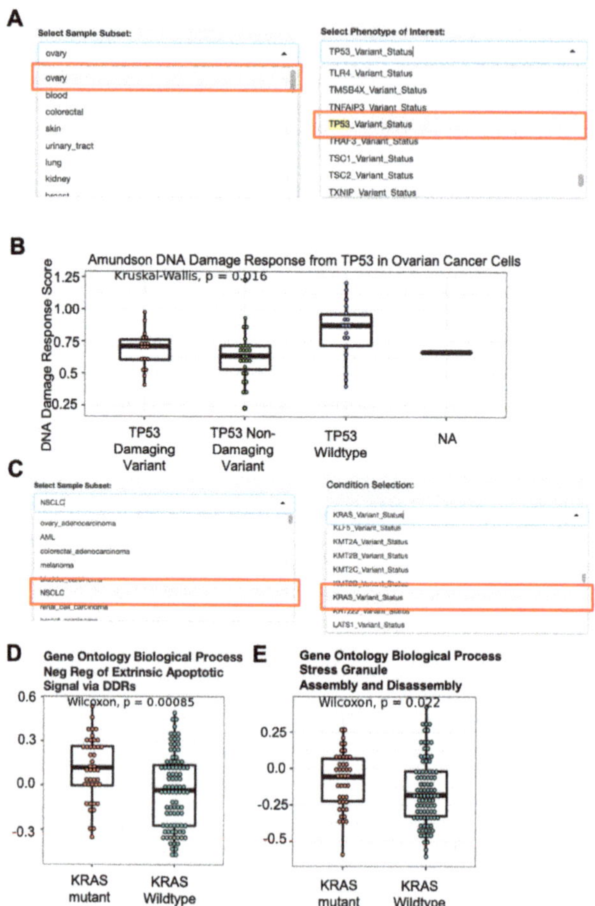

Figure 4. Use case analysis example of CCLE Expression data. (**A**) Drop-down menu selection of sample cohort and sample phenotype characteristic. CCLE ovary samples and TP53 mutation status were selected from the drop-down menu option. (**B**) Single-sample GSEA analysis of genes defining the DNA damage response by Amundson et al. Analyzed samples were selected from the drop-down menu from (**A**). (**C**) Drop-down menu selection of sample cohort and sample phenotype characteristic. CCLE non-small cell lung cancer samples and phenotype associated with the KRAS mutation status were selected from the drop-down menu option. (**D**) Single sample GSEA analysis of genes negatively regulating the DNA damage response. (**E**) Single sample GSEA of genes defining the stress granule assembly and disassembly. Gene sets were compiled from Biological Pathways from the Gene Ontology database (GOBP). Analyzed samples were selected from the drop-down menu from (**C**).

4. Discussion

An effective method for visualization and data analysis is key to the analysis of multi-omics data that captures the molecular processes of cancer initiation and progression. Several Shiny apps have been published to date and can be categorized into the following three categories: (1) tools that focus on pairwise differential expression and biomarker discovery (e.g., POMAShiny 10], TCC-GUI [11], and START App [12]), (2) tools that perform pathway and network analysis (e.g., iOmics [14] and JUMPn [15]), and (3) tools that facilitate the query of large datasets, such as from public repositories or consortium deposited datasets and deposited expression data (e.g., shinyGEO [16], ImaGEO [17], and GENAVi [13]). While numerous web tools have been developed thus far, there is a lack of tools that directly address challenges associated with multi-data integration, such as evaluating the consistency between omics datasets.

Here, we developed an interactive software tool, DRPPM-EASY, that allows users to perform complex omics data integration in both small (pairwise comparison) and large (consortium) projects. DRPPM-EASY puts together an interactive flexible interface that enables the exploration of biomarkers and enriched pathways across multiple datasets. DRPPM-EASY can perform routine gene analysis, such as hierarchical clustering, differential gene expression, pathway analysis, GSEA, and ssGSEA. Additionally, DRPPM-EASY can perform a joint analysis of two expression datasets. As an example, we have highlighted the application's ability to evaluate the consistency between transcriptome and protein datasets. This is made possible by deriving a gene set feature in one dataset (i.e., transcriptomics), which is applied in the GSEA analysis of the other dataset (i.e., proteomics). DRPPM-EASY can be easily adapted for large consortium data, which we highlight as an example in CCLE cancer cell lines and lung squamous cell carcinoma CPTAC proteome data. Finally, to further expand the utility of our tool, the user can upload their own expression data and use it to compare against CCLE cell lines and lung squamous cell carcinoma proteome data. One major limitation of our application requires the user to normalize their gene expression matrix prior to using our application. Existing pipelines are available to streamline the normalization procedure, such as Shiny-Seq [36]. A normalization procedure will be included in future updates of our application.

Finally, the ability to run the application with a user interface on a local desktop reduces the need for computational domain knowledge of expression analysis. The DRPPM-EASY application can be set up on the server in real-time, enabling collaborative discussion on potential hypotheses derived from the high-throughput data. Our tool also ensures reproducibility of the data analysis, which is one of the most significant issues in omics research [37]. While the current application is highlighted to work in RNA-seq and proteomics data, our framework could easily be adapted to incorporate drug response, genetic screening, or splicing associated features in future versions of our application. Thus, we believe DRPPM-EASY will be a useful and valuable tool for the biomedical research community.

Supplementary Materials: The following supporting information can be downloaded at: https://www.mdpi.com/article/10.3390/biology11020260/s1, Supplementary Method. Supplementary Table S1. KEGG Pathways Jointly Enriched in the Transcriptome and Proteome. Supplementary Table S2. Hallmark Pathways Jointly Enriched in the Transcriptome and Proteome. Supplementary Table S3. CCLE Sample Meta-Information. Supplementary Figure S1. Schematic of the GSEA pre-processing. Supplementary Figure S2. Unsupervised hierarchical clustering of Jurkat samples after USP7 knockdown. Supplementary Figure S3. Experimental design of the total proteome profiling of the USP7 knockdown experiment. Supplementary Figure S4. Pathway enrichment analysis of genes differentially upregulated in KRAS mutated samples in NSCLC. Supplementary Figure S5. Screen shot showing the user option to upload user data in the DRPPM-Large Project Integration.

Author Contributions: Conceptualization, T.I.S. and A.-C.T.; methodology, T.I.S., A.O.; software, T.I.S., A.O., Q.H. and J.D.N.; validation, L.D.; formal analysis, T.I.S.; data curation, M.G.; writing—original draft preparation, T.I.S., A.-C.T.; writing—review and editing, T.I.S., A.-C.T., P.R., T.J.R. and M.T.; supervision, T.I.S.; funding acquisition, T.I.S. All authors have read and agreed to the published version of the manuscript.

Funding: Moffitt Cancer Center (NCI P30 CA076292) and the Moffitt Cancer Center Department of Biostatistics and Bioinformatics Pilot Projects (PI: T.S.).

Institutional Review Board Statement: Not applicable.

Informed Consent Statement: Not applicable.

Data Availability Statement: The developed software and processed data can be downloaded from the following GitHub page https://github.com/shawlab-moffitt/DRPPM-EASY-ExprAnalysisShinY (accessed on 1 February 2022).

Acknowledgments: This work has been supported in part by the Biostatistics and Bioinformatics Shared Resource at the Moffitt Cancer Center (NCI P30 CA076292) and the Moffitt Cancer Center Department of Biostatistics and Bioinformatics Pilot Project (PI: T.S.). We would like to thank Rodrigo Carvajal and Guillermo Gonzalez-Calderon for their help in setting up the internal web server. We would like to thank Dongliang Du and Ling Cen for their help in the USP7 RNAseq analysis.

Conflicts of Interest: The funders had no role in the design of the study; in the collection, analyses, or interpretation of data; in the writing of the manuscript, or in the decision to publish the results.

References

1. De Anda-Jauregui, G.; Hernandez-Lemus, E. Computational Oncology in the Multi-Omics Era: State of the Art. *Front. Oncol.* **2020**, *10*, 423. [CrossRef]
2. Menyhart, O.; Gyorffy, B. Multi-omics approaches in cancer research with applications in tumor subtyping, prognosis, and diagnosis. *Comput. Struct. Biotechnol. J.* **2021**, *19*, 949–960. [CrossRef]
3. Chai, A.W.Y.; Tan, A.C.; Cheong, S.C. Uncovering drug repurposing candidates for head and neck cancers: Insights from systematic pharmacogenomics data analysis. *Sci. Rep.* **2021**, *11*, 23933. [CrossRef]
4. Edgar, R.; Domrachev, M.; Lash, A.E. Gene Expression Omnibus: NCBI gene expression and hybridization array data repository. *Nucleic Acids Res.* **2002**, *30*, 207–210. [CrossRef]
5. Barrett, T.; Wilhite, S.E.; Ledoux, P.; Evangelista, C.; Kim, I.F.; Tomashevsky, M.; Marshall, K.A.; Phillippy, K.H.; Sherman, P.M.; Holko, M.; et al. NCBI GEO: Archive for functional genomics data sets—Update. *Nucleic Acids Res.* **2013**, *41*, D991–D995. [CrossRef]
6. Wilks, C.; Zheng, S.C.; Chen, F.Y.; Charles, R.; Solomon, B.; Ling, J.P.; Imada, E.L.; Zhang, D.; Joseph, L.; Leek, J.T.; et al. Recount3: Summaries and queries for large-scale RNA-seq expression and splicing. *Genome Biol.* **2021**, *22*, 323. [CrossRef]
7. Edwards, N.J.; Oberti, M.; Thangudu, R.R.; Cai, S.; McGarvey, P.B.; Jacob, S.; Madhavan, S.; Ketchum, K.A. The CPTAC Data Portal: A Resource for Cancer Proteomics Research. *J. Proteome Res.* **2015**, *14*, 2707–2713. [CrossRef]
8. Ghandi, M.; Huang, F.W.; Jane-Valbuena, J.; Kryukov, G.V.; Lo, C.C.; McDonald, E.R., 3rd; Barretina, J.; Gelfand, E.T.; Bielski, C.M.; Li, H.; et al. Next-generation characterization of the Cancer Cell Line Encyclopedia. *Nature* **2019**, *569*, 503–508. [CrossRef]
9. Davis-Turak, J.; Courtney, S.M.; Hazard, E.S.; Glen, W.B., Jr.; da Silveira, W.A.; Wesselman, T.; Harbin, L.P.; Wolf, B.J.; Chung, D.; Hardiman, G. Genomics pipelines and data integration: Challenges and opportunities in the research setting. *Expert Rev. Mol. Diagn.* **2017**, *17*, 225–237. [CrossRef]
10. Castellano-Escuder, P.; Gonzalez-Dominguez, R.; Carmona-Pontaque, F.; Andres-Lacueva, C.; Sanchez-Pla, A. POMAShiny: A user-friendly web-based workflow for metabolomics and proteomics data analysis. *PLoS Comput. Biol.* **2021**, *17*, e1009148. [CrossRef]
11. Su, W.; Sun, J.; Shimizu, K.; Kadota, K. TCC-GUI: A Shiny-based application for differential expression analysis of RNA-Seq count data. *BMC Res. Notes* **2019**, *12*, 133. [CrossRef] [PubMed]
12. Nelson, J.W.; Sklenar, J.; Barnes, A.P.; Minnier, J. The START App: A web-based RNAseq analysis and visualization resource. *Bioinformatics* **2017**, *33*, 447–449. [CrossRef]
13. Reyes, A.L.P.; Silva, T.C.; Coetzee, S.G.; Plummer, J.T.; Davis, B.D.; Chen, S.; Hazelett, D.J.; Lawrenson, K.; Berman, B.P.; Gayther, S.A.; et al. GENAVi: A shiny web application for gene expression normalization, analysis and visualization. *BMC Genom.* **2019**, *20*, 745. [CrossRef]
14. Zoppi, J.; Guillaume, J.F.; Neunlist, M.; Chaffron, S. MiBiOmics: An interactive web application for multi-omics data exploration and integration. *BMC Bioinform.* **2021**, *22*, 6. [CrossRef]
15. Vanderwall, D.; Suresh, P.; Fu, Y.; Cho, J.H.; Shaw, T.I.; Mishra, A.; High, A.A.; Peng, J.; Li, Y. JUMPn: A Streamlined Application for Protein Co-Expression Clustering and Network Analysis in Proteomics. *J. Vis. Exp.* **2021**, *176*. [CrossRef]
16. Dumas, J.; Gargano, M.A.; Dancik, G.M. shinyGEO: A web-based application for analyzing gene expression omnibus datasets. *Bioinformatics* **2016**, *32*, 3679–3681. [CrossRef] [PubMed]
17. Toro-Dominguez, D.; Martorell-Marugan, J.; Lopez-Dominguez, R.; Garcia-Moreno, A.; Gonzalez-Rumayor, V.; Alarcon-Riquelme, M.E.; Carmona-Saez, P. ImaGEO: Integrative gene expression meta-analysis from GEO database. *Bioinformatics* **2019**, *35*, 880–882. [CrossRef]
18. Shimada, K.; Bachman, J.A.; Muhlich, J.L.; Mitchison, T.J. ShinyDepMap, a tool to identify targetable cancer genes and their functional connections from Cancer Dependency Map data. *Elife* **2021**, *10*, e57116. [CrossRef]
19. Ritchie, M.E.; Phipson, B.; Wu, D.; Hu, Y.; Law, C.W.; Shi, W.; Smyth, G.K. Limma powers differential expression analyses for RNA-sequencing and microarray studies. *Nucleic Acids Res.* **2015**, *43*, e47. [CrossRef] [PubMed]
20. Hanzelmann, S.; Castelo, R.; Guinney, J. GSVA: Gene set variation analysis for microarray and RNA-seq data. *BMC Bioinform.* **2013**, *14*, 7. [CrossRef] [PubMed]

21. Shaw, T.I.; Dong, L.; Tian, L.; Qian, C.; Liu, Y.; Ju, B.; High, A.; Kavdia, K.; Pagala, V.R.; Shaner, B.; et al. Integrative network analysis reveals USP7 haploinsufficiency inhibits E-protein activity in pediatric T-lineage acute lymphoblastic leukemia (T-ALL). *Sci. Rep.* **2021**, *11*, 5154. [CrossRef]
22. Dobin, A.; Davis, C.A.; Schlesinger, F.; Drenkow, J.; Zaleski, C.; Jha, S.; Batut, P.; Chaisson, M.; Gingeras, T.R. STAR: Ultrafast universal RNA-seq aligner. *Bioinformatics* **2013**, *29*, 15–21. [CrossRef] [PubMed]
23. Anders, S.; Pyl, P.T.; Huber, W. HTSeq—A Python framework to work with high-throughput sequencing data. *Bioinformatics* **2015**, *31*, 166–169. [CrossRef]
24. Xu, P.; Duong, D.M.; Peng, J. Systematical optimization of reverse-phase chromatography for shotgun proteomics. *J. Proteome Res.* **2009**, *8*, 3944–3950. [CrossRef] [PubMed]
25. Pagala, V.R.; High, A.A.; Wang, X.; Tan, H.; Kodali, K.; Mishra, A.; Kavdia, K.; Xu, Y.; Wu, Z.; Peng, J. Quantitative protein analysis by mass spectrometry. *Methods Mol. Biol.* **2015**, *1278*, 281–305. [CrossRef]
26. Wang, X.; Li, Y.; Wu, Z.; Wang, H.; Tan, H.; Peng, J. JUMP: A tag-based database search tool for peptide identification with high sensitivity and accuracy. *Mol. Cell. Proteom.* **2014**, *13*, 3663–3673. [CrossRef]
27. Hao, Y.H.; Fountain, M.D., Jr.; Fon Tacer, K.; Xia, F.; Bi, W.; Kang, S.H.; Patel, A.; Rosenfeld, J.A.; Le Caignec, C.; Isidor, B.; et al. USP7 Acts as a Molecular Rheostat to Promote WASH-Dependent Endosomal Protein Recycling and Is Mutated in a Human Neurodevelopmental Disorder. *Mol. Cell* **2015**, *59*, 956–969. [CrossRef]
28. Jin, Q.; Martinez, C.A.; Arcipowski, K.M.; Zhu, Y.; Gutierrez-Diaz, B.T.; Wang, K.K.; Johnson, M.R.; Volk, A.G.; Wang, F.; Wu, J.; et al. USP7 Cooperates with NOTCH1 to Drive the Oncogenic Transcriptional Program in T-Cell Leukemia. *Clin. Cancer Res.* **2019**, *25*, 222–239. [CrossRef] [PubMed]
29. Lamb, J. The Connectivity Map: A new tool for biomedical research. *Nat. Rev. Cancer* **2007**, *7*, 54–60. [CrossRef]
30. Lamb, J.; Crawford, E.D.; Peck, D.; Modell, J.W.; Blat, I.C.; Wrobel, M.J.; Lerner, J.; Brunet, J.P.; Subramanian, A.; Ross, K.N.; et al. The Connectivity Map: Using gene-expression signatures to connect small molecules, genes, and disease. *Science* **2006**, *313*, 1929–1935. [CrossRef] [PubMed]
31. Bailey, M.H.; Tokheim, C.; Porta-Pardo, E.; Sengupta, S.; Bertrand, D.; Weerasinghe, A.; Colaprico, A.; Wendl, M.C.; Kim, J.; Reardon, B.; et al. Comprehensive Characterization of Cancer Driver Genes and Mutations. *Cell* **2018**, *173*, 371–385. [CrossRef] [PubMed]
32. Tsherniak, A.; Vazquez, F.; Montgomery, P.G.; Weir, B.A.; Kryukov, G.; Cowley, G.S.; Gill, S.; Harrington, W.F.; Pantel, S.; Krill-Burger, J.M.; et al. Defining a Cancer Dependency Map. *Cell* **2017**, *170*, 564–576. [CrossRef] [PubMed]
33. Adderley, H.; Blackhall, F.H.; Lindsay, C.R. KRAS-mutant non-small cell lung cancer: Converging small molecules and immune checkpoint inhibition. *EBioMedicine* **2019**, *41*, 711–716. [CrossRef]
34. Wheeler, J.R.; Matheny, T.; Jain, S.; Abrisch, R.; Parker, R. Distinct stages in stress granule assembly and disassembly. *Elife* **2016**, *5*, e18413. [CrossRef]
35. Grabocka, E.; Bar-Sagi, D. Mutant KRAS Enhances Tumor Cell Fitness by Upregulating Stress Granules. *Cell* **2016**, *167*, 1803–1813. [CrossRef]
36. Sundararajan, Z.; Knoll, R.; Hombach, P.; Becker, M.; Schultze, J.L.; Ulas, T. Shiny-Seq: Advanced guided transcriptome analysis. *BMC Res. Notes* **2019**, *12*, 432. [CrossRef] [PubMed]
37. Krassowski, M., Das, V., Sahu, S.K., Misra, B.B. State of the Field in Multi-Omics Research: From Computational Needs to Data Mining and Sharing. *Front. Genet.* **2020**, *11*, 610798. [CrossRef]

Article

Identification and Validation of an Annexin-Related Prognostic Signature and Therapeutic Targets for Bladder Cancer: Integrative Analysis

Xitong Yao [1,†], Xinlei Qi [1,†], Yao Wang [1,†], Baokun Zhang [2], Tianshuai He [1], Taoning Yan [1], Lu Zhang [1], Yange Wang [1], Hong Zheng [1], Guosen Zhang [1,*] and Xiangqian Guo [1,*]

[1] Cell Signal Transduction Laboratory, Department of Predictive Medicine, Institute of Biomedical Informatics, Bioinformatics Center, Henan Provincial Engineering Center for Tumor Molecular Medicine, School of Basic Medical Sciences, Academy for Advanced Interdisciplinary Studies, Henan University, Kaifeng 475004, China; 104753190854@henu.edu.cn (X.Y.); 104753190845@henu.edu.cn (X.Q.); 104753201156@henu.edu.cn (Y.W.); hts@henu.edu.cn (T.H.); 104753211246@henu.edu.cn (T.Y.); 10190146@vip.henu.edu.cn (L.Z.); wangyange3@henu.edu.cn (Y.W.); zhhong@henu.edu.cn (H.Z.)

[2] Beijing Key Laboratory of New Molecular Diagnosis Technologies for Infectious Diseases, Department of Biotechnology, Beijing Institute of Radiation Medicine, Beijing 100850, China; kunbzh0201@163.com

*   Correspondence: zhangguosen1989@126.com or 40060006@vip.henu.edu.cn (G.Z.); xqguo@henu.edu.cn (X.G.); Tel.: +86-18237808750 (G.Z.)

†   These authors contributed equally to this work.

Citation: Yao, X.; Qi, X.; Wang, Y.; Zhang, B.; He, T.; Yan, T.; Zhang, L.; Wang, Y.; Zheng, H.; Zhang, G.; et al. Identification and Validation of an Annexin-Related Prognostic Signature and Therapeutic Targets for Bladder Cancer: Integrative Analysis. *Biology* **2022**, *11*, 259. https://doi.org/10.3390/biology11020259

Academic Editors: Shibiao Wan, Yiping Fan, Chunjie Jiang and Shengli Li

Received: 15 December 2021
Accepted: 3 February 2022
Published: 7 February 2022

Publisher's Note: MDPI stays neutral with regard to jurisdictional claims in published maps and institutional affiliations.

Simple Summary: Identification of new prognostic biomarkers and therapeutic targets could be essential ways to improve the outcome of bladder cancer (BC) patients. In this study, we comprehensively analyzed the mRNA expression and prognosis of Annexin family members (ANXA1-11, 13) in BC through public analysis tools, including Oncomine, GEPIA2 and our in-house OSblca web server, and found that several Annexins were aberrantly expressed and associated with prognosis in BC. Then, we constructed and validated an Annexin-related prognostic signature (ARPS) in four individual BC cohorts through LASSO and COX regression, indicating that ARPS was an independent prognostic factor for BC. Briefly, our study was to determine the clinical significance of Annexins and provided a potential prognostic model and potential therapeutic targets for BC.

Abstract: Abnormal expression and dysfunction of Annexins (ANXA1-11, 13) have been widely found in several types of cancer. However, the expression pattern and prognostic value of Annexins in bladder cancer (BC) are currently still unknown. In this study, survival analysis by our in-house OSblca web server revealed that high *ANXA1/2/3/5/6* expression was significantly associated with poor overall survival (OS) in BC patients, while higher *ANXA11* was associated with increased OS. Through Oncomine and GEPIA2 database analysis, we found that *ANXA2/3/4/13* were up-regulated, whereas *ANXA1/5/6* were down-regulated in BC compared with normal bladder tissues. Further LASSO analysis built an Annexin-Related Prognostic Signature (ARPS, including four members *ANXA1/5/6/10*) in the TCGA BC cohort and validated it in three independent GEO BC cohorts (GSE31684, GSE32548, GSE48075). Multivariate COX analysis demonstrated that ARPS is an independent prognostic signature for BC. Moreover, GSEA results showed that immune-related pathways, such as epithelial–mesenchymal transition and IL6/JAK/STAT3 signaling were enriched in the high ARPS risk groups, while the low ARPS risk group mainly regulated metabolism-related processes, such as adipogenesis and bile acid metabolism. In conclusion, our study comprehensively analyzed the mRNA expression and prognosis of Annexin family members in BC, constructed an Annexin-related prognostic signature using LASSO and COX regression, and validated it in four independent BC cohorts, which might help to improve clinical outcomes of BC patients, offer insights into the underlying molecular mechanisms of BC development and suggest potential therapeutic targets for BC.

Keywords: bladder cancer; Annexin family; survival analysis; prognostic signature; therapeutic target

1. Introduction

Bladder cancer (BC) is one of the most common malignancies with high risk of tumor recurrence and fatality in the urinary system. According to Global Cancer Statistics 2020, there were about 573,000 new cases and 212,000 deaths of BC around the world [1]. Although the significant advances in understanding of the underlying biology of BC have improved the accuracy and effect of diagnosing and treating this disease in recent years, BC still represents a spectra of diseases from recurrent noninvasive tumors to aggressive or advanced-stage disease that requires multimodal and invasive treatment [2,3]. Frequent postoperative recurrence and distant metastasis lead to the poor prognosis in BC patients [4,5]. Identification of efficient therapeutic targets, as well as new prognostic biomarkers are needed to improve the outcomes of BC patients.

Annexins belongs to a superfamily of calcium-dependent phospholipid-binding proteins and contains 12 members (ANXA1-11, 13) [6]. In eukaryotic cells, Annexins are involved in membrane trafficking and organization, such as vesicle transport, signal transduction, cell proliferation, cell differentiation and apoptosis [7,8]. Recent studies found that abnormal expression and dysfunction of Annexin proteins commonly occurred in tumor tissue and indicated that the disordered Annexin proteins may play important roles in tumorigenesis and progression, as well as chemoresistance in several types of cancer [9,10]. However, few studies reported the roles of Annexins in the carcinogenesis and prognosis in BC. Yu et al. found that the expression of ANXA1 was related to disease-free survival in BC patients and can be used as a recurrence biomarker for BC [11]. In addition, ANXA2 has also been found to play a key role in the formation, progression and recurrence of BC [12], and high expression of ANXA10 is significantly correlated with poor progression-free survival in BC patients [13]. However, the roles and mechanisms of most Annexins in BC remain unclear.

In this study, we comprehensively analyzed the mRNA expression and prognosis of Annexin family members in BC through public analysis tools, including Oncomine, GEPIA2 and our in-house OSblca web server, and found that several Annexins were aberrantly expressed and associated with prognosis in BC. Then, we collected four BC datasets, including 703 BC samples with survival information from The Cancer Genome Atlas (TCGA) and Gene Expression Omnibus (GEO), and constructed and validated an Annexin-related prognostic signature (ARPS) in four individual BC cohorts through LASSO and COX regression, indicating that ARPS was an independent prognostic factor for BC. In addition, we further explored the biological functions and relevant pathways of ARPS through gene set enrichment analysis (GSEA), and analyzed the correlation between ARPS and the infiltrating immune cells using ssGSEA. Briefly, our study was to determine the clinical significance of Annexins and provided a potential prognostic model and potential therapeutic targets for BC.

2. Materials and Methods

2.1. Survival Analysis of Annexin Family Members in OSblca

OSblca (http://bioinfo.henu.edu.cn/BLCA/BLCAList.jsp, accessed on 2 December 2020) [14] is our in-house online survival analysis tool providing 1075 BC gene expression profiles and accompanied patient clinical follow-up information from TCGA and GEO databases. In OSblca, four types of survival endpoints, including overall survival (OS), disease-specific survival (DSS), disease-free interval (DFI), and progression-free interval (PFI) were provided for prognosis analysis. Each member of the Annexin family was analyzed for the relationship between their mRNA expression and BC outcomes in OSblca prognostic values of these genes were evaluated in all cohorts and survival terms, and, all cutoff values in 'splitting the patients' were tested in each cohort to get the best cutoff value.

2.2. Differential Expression Analysis of Annexin Family Members between BC and Adjacent Tissue by Oncomine and GEPIA2

Oncomine (www.oncomine.org, accessed on 2 December 2020) [15] and GEPIA2 (http://gepia2.cancer-pku.cn/#index, accessed on 2 December 2020) [16] databases were used to analyze the differential expression of Annexins between cancer and adjacent normal tissues. Oncomine is an online database that provides differentially expressed gene analysis using public microarray datasets. In Oncomine, mRNA expression of Annexin members between cancer tissue and adjacent normal tissue were compared with the thresholds of p-value < 0.05, $|$log2 (fold-change)$| > 1$, and the gene rank percentage $< 10\%$. GEPIA2 provided the gene expression analysis based on TCGA and GTEx data. In GEPIA2, the expression of Annexins were compared between 404 bladder cancer samples and 28 normal samples with the threshold for p-value < 0.05 and $|$log2(fold-change)$| > 1$. In addition, differential expression of Annexins members in distinct clinical stages was also analyzed in TCGA bladder cancer samples using TISIDB database (http://cis.hku.hk/TISIDB/, accessed on 3 December 2020) [17].

2.3. Construction and Validation of the Annexin-Related Prognostic Signature through LASSO

Four individual BC cohorts with both gene expression data and related clinical follow-up information were downloaded from TCGA and GEO databases, including one TCGA BC dataset [18] (Discovery cohort) and three GEO BC datasets (Validation cohorts, GSE31684 [19], GSE32548 [20], GSE48075 [21]). Detailed information of each BC cohort was summarized in Supplementary Table S1. The work-flow was illustrated in Supplementary Figure S1. The ARPS was constructed using least absolute shrinkage and selection operator (LASSO) Cox regression through R package "glmnet". The optimal parameter was determined through 10-fold cross validation with "family = cox, alpha = 1", and with all other parameters set to default. Ultimately, ARPS is developed according to the following risk score formula:

$$\text{risk score} = \sum_{i}^{n} (\text{Coefi} * \text{Expri})$$

where Coef_i is the coefficient of gene i in LASSO and Expr_i is the FPKM value of the included gene i.

Best cut-off risk score was calculated by using the "surv_cutpoint" function of R package "survminer" (https://CRAN.R-project.org/package=survminer, accessed on 24 December 2020). According to the best cut-off risk score, TCGA BC patients were divided into high- and low-risk groups and the prognosis between the two groups was evaluated through Kaplan-Meier survival analysis with the log-rank test. In addition, the expression heatmap of each Annexin member in ARPS and the risk score distribution and survival of patients were visualized through "pheatmap" package. Similar analyses were performed in three individual BC cohorts (GSE31684, GSE32548, GSE48075) to validate the prognosis performance of ARPS in BC.

2.4. Independent Prognostic Performance Analysis of ARPS in BC Cohorts

In the TCGA BC cohort, univariate Cox regression models were used to identify the prognostic clinical characteristics related to prognosis, and subsequently these significantly prognostic factors were further tested their independent prognostic performance through multivariate Cox regression models. Similar analyses were performed in other three individual BC cohorts (GSE31684, GSE32548, GSE48075) to validate the independent prognostic performance of ARPS in BC.

2.5. Association between ARPS and Clinicopathology

The chi-squared test was performed to determine the association of clinical features between and ARPS in BC patients, where a p-value less than 0.05 indicates statistical significance. In addition, to verify the predictive effectiveness of ARPS in different subgroups, Kaplan-Meier survival analysis was used to compare the prognostic capability between subgroups in certain clinical features including age, gender, grade, lymph invasion status, T status, M status, and N status, TNM stage, and race.

2.6. Gene Interaction and Biological Functions of ARPS

GeneMANIA was used to analyze the protein interactions between Annexin members in ARPS. To evaluate the biological functions of ARPS, differentially expressed genes (DEGs) between the two ARPS risk groups were identified through limma R package, and then were analyzed in the DAVID database to predict the gene ontology (GO) function and KEGG pathway. Furthermore, GSEA was implemented to reveal the potential mechanism that ARPS was involved in.

2.7. Correlation between ARPS Risk Score and Imune Cell Infiltration and Immune Checkpoint Genes

In order to characterize the immune cell infiltration in the tumor microenvironment of two ARPS risk groups, the immune cell abundance of the TCGA BC cohort was calculated by estimate, timer, MCPcounter and xCell algorithm, and visualized through the "pheatmap" package of R software. Correlations between ARPS risk score and different immune cell abundances and immune checkpoint genes were analyzed through Pearson coefficient analysis. Then, the levels of immune cell infiltration and immune checkpoint gene expression between high- and low-risk groups were compared using the 'limma' package, which revealed the effect of ARPS risk score on BC immune microenvironment.

2.8. Statistical Analysis

Statistical analysis was performed using SPSS 16.0 and GraphPad Prism 5.0 software. Differences were compared by the Student's t test or one-way analysis of variance (ANOVA) where appropriate. Statistical significance was determined by p-value less than 0.05.

3. Results

3.1. Survival and Differential Analysis of Annexins in Bladder Cancer

Survival analysis results revealed that the mRNA expression of more than half of Annexin members were related to the prognosis of BC patients. As shown in Figure 1, BC patients with high expression of *ANXA1/2/3/5/6* had a shorter OS time in comparison to those with low expression of *ANXA1/2/3/5/6*, while BC patients with high expressed *ANXA11* had a longer OS time. In particular, upregulated *ANXA1/2/5* were significantly associated with poor prognosis of BC in three or more datasets. In addition, the mRNA expressions of Annexins were also related to DSS in BC patients (Figure 2). The results indicated that BC patients with high expression of *ANXA1/2/5/6/7/13* showed a shorter DSS time than those with low expression, as opposed to the patients with high expression of *ANXA11*. Moreover, high expression of *ANXA5* and *ANXA13* were found to be associated with poor DFI and PFI in the BC cohort (Figure 3), indicating that these two genes might be involved in recurrence and progression of BC.

Using the Oncomine database (Table 1), we found that most Annexins were significantly differentially expressed between BC and adjacent normal tissues. Markedly lower expressions of *ANXA1/5/6* were found in BC tissues consistently, while the expression levels of *ANXA2/3/4/13* were significantly increased in multiple BC cohorts. In the GEPIA2 database, *ANXA6* was significantly downregulated in TCGA BC samples compared to normal samples (p-value < 0.05), while *ANXA8* was significantly upregulated in TCGA BC samples, and no significant differences were found for other Annexins in BC (Supplementary Figure S1).

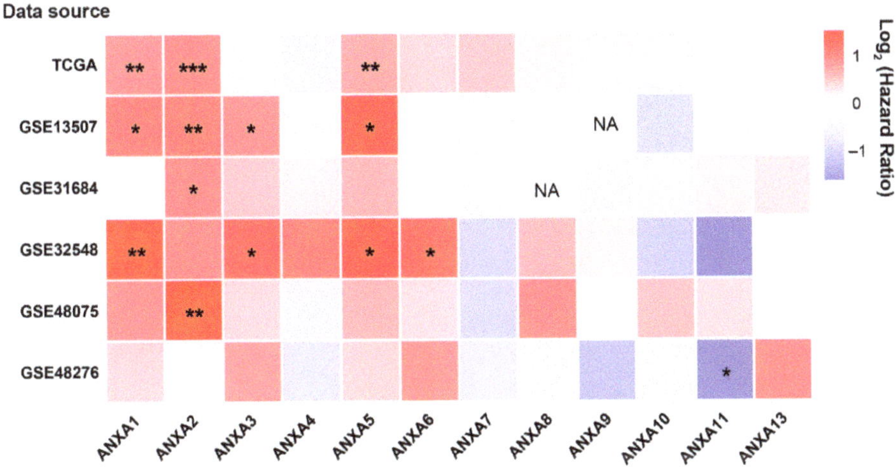

Figure 1. The hot map of prognostic value of Annexin family members regarding overall survival (OS) in BC patients by OSblca web server. Where * $p < 0.05$, ** $p < 0.01$ and *** $p < 0.001$, NA means the queried gene was non-existent in the dataset.

Disease-Specific Survival

Gene	Data source	Cutoff (Upper)		p value	HR (95% CI)
ANXA1					
	TCGA	50%		0.0014	1.88 (1.28–2.76)
	GSE19915	25%		0.0088	2.93 (1.31–6.54)
ANXA2					
	TCGA	25%		0.0018	1.89 (1.27–2.82)
	GSE31684	25%		0.0412	1.81 (1.02–3.20)
ANXA5					
	TCGA	25%		0.0099	1.70 (1.14–2.55)
	GSE19915	25%		0.003	3.36 (1.51–7.50)
	GSE31684	50%		0.0291	1.74 (1.06–2.85)
ANXA6					
	TCGA	50%		0.0181	1.59 (1.08–2.34)
	GSE48276	50%		0.0093	3.45 (1.36–8.75)
ANXA7					
	TCGA	50%		0.0153	1.61 (1.10–2.38)
ANXA11					
	GSE19915	50%		0.0206	0.35 (0.15–0.85)
	GSE48276	25%		0.0364	0.21 (0.05–0.91)
ANXA13					
	GSE31684	50%		0.0177	1.84 (1.11–3.05)

Figure 2. Forest plots displayed prognostic value of Annexin family members regarding disease-specific survival (DSS) in BC patients using OSblca web server.

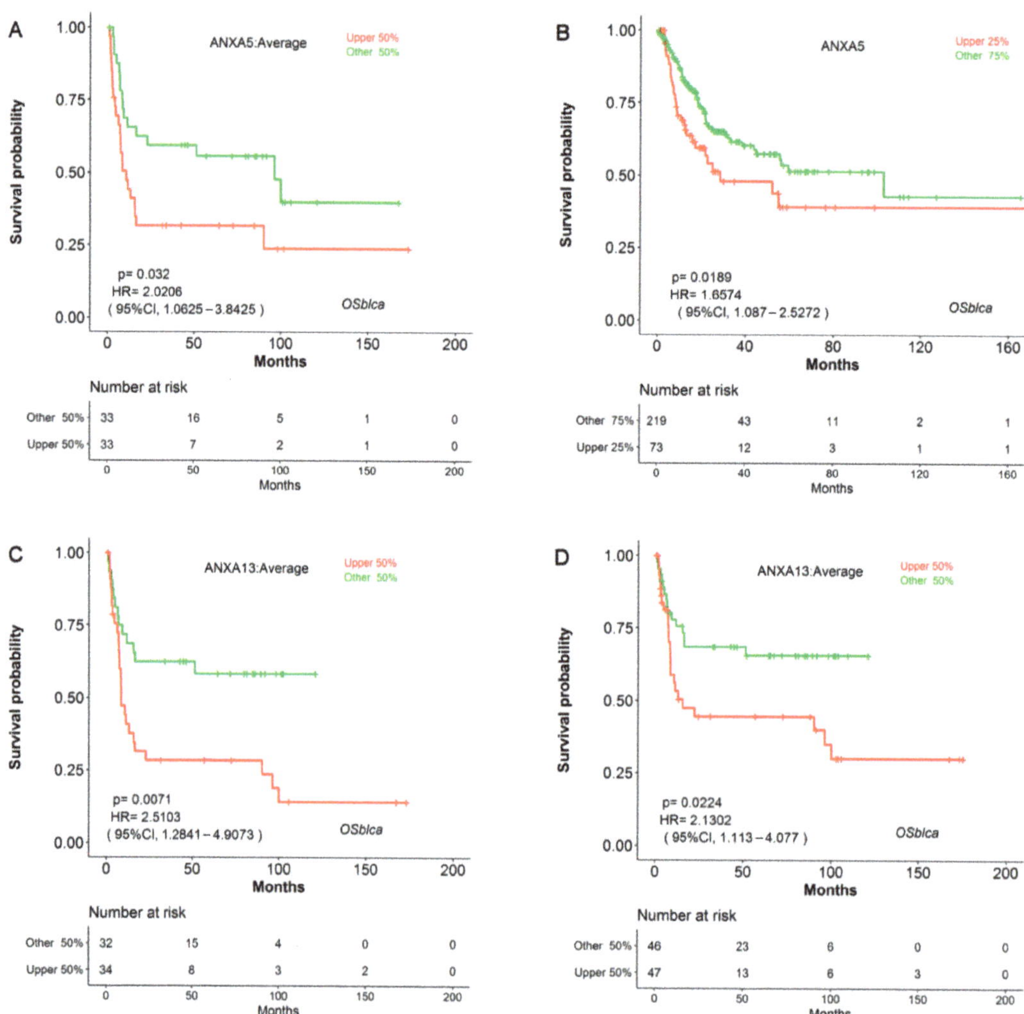

Figure 3. Survival analysis of *ANXA5* and *ANXA13* in disease-free interval (DFI) and progression-free interval (PFI) in BC patients using OSblca web server. Kaplan-Meier plotter of *ANXA5* with DFI (**A**) and PFI (**B**); Kaplan-Meier plotter of *ANXA13* with DFI (**C**) and PFI (**D**) in OSblca.

Correlation between Annexin expression and clinical stage of BC are shown in Figure 4A. The results showed that the expression levels of *ANXA1/2/5/6* were positively correlated with clinical stage, and *ANXA10* showed negative correlation with clinical stage, while no significant correlation was found in other Annexins. Additionally, high expression of *ANXA1/2/5/6* were found in BC patients with stage III/IV compared to those in BC patients with stage I/II (Figure 4B–E), whereas low expression of *ANXA10* were found in stage III/IV BC patients (Figure 4F).

3.2. Construction and Validation of the Annexin-Related Prognostic Signature

Through LASSO Cox regression, four Annexin members including ANXA1/5/6/10 were identified and used to construct Annexin-related prognostic signature (ARPS) (Figure 5A–C). Risk score of ARPS was calculated according to the formula, Risk score = $0.00083 \times \mathrm{Exp}_{\mathrm{ANXA1}}$

$$+\ 0.0017 \times \mathrm{Exp_{ANXA5}} + 0.00016 \times \mathrm{Exp_{ANXA6}} - 0.00012 \times \mathrm{Exp_{ANXA10}}.$$ Then, BC samples were divided into high/low-risk groups by ARPS according to the best cut-off of risk score.

Table 1. Comparison of mRNA expression of Annexin family members between bladder cancer and adjacent normal tissues (Oncomine database).

Gene	Datasets	Tumor (Cases Number)	Normal (Cases Number)	Fold Change	p-Value
ANXA1	Lee et al.	Superficial Bladder Cancer (126)	Bladder Mucosa (68)	−2.916	9.51×10^{-14}
		Infiltrating Bladder Urothelial Carcinoma (62)	Bladder Mucosa (68)	−1.466	1.90×10^{-2}
	Sanchez et al.	Superficial Bladder Cancer (28)	Bladder (48)	−2.131	7.07×10^{-4}
ANXA2	Sanchez et al.	Infiltrating Bladder Urothelial Carcinoma (81)	Bladder (48)	1.493	6.23×10^{-5}
		Superficial Bladder Cancer (28)	Bladder (48)	1.371	2.00×10^{-3}
	Dyrskjot et al.	Infiltrating Bladder Urothelial Carcinoma (13)	Bladder (9) Bladder Mucosa (5)	2.085	4.41×10^{-4}
ANXA3	Sanche et al.	Infiltrating Bladder Urothelial Carcinoma (81)	Bladder (48)	2.323	5.55×10^{-8}
	Dyrskjot et al.	Infiltrating Bladder Urothelial Carcinoma (13)	Bladder (9) Bladder Mucosa (5)	2.607	2.00×10^{-3}
ANXA4	Sanchez et al.	Infiltrating Bladder Urothelial Carcinoma (81)	Bladder (48)	1.731	3.28×10^{-8}
		Superficial Bladder Cancer (28)	Bladder (48)	2.506	2.28×10^{-13}
	Dyrskjot et al.	Superficial Bladder Cancer (28)	Bladder (9) Bladder Mucosa (5)	2.770	2.84×10^{-5}
		Infiltrating Bladder Urothelial Carcinoma (13)	Bladder (9) Bladder Mucosa (5)	1.915	2.00×10^{-3}
ANXA5	Lee et al.	Superficial Bladder Cancer (126)	Bladder Mucosa (68)	−2.392	1.01×10^{-13}
		Infiltrating Bladder Urothelial Carcinoma (62)	Bladder Mucosa (68)	−1.417	4.00×10^{-3}
	Sanchez et al.	Superficial Bladder Cancer (28)	Bladder (48)	−2.428	4.70×10^{-10}
		Infiltrating Bladder Urothelial Carcinoma (81)	Bladder (48)	−1.473	6.43×10^{-7}
	Blaveri et al.	Superficial Bladder Cancer (26)	Bladder (3)	−4.211	3.00×10^{-3}
ANXA6	Sanchez et al.	Superficial Bladder Cancer (28)	Bladder (48)	−8.011	5.24×10^{-25}
		Infiltrating Bladder Urothelial Carcinoma (81)	Bladder (48)	−2.846	3.69×10^{-14}
	Dyrskjot et al.	Stage 0is Bladder Urothelial Carcinoma (5)	Bladder (9) Bladder Mucosa (5)	−1.295	4.60×10^{-2}
		Superficial Bladder Cancer (28)	Bladder (9) Bladder Mucosa (5)	−1.558	1.00×10^{-3}
ANXA13	Lee et al.	Infiltrating Bladder Urothelial Carcinoma (62)	Bladder Mucosa (68)	1.033	2.70×10^{-2}
	Blaveri et al.	Infiltrating Bladder Urothelial Carcinoma (41)	Bladder (2)	2.374	8.64×10^{-4}
		Superficial Bladder Cancer (21)	Bladder (2)	2.239	1.00×10^{-3}

The prognostic performance of ARPS in the BC cohort was evaluated in the TCGA dataset (Discovery cohort) and validated in three independent GEO datasets (Validation cohorts, GSE31684, GSE32548, GSE48075). As shown in Figure 5D, Kaplan-Meier plot showed that BC patients in the high ARPS risk group had shorter OS time than those in the low ARPS risk group ($p < 0.0001$, HR = 2.232). The gene expression heat map indicated that high expression of *ANXA1, ANXA5,* and *ANXA6* but low expression of *ANXA10* were shown in the high-risk group in comparison to the low-risk group. In addition, high ARPS risk score were consistently related to short OS in GSE31684 ($p = 0.0079$, HR = 1.987, Figure 5E), GSE32548 ($p = 0.0005$, HR = 4.255, Figure 5F) and GSE48075 ($p = 0.0296$, HR = 1.999, Figure 5G). In addition, the high ARPS risk group had a shorter DSS and PFI than those with low ARPS risk in the TCGA BC cohort and GSE31684 BC cohort (Supplementary Figure S2).

Univariate and multivariate COX regression were performed to explore whether the ARPS was an independent prognostic predictor for BC. In the univariate analysis of the TCGA dataset (Discovery cohort), risk score, grade and age were all correlated with OS, and then included in subsequently multivariate analysis. Multivariate analysis showed that high ARPS risk score was associated with poor prognosis in both discovery BC cohort [($p < 0.0001$, HR= 2.045 (1.485–2.817), Table 2] and three independent validated BC cohorts GSE31684 [$p = 0.0010$, HR= 2.259 (1.375–3.711), Table 3], GSE3254 [($p = 0.0060$, HR = 3.591(1.453–8.872)] and GSE48075 [($p = 0.0100$, HR = 2.291 (1.224–4.286)]. Overall, these results all confirmed that the ARPS risk score is an independent survival predictor for BC patients.

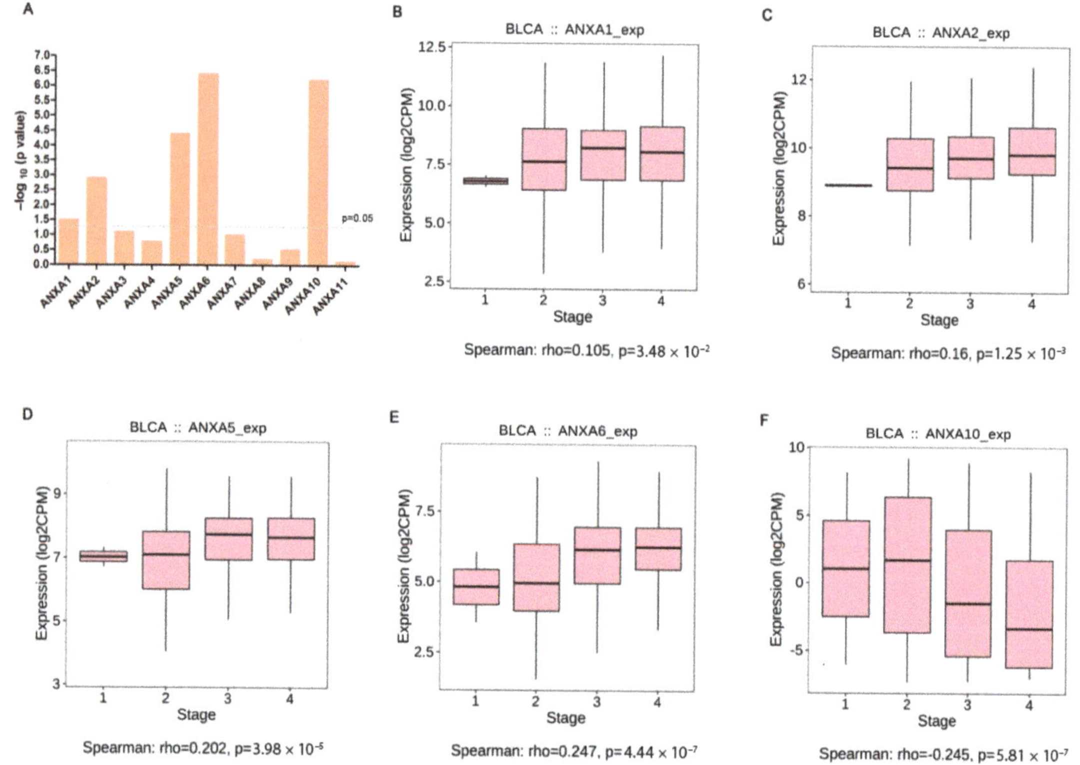

Figure 4. Correlation analysis of Annexin family members and clinical stages of bladder cancer in TISIDB. Summary of spearman's correlation between Annexin family members and clinical stages (**A**); Different expression of ANXA1 (**B**), ANXA2 (**C**), ANXA5 (**D**), ANXA6 (**E**), ANXA10 (**F**) between clinical stages.

3.3. Associations of ARPS with Clinicopathological Features of BC

In order to better understand the role of ARPS in clinical outcomes of BC, we further investigated the relationships between ARPS and the pathological features of BC, including age, gender, grade, lymph invasion status, pT stage, pN stage, pM stage, TNM stage and race. Chi-squared test (Table 4) demonstrated that the clinicopathological features including gender, grade, pT stage, pN stage, TNM stage and race showed significant association with ARPS risk score. Further subgroup analyses were performed to determine whether ARPS could predict prognosis of BC patients under certain clinicopathological circumstances. Kaplan-Meier survival analysis (Figure 6) revealed that worse OS was noted in the high-risk ARPS groups regardless of age (Figure 6A), gender (Figure 6B), and pT stage (Figure 6D).

However, ARPS is more potent to predict the outcome for higher TNM stages (Figure 6C), pN0 stage (Figure 6E), pM0 stage (Figure 6F), high grade (Figure 6G), and white (Figure 6H) than lower TNM stages, pN 1/2/3, pM 1, low grade and non-white, respectively.

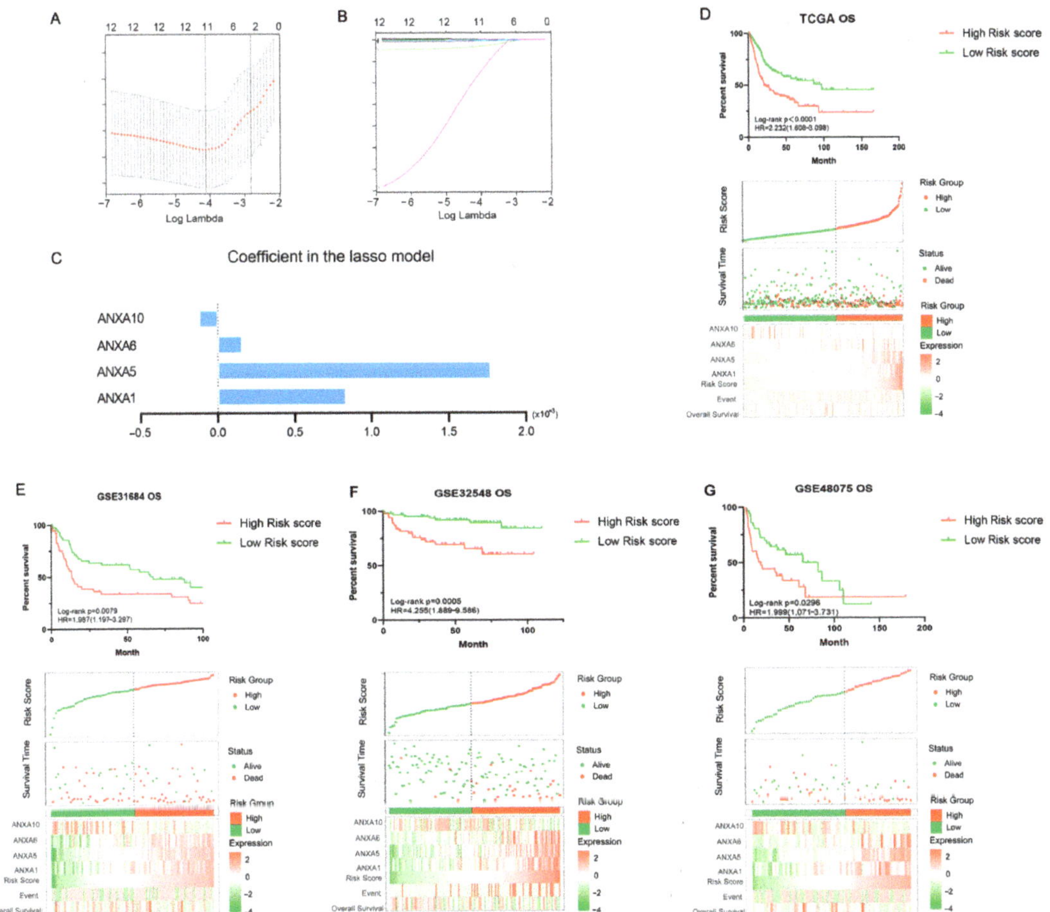

Figure 5. Construction and validation of Annexin-Related Prognostic Signature (ARPS). LASSO algorithm was used to construct a prognosis model (**A–C**); Kaplan-Meier curve, distribution diagram of risk score and survival status in TCGA BC patients (discovery cohort) between high/low-risk groups (**D**); Kaplan-Meier curve, distribution diagram of risk score and survival status in three BC validation cohorts GSE31684 (**E**), GSE32548 (**F**) and GSE48075 (**G**).

Table 2. Univariate and multivariate Cox analyses of ARPS risk score with OS in TCGA.

Covariates	Univariate Cox Analysis			Multivariate Cox Analysis		
	p Value	HR	95% CI	*p* Value	HR	95% CI
Age (>65 vs. ≤65 years)	<0.0001 ****	2.039	1.426–2.916	<0.0001 ****	1.981	1.384–2.837
Gender (Male vs. Female)	0.3460	0.846	0.598–1.198	-	-	-
Stage (III/IV vs. I/II)	<0.0001 ****	2.531	1.670–3.836	0.0010 **	2.084	1.366–3.180
Grade (High vs. Low)	0.1310	21.473	0.400–1151.678	-	-	-
Lymph (Yes vs. No)	<0.0001 ****	1.931	1.366–2.792	0.0010 **	1.843	1.271–2.671
Race (White vs. Non-white)	0.3230	1.291	0.778–2.140	-	-	-
Risk score (High vs. Low)	<0.0001 ****	2.144	1.559–2.950	<0.0001 ****	2.045	1.485–2.817

Note: Where **, $p < 0.01$ and ****, $p < 0.0001$.

Table 3. Univariate and multivariate Cox analyses of ARPS risk score with OS in three BC validation datasets.

Covariates	Univariate Cox Analysis						Multivariate Cox Analysis					
	GSE31684		GSE32548		GSE48075		GSE31684		GSE32548		GSE48075	
	p Value	HR (95% CI)	*p* Value	HR (95% CI)	*p* Value	HR (95% CI)	*p* Value	HR (95% CI)	*p* Value	HR (95% CI)	*p* Value	HR (95% CI)
Age (>65 vs. ≤65 years)	0.8710	1.047 (0.605–1.811)	0.3750	1.487 (0.619–3.569)	0.0230 *	2.371 (1.129–4.979)	-	-	-	-	0.0080 **	2.786 (1.304–5.956)
Gender (Male vs. Female)	0.9660	0.988 (0.560–1.741)	0.6300	1.273 (0.477–3.393)	-	-	-	-	-	-	-	-
Grade (High vs. Low)	0.1230	3.035 (0.741–12.43)	0.0290 *	2.790 (1.112–6.999)	-	-	-	-	0.1460	2.014 (0.783–5.182)	-	-
Stage (III/IV vs. I/II)	0.0110	2.162 (1.192–3.922)	-	-	-	-	0.0080 **	2.237 (1.230–4.068)	-	-	-	-
Risk score (High vs. Low)	0.0020 **	2.200 (1.340–3.613)	0.0010 **	4.248 (1.761–10.250)	0.0330 *	1.939 (1.056–3.559)	0.0010 **	2.259 (1.375–3.711)	0.0060 **	3.591 (1.453–8.872)	0.0100 *	2.291 (1.224–4.286)

Note: Where *, $p < 0.05$ and **, $p < 0.01$.

Table 4. Association of ARPS risk score with clinicopathological features in TCGA BC cohort.

Characteristics	Sample (*n* = 406)	Risk Score		χ^2	*p* Value
		High Risk Score (*n* = 173)	Low Risk Score (*n* = 233)		
Age				2.173	0.1410
>65 years	246	112	134		
≤65 years	160	61	99		
Gender				4.592	0.0320 *
Male	299	118	181		
Female	107	55	52		
Grade				12.360	0.0004 ***
High	383	172	211		
Low	20	1	19		
Lymph invasion				0.582	0.4450
Yes	149	62	87		
No	130	60	70		
TNM Stage				6.403	0.0110 *
I-II	273	128	145		
III-IV	131	44	87		
pT Stage				9.258	0.0020 **
T0-T2	122	49	73		
T3-T4	251	143	108		
pN Stage				4.166	0.0410 *
N0	236	110	126		
N1-N3	128	74	54		
pM Stage				0.155	0.6940
M0	195	77	118		
M1	11	5	6		
Race				6.265	0.0120 *
White	323	147	176		
Non-White	66	19	47		

Note: Where *, $p < 0.05$, **, $p < 0.01$, and ***, $p < 0.001$.

3.4. Gene-Gene Interaction Network and Function Analysis of ARPS in BC

A gene-gene interaction network of ARPS was constructed using the GeneMANIA database. As shown in Figure 7A, the top five genes displaying the greatest correlations with ARPS included *U2AF2*, *RASA1*, *ANXA4*, *COL10A1* and *DYSF*. Functional analysis revealed that these genes showed the greatest correlation with calcium-dependent phospholipid binding, lipase inhibitor activity, phospholipid binding S100 protein binding and enzyme-inhibitor activity. The predictive power of ARPS in predicting recurrence risk of BC patients could be attributed to their crucial roles in tumor development or metastases. Therefore, we further explored the underlying biological functions of ARPS through GO, KEGG, and GSEA pathway enrichment analyses. Gene differential analysis identified that there were 2439 differentially expressed genes (DEGs) between these two groups with high/low ARPS risk, including 1604 upregulated genes and 1710 downregulated genes. GO analysis (Figure 7B) showed that DEGs were mainly involved in cell-cell signaling (GO:0007267), immune response (GO:0006955) and chemokine-mediated signaling pathway (GO:0070098), and KEGG pathway enrichment (Figure 7C) revealed that the DEGs were mainly enriched in cytokine-cytokine receptor interaction (hsa04060), chemokine signaling pathway (hsa04062), and drug metabolism (hsa00982). Moreover, GSEA enrichment results (Figure 8A–C, Supplementary Figure S4) showed that immune related pathways such as epithelial–mesenchymal transition, IL6/JAK/STAT3 signaling, inflammatory response and TNFA signaling via NFKB were enriched in the high-risk groups (Figure 8B), while the low-risk group mainly regulated metabolism-related processes, such as adipogenesis, bile acid metabolism, oxidative phosphorylation and peroxisome (Figure 8C).

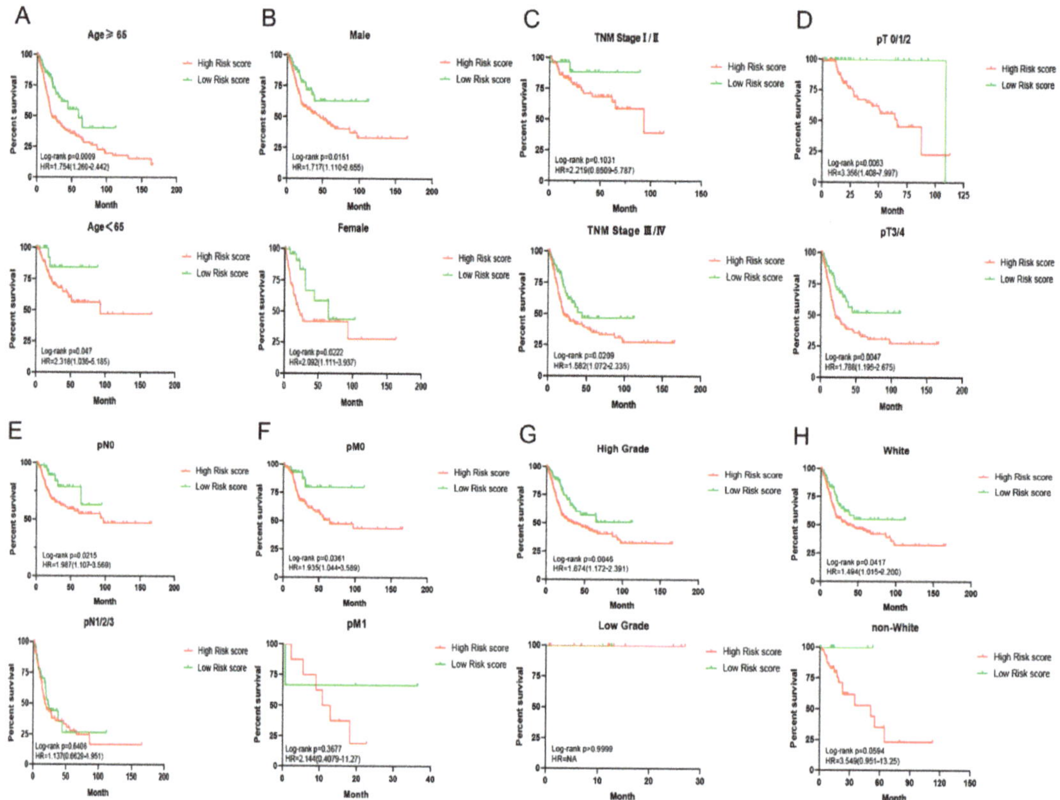

Figure 6. Survival analyses of ARPS in subgroup of BC patients stratified by different clinical characteristics. (**A**) Age; (**B**) gender; (**C**) TNM stage; (**D**) T stage; (**E**) M stage; (**F**) N stage, (**G**) grade and (**H**) race.

3.5. Relation between ARPS and the Degree of Immune Cell Infiltration

Immune cell infiltration of BC cases with high/low ARPS risk were estimated and compared by estimate algorithm (Figure 9A–D). The result showed that the risk score of ARPS was significantly positively correlated with immune infiltration level, and BC cases with high ARPS risk score had greater ESTIMATE score (Figure 9A), immune score (Figure 9B) and stromal score (Figure 9C), but unsurprisingly lower purity than those in the low ARPS risk group (Figure 9D). Through Timer, MCPcounter and xCell algorithm, we compared the immune cell abundance between the high- and low-risk groups and found that several types of immune cells, including $CD8^+$ T cells, neutrophils, macrophages, myeloid dendritic cell, Tregs, and cancer-associated fibroblasts were significantly more abundant in the high-risk group than those in the low-risk group (Figure 9E). Moreover, we compared the different expression of several immune check genes between the high- and low-risk groups. The results revealed that elevated expression of most immune check genes, including *CD274, CD276, CD28, CD80, CD86, ICOS, ICOSLG, LAG3, PDCD1* and *PDCDLG2*, were found in the high-risk group compared to those in the low-risk group (Figure 9F).

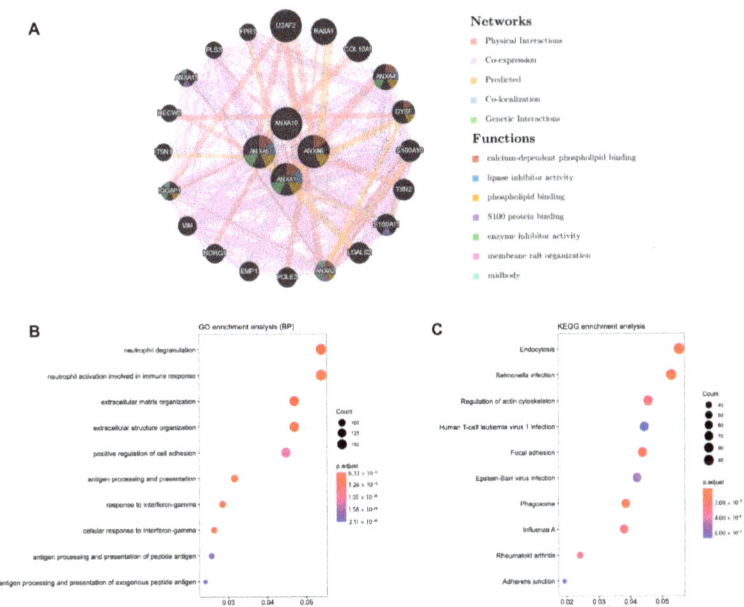

Figure 7. Results of gene-gene interaction network (GeneMANIA) and functional enrichment analyses between the high/low risk groups. (**A**) Gene-gene interaction network analysis of ARPS members (GeneMANIA); (**B**) GO enrichment analysis (Biological Process); (**C**) KEGG pathway analysis.

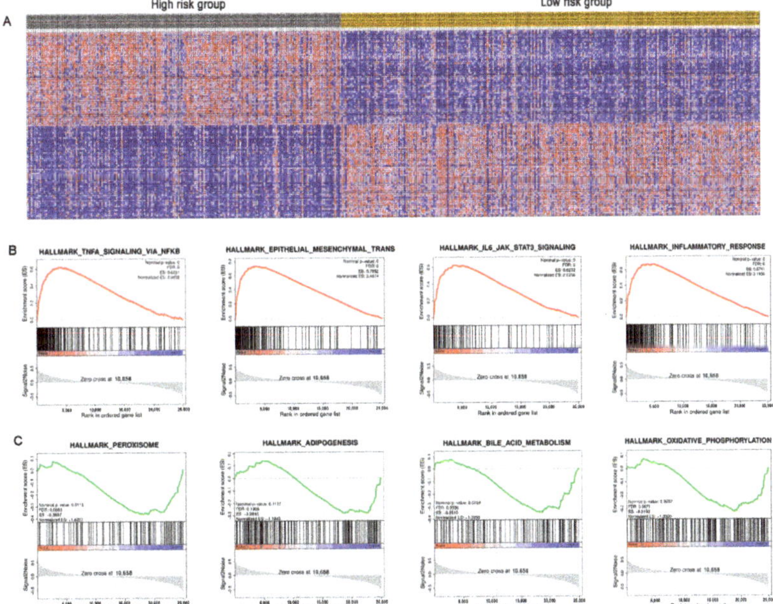

Figure 8. Gene-set enrichment analysis (GSEA) between the high/low-risk groups. (**A**) Heat map of differential expression genes; (**B**) significantly enriched pathways in the high-risk group; (**C**) significantly enriched pathways in the low-risk group.

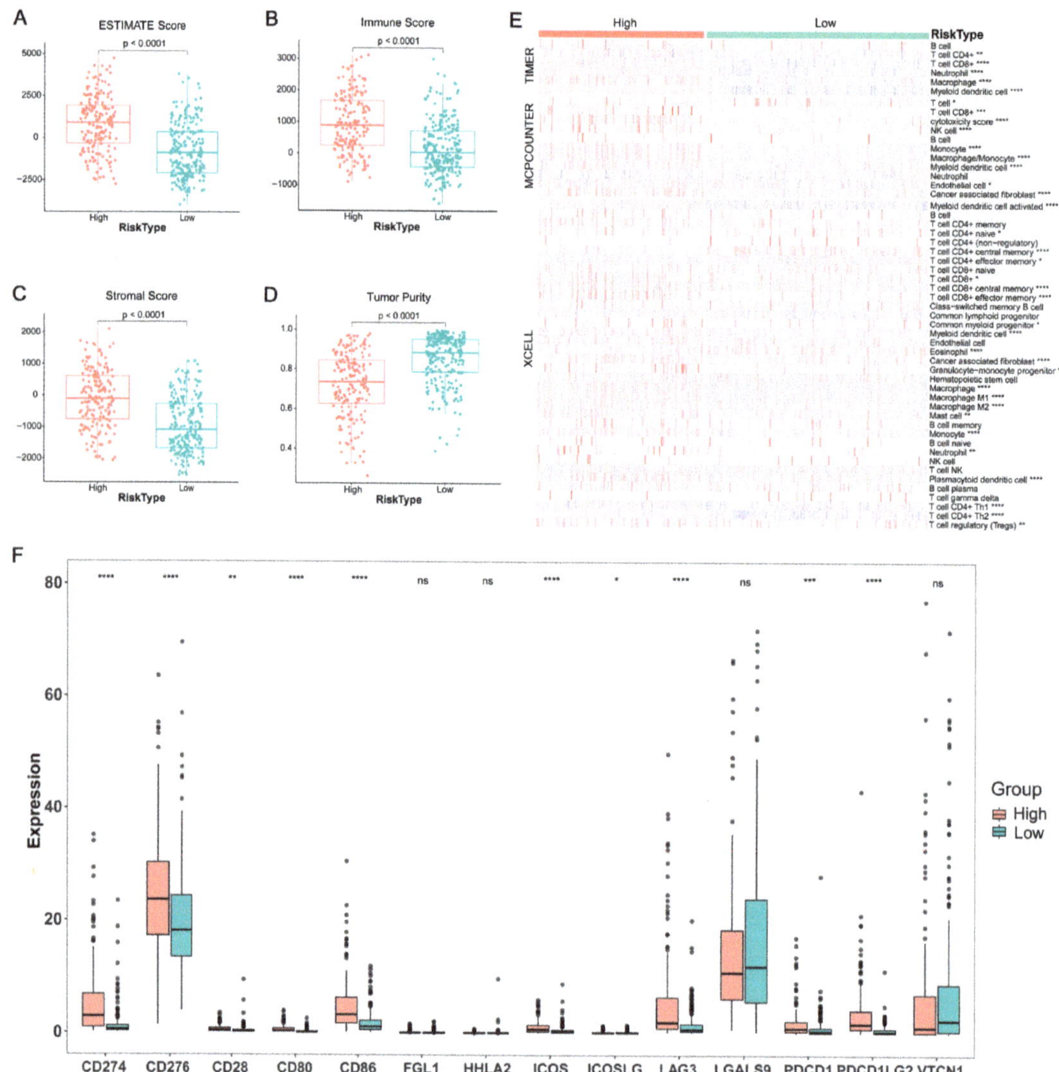

Figure 9. Tumor immune infiltrating and the expression of immune check gene between the high/low-risk groups. Tumor immune infiltrating level between the high/low-risk groups (**A–D**); proportions of tumor-infiltrating cells between the high/low-risk groups (**E**); different expression levels of immune check genes between the high/low-risk groups (**F**). Where *, $p < 0.05$, **, $p < 0.01$, ***, $p < 0.001$ and ****, $p < 0.0001$, ns not significant.

4. Discussion

Although the advance of surgical methods and medical therapies have improved the treatment of bladder cancer, high rate of recurrence after operation and frequent metastasis lead to poor prognosis of BC patients. Identification of new prognostic biomarkers and therapeutic targets could be essential ways to improve the outcome of BC patients. In this study, we comprehensively analyzed the gene expression and prognosis of Annexin family members in BC, and constructed and validated an ARPS, which could be an independent prognostic biomarker in four individual BC cohorts.

During our evaluation of the gene expression and prognostic value of Annexins in BC, we found that several Annexins were aberrantly expressed and associated to prognosis in BC. For example, high expression of *ANXA2/3/13* were found in BC compared to normal tissue (Table 1) and related to poor prognosis in BC patients (Figures 2–4). ANXA2 is mainly distributed in the nucleus and cytoplasm, and important role in cancer progression and invasion has been reported [22]. Previous studies reported that ANXA2 was significantly elevated in tumor issues and related to poor prognosis in breast cancer [23], glioma [24], gastric cancer [25] and liver cancer [26]. ANXA3 was also reported as an important role in a variety of tumor development processes [27]. Overexpressed ANXA3 could promote tumor proliferation and metastasis in breast, lung, liver, and ovarian cancer, and was associated with chemotherapy resistance [28,29]. In addition, increased expression of ANXA13 could promote the proliferation and migration of lung cancer cells in vitro and was associated with poor survival in lung adenocarcinoma patients [30]. Moreover, Wu et al. (2021) recently reported that the expression of Annexins were related to the molecular subtypes of MIBC [31]. They found that *ANXA1/2/3/5/6/7/8* were highly expressed in basal-subtype MIBC, while *ANXA4/9/10/11* were mainly expressed in luminal-subtype MIBC, which might be used as potential markers for subtype classification of BC. Their results could show that the abnormal expression of Annexin members were common in several types of cancer and might play key roles in carcinogenesis and cancer progression, including BC.

We then constructed an ARPS using the machine learning algorithm LASSO and demonstrated that BC patients in the high ARPS risk group had a shorter OS/DSS/PFI in BC cohorts than those with low risk through KM-survival analysis (Figure 5). Additionally, Cox regression analysis showed that ARPS was an independent prognostic predictor in both the discovery BC cohort and three independent validation cohorts, respectively. Moreover, ARPS can even predict the prognosis of BC patients within different subgroups stratified by clinical characteristics, including age, gender and T stage. Overall, these results all confirmed that the risk score derived from ARPS could accurately and stably predict the survival outcome of BC patients independently.

KEGG pathway and GSEA analysis revealed that EMT and its regulators pathways (TGF-β signaling pathway, TNF-alpha/NF-kappaB, PI3K/AKT/mTOR) were found to be differentially enriched between the high- and low-risk groups. EMT is a process by which epithelial cells lose their epithelial properties and obtain a mesenchymal phenotype, and could transform tumor cells from inactive cancer to malignant phenotypes [27,28]. Previous studies have indicated that EMT was a key controller in tumor progression and metastasis of BC [32–34]. Upregulation of EMT transcription factors, such as TWIST1, ZEB1/2 and SNAI1/2 have been reported to promote migration and invasion of tumor cells in many types of tumors [35–38]. In addition, several EMT regulatory pathways, such as TNF-alpha/NF-kappaB and TGF-β were significantly highly enriched in the high-risk ARPS group. Li et al. revealed that activation of TNF-alpha/NF-kappaB could induce EMT through upregulation of EMT transcription factor Twist1 and contribute to metastatic BC [39]. Upregulation of TGF-β can activate Wnt signaling pathways and play a synergistic role to start the EMT process [40]. Moreover, the PI3K/AKT/mTOR pathway participated in numerous cell biological processes. Activated AKT and mTOR can increase E-cadherin expression and promote EMT activation [41]. Therefore, the cross-talk of these signaling pathways may contribute to the poor prognosis of the high ARPS risk group through promoting tumor recurrence and metastasis by the EMT process.

In order to escape the anti-tumor immune response, tumor cells could secrete immuno-suppressive and anti-apoptotic factors or recruit suppressive immune cells to generate a highly immunosuppressive microenvironment through different mechanisms [42,43]. In BC TME, accumulated immunosuppressive cells (e.g., myeloid-derived suppressor cells (MDSCs), tumor-associated macrophages (TAMs) and regulatory T cells (T regs) and evaluated expression of immune checkpoints (e.g., CTLA-4 and PD-1) were reported to induce immune evasion of tumor cells [44,45]. Therefore, we evaluated the landscape of immune cell infiltration for the high and low ARPS risk groups by ESTIMATE, Timer,

MCPcounter and xCell algorithm, which revealed that a higher degree of immune cell infiltration and greater abundance of immunosuppressive cells including Tregs, TAMs and MDSCs were found in the high ARPS risk group than these in the low ARPS risk group. Previous studies have proved that increased infiltration of Tregs, TAMs and MDSCs were found in BC tissue and were associated with poor prognosis of BC patients [46–48]. As key cellular components of TME, Tregs could facilitate immune evasion of cancer cells through secreting inhibitory cytokines [49], and TAMs could greatly contribute to form a tolerogenic TME by directly exhausting CD8 T cells, and supporting to traffic Tregs [50]. Additionally, MDSCs can also inhibit the immune response by suppressing CD4 T-cells, CD8 T-cells, and NK cells, inducing Tregs and facilitating TAMs polarizing into M2 phenotype [46]. Notably, MDSC-induced immunosuppression has been demonstrated to accelerate the tumor progression and enhance the formation of metastatic lesions through promoting the EMT process [51,52]. Moreover, our study suggested that the high-risk ARPS prognostic group showed high expression of *CD274, CD276, CD28, CD86, LAG3, PDCD1* and *PDCDLG2*, and may be more sensitive to anti-PD1 treatment. Based on above findings, we deliberate that the high-risk group might be related to a high degree of immunosuppression and low immunoreactivity in TME, thereby promoting tumor recurrence and metastasis through EMT-related pathways. As a result, the high-risk group might get more benefits from immunotherapy.

5. Conclusions

In conclusion, we found that several Annexins were aberrantly expressed and associated with prognosis in BC through public tools and identified and validated an ARPS comprised of four members, ANXA1/5/6/10, proving that ARPS was an independent prognostic factor in four individual BC cohorts. This model might be helpful for clinicians to guide the treatment strategy and eventually benefit BC patients. These results could also provide insights into the underlying molecular mechanisms of development and progression of BC and offer potential therapeutic targets for BC.

Supplementary Materials: The following are available online at https://www.mdpi.com/article/10.3390/biology11020259/s1. Figure S1: The work-flow of our study, Figure S2: Comparison of mRNA expression of Annexins family members between bladder cancer and adjacent normal tissues (GEPIA2). ANXA1-11 (A-K) and ANXA13 (L); where * $p < 0.05$, Figure S3: Survival analyses of ARPS regarding Disease Free Interval (DSS) and Progression Free Interval (PFI) in TCGA (A-B) and GSE31684 (C-D) BC cohorts, Figure S4: Primary figure of Heat map of differential expression genes between the high/low risk groups (Figure 8A), Table S1: Clinical characteristics of the BC patients collected in this study.

Author Contributions: Conceptualization, G.Z. and X.G.; methodology, X.Y., X.Q. and Y.W. (Yao Wang); software, X.Q., Y.W. (Yao Wang) and B.Z.; validation, T.H., T.Y. and L.Z.; formal analysis, X.Y. and X.Q.; investigation, G.Z. and X.G.; resources, X.G.; data curation, X.Y., X.Q., Y.W. (Yao Wang) and B.Z.; writing—original draft preparation, G.Z., X.G. and X.Y.; writing—review and editing, X.Q., Y.W. (Yao Wang), B.Z., T.H., T.Y., L.Z., Y.W. (Yange Wang) and H.Z.; visualization, X.Q.; supervision, G.Z. and X.G.; project administration, G.Z. and X.G.; funding acquisition, X.G. All authors have read and agreed to the published version of the manuscript.

Funding: This research was funded by "SUPPORTING PROGRAM FOR CENTRAL PLAIN YOUNG TOP TALENTS IN HENAN PROVINCE, grant number ZYQR201912176", "PROGRAM FROM ACADEMY FOR ADVANCED INTERDISCIPLINARY STUDIES OF HENAN UNIVERSITY, grant number Y21008L", "PROGRAM FOR SCIENCE AND TECHNOLOGY DEVELOPMENT IN HENAN PROVINCE, grant number 212102310150", "KAIFENG SCIENCE AND TECHNOLOGY PROJECT, grant number No.1908001", and The APC was funded by SUPPORTING PROGRAM FOR CENTRAL PLAIN YOUNG TOP TALENTS IN HENAN PROVINCE".

Institutional Review Board Statement: Not applicable.

Informed Consent Statement: Not applicable.

Data Availability Statement: All of the data in this manuscript are available at TCGA (https://portal.gdc.cancer.gov/, accessed on 20 April 2018) AND GEO (https://www.ncbi.nlm.nih.gov/gds/, accessed on 20 April 2018) databases.

Conflicts of Interest: The authors declare no conflict of interest.

References

1. Sung, H.; Ferlay, J.; Siegel, R.L.; Laversanne, M.; Soerjomataram, I.; Jemal, A.; Bray, F. Global Cancer Statistics 2020: Globocan Estimates of Incidence and Mortality Worldwide for 36 Cancers in 185 Countries. *CA Cancer J. Clin.* **2021**, *71*, 209–249. [CrossRef] [PubMed]
2. Antoni, S.; Ferlay, J.; Soerjomataram, I.; Znaor, A.; Jemal, A.; Bray, F. Bladder Cancer Incidence and Mortality: A Global Overview and Recent Trends. *Eur. Urol.* **2017**, *71*, 96–108. [CrossRef] [PubMed]
3. Alifrangis, C.; McGovern, U.; Freeman, A.; Powles, T.; Linch, M. Molecular and histopathology directed therapy for advanced bladder cancer. *Nat. Rev. Urol.* **2019**, *16*, 465–483. [CrossRef] [PubMed]
4. Nadal, R.; Bellmunt, J. Management of metastatic bladder cancer. *Cancer Treat. Rev.* **2019**, *76*, 10–21. [CrossRef] [PubMed]
5. Lenis, A.T.; Lec, P.M.; Chamie, K. Bladder Cancer: A Review. *JAMA* **2020**, *324*, 1980–1991. [CrossRef] [PubMed]
6. Gerke, V.; Creutz, C.E.; Moss, S.E. Annexins: Linking Ca^{2+} signalling to membrane dynamics. *Nat. Rev. Mol. Cell Biol.* **2005**, *6*, 449–461. [CrossRef] [PubMed]
7. Boye, T.L.; Nylandsted, J. Annexins in plasma membrane repair. *Biol. Chem.* **2016**, *397*, 961–969. [CrossRef] [PubMed]
8. Grewal, T. Annexins in cell migration and adhesion. *Cell Adh. Migr.* **2017**, *11*, 245–246. [CrossRef]
9. Mussunoor, S.; Murray, G.I. The role of annexins in tumour development and progression. *J. Pathol.* **2008**, *216*, 131–140. [CrossRef]
10. Schloer, S.; Pajonczyk, D.; Rescher, U. Annexins in Translational Research: Hidden Treasures to Be Found. *Int. J. Mol. Sci.* **2018**, *19*, 1781. [CrossRef]
11. Yu, S.; Meng, Q.; Hu, H.; Zhang, M. Correlation of ANXA1 expression with drug resistance and relapse in bladder cancer. *Int. J. Clin. Exp. Pathol.* **2014**, *7*, 5538–5548. [PubMed]
12. Hu, H.; Zhao, J.; Zhang, M. Expression of Annexin A2 and Its Correlation With Drug Resistance and Recurrence of Bladder Cancer. *Technol. Cancer Res. Treat.* **2016**, *15*, NP61–NP68. [CrossRef] [PubMed]
13. Munksgaard, P.P.; Mansilla, F.; Brems Eskildsen, A.S.; Fristrup, N.; Birkenkamp-Demtröder, K.; Ulhøi, B.P.; Borre, M.; Agerbæk, M.; Hermann, G.G.; Ørntoft, T.F.; et al. Low ANXA10 expression is associated with disease aggressiveness in bladder cancer. *Br. J. Cancer* **2011**, *105*, 1379–1387. [CrossRef]
14. Zhang, G.; Wang, Q.; Yang, M.; Yuan, Q.; Dang, Y.; Sun, X.; An, Y.; Dong, H.; Xie, L.; Zhu, W.; et al. OSblca: A Web Server for Investigating Prognostic Biomarkers of Bladder Cancer Patients. *Front. Oncol.* **2019**, *9*, 466. [CrossRef] [PubMed]
15. Rhodes, D.R.; Kalyana-Sundaram, S.; Mahavisno, V.; Varambally, R.; Yu, J.; Briggs, B.B.; Barrette, T.R.; Anstet, M.J.; Kincead-Beal, C.; Kulkarni, P.; et al. Oncomine 3.0: Genes, pathways, and networks in a collection of 18,000 cancer gene expression profiles. *Neoplasia* **2007**, *9*, 166–180. [CrossRef]
16. Tang, Z.; Kang, B.; Li, C.; Chen, T.; Zhang, Z. GEPIA2: An enhanced web server for large-scale expression profiling and interactive analysis. *Nucleic Acids Res.* **2019**, *47*, W556–W560. [CrossRef]
17. Ru, B.; Wong, C.N.; Tong, Y.; Zhong, J.Y.; Zhong, S.S.W.; Wu, W.C.; Chu, K.C.; Wong, C.Y.; Lau, C.Y.; Chen, I.; et al. TISIDB: An integrated repository portal for tumor-immune system interactions. *Bioinformatics* **2019**, *35*, 4200–4202.
18. Cancer Genome Atlas Research Network. Comprehensive molecular characterization of urothelial bladder carcinoma. *Nature* **2014**, *507*, 315–322. [CrossRef]
19. Riester, M.; Taylor, J.M.; Feifer, A.; Koppie, T.; Rosenberg, J.E.; Downey, R.J.; Bochner, B.H.; Michor, F. Combination of a novel gene expression signature with a clinical nomogram improves the prediction of survival in high-risk bladder cancer. *Clin. Cancer Res.* **2012**, *18*, 1323–1333.
20. Lindgren, D.; Sjödahl, G.; Lauss, M.; Staaf, J.; Chebil, G.; Lövgren, K.; Gudjonsson, S.; Liedberg, F.; Patschan, O.; Månsson, W.; et al. Integrated genomic and gene expression profiling identifies two major genomic circuits in urothelial carcinoma. *PLoS ONE* **2012**, *7*, e38863.
21. Choi, W.; Porten, S.; Kim, S.; Willis, D.; Plimack, E.R.; Hoffman-Censits, J.; Roth, B.; Cheng, T.; Tran, M.; Lee, I.L.; et al. Identification of distinct basal and luminal subtypes of muscle-invasive bladder cancer with different sensitivities to frontline chemotherapy. *Cancer Cell.* **2014**, *25*, 152–165. [CrossRef] [PubMed]
22. Wang, T.; Wang, Z.; Niu, R.; Wang, L. Crucial role of Anxa2 in cancer progression: Highlights on its novel regulatory mechanism. *Cancer Biol. Med.* **2019**, *16*, 671–687. [PubMed]
23. Gibbs, L.D.; Vishwanatha, J.K. Prognostic impact of AnxA1 and AnxA2 gene expression in triple-negative breast cancer. *Oncotarget* **2018**, *9*, 2697–2704. [CrossRef] [PubMed]
24. Gao, H.; Yu, B.; Yan, Y.; Shen, J.; Zhao, S.; Zhu, J.; Qin, W.; Gao, Y. Correlation of expression levels of ANXA2, PGAM1, and CALR with glioma grade and prognosis. *J. Neurosurg.* **2013**, *118*, 846–853. [CrossRef]
25. Xie, R.; Liu, J.; Yu, X.; Li, C.; Wang, Y.; Yang, W.; Hu, J.; Liu, P.; Sui, H.; Liang, P.; et al. ANXA2 Silencing Inhibits Proliferation, Invasion, and Migration in Gastric Cancer Cells. *J. Oncol.* **2019**, *2019*, 4035460. [CrossRef] [PubMed]

26. Tang, L.; Liu, J.X.; Zhang, Z.J.; Xu, C.Z.; Zhang, X.N.; Huang, W.R.; Zhou, D.H.; Wang, R.R.; Chen, X.D.; Xiao, M.B.; et al. High expression of Anxa2 and Stat3 promote progression of hepatocellular carcinoma and predict poor prognosis. *Pathol. Res. Pract.* **2019**, *215*, 152386. [CrossRef]

27. Wu, N.; Liu, S.; Guo, C.; Hou, Z.; Sun, M.Z. The role of annexin A3 playing in cancers. *Clin. Transl. Oncol.* **2013**, *15*, 106–110. [CrossRef] [PubMed]

28. Wang, L.; Li, X.; Ren, Y.; Geng, H.; Zhang, Q.; Cao, L.; Meng, Z.; Wu, X.; Xu, M.; Xu, K. Cancer-associated fibroblasts contribute to cisplatin resistance by modulating ANXA3 in lung cancer cells. *Cancer Sci.* **2019**, *110*, 1609–1620. [CrossRef]

29. Tong, M.; Fung, T.M.; Luk, S.T.; Ng, K.Y.; Lee, T.K.; Lin, C.H.; Yam, J.W.; Chan, K.W.; Ng, F.; Zheng, B.J.; et al. ANXA3/JNK Signaling Promotes Self-Renewal and Tumor Growth, and Its Blockade Provides a Therapeutic Target for Hepatocellular Carcinoma. *Stem Cell Rep.* **2015**, *5*, 45–59. [CrossRef]

30. Xue, G.L.; Zhang, C.; Zheng, G.L.; Zhang, L.J.; Bi, J.W. Annexin A13 predicts poor prognosis for lung adenocarcinoma patients and accelerates the proliferation and migration of lung adenocarcinoma cells by modulating epithelial-mesenchymal transition. *Fundam. Clin. Pharmacol.* **2020**, *34*, 687–696. [CrossRef]

31. Wu, W.; Jia, G.; Chen, L.; Liu, H.; Xia, S. Analysis of the Expression and Prognostic Value of Annexin Family Proteins in Bladder Cancer. *Front. Genet.* **2021**, *12*, 731625. [CrossRef] [PubMed]

32. Heerboth, S.; Housman, G.; Leary, M.; Longacre, M.; Byler, S.; Lapinska, K.; Willbanks, A.; Sarkar, S. EMT and tumor metastasis. *Clin. Transl. Med.* **2015**, *4*, 6. [CrossRef]

33. Guo, F.; Parker Kerrigan, B.C.; Yang, D.; Hu, L.; Shmulevich, I.; Sood, A.K.; Xue, F.; Zhang, W. Post-transcriptional regulatory network of epithelial-to-mesenchymal and mesenchymal-to-epithelial transitions. *J. Hematol. Oncol.* **2014**, *7*, 19. [CrossRef]

34. Ashrafizadeh, M.; Hushmandi, K.; Hashemi, M.; Akbari, M.E.; Kubatka, P.; Raei, M.; Koklesova, L.; Shahinozzaman, M.; Mohammadinejad, R.; Najafi, M.; et al. Role of microRNA/Epithelial-to-Mesenchymal Transition Axis in the Metastasis of Bladder Cancer. *Biomolecules* **2020**, *10*, 1159. [CrossRef] [PubMed]

35. Geradts, J.; de Herreros, A.G.; Su, Z.; Burchette, J.; Broadwater, G.; Bachelder, R.E. Nuclear Snail1 and nuclear ZEB1 protein expression in invasive and intraductal human breast carcinomas. *Hum. Pathol.* **2011**, *42*, 1125–1131. [CrossRef] [PubMed]

36. Spoelstra, N.S.; Manning, N.G.; Higashi, Y.; Darling, D.; Singh, M.; Shroyer, K.R.; Broaddus, R.R.; Horwitz, K.B.; Richer, J.K. The transcription factor ZEB1 is aberrantly expressed in aggressive uterine cancers. *Cancer Res.* **2006**, *66*, 3893–3902. [CrossRef]

37. Fang, Y.; Wei, J.; Cao, J.; Zhao, H.; Liao, B.; Qiu, S.; Wang, D.; Luo, J.; Chen, W. Protein expression of ZEB2 in renal cell carcinoma and its prognostic significance in patient survival. *PLoS ONE* **2013**, *8*, e62558.

38. Galván, J.A.; Zlobec, I.; Wartenberg, M.; Lugli, A.; Gloor, B.; Perren, A.; Karamitopoulou, E. Expression of E-cadherin repressors SNAIL, ZEB1 and ZEB2 by tumour and stromal cells influences tumour-budding phenotype and suggests heterogeneity of stromal cells in pancreatic cancer. *Br. J. Cancer* **2015**, *112*, 1944–1950. [CrossRef]

39. Li, C.W.; Xia, W.; Huo, L.; Lim, S.O.; Wu, Y.; Hsu, J.L.; Chao, C.H.; Yamaguchi, H.; Yang, N.K.; Ding, Q.; et al. Epithelial-mesenchymal transition induced by TNF-alpha requires NF-kappaB-mediated transcriptional upregulation of Twist1. *Cancer Res.* **2012**, *72*, 1290–1300. [CrossRef]

40. Wang, H.; Wang, H.S.; Zhou, B.H.; Li, C.L.; Zhang, F.; Wang, X.F.; Zhang, G.; Bu, X.Z.; Cai, S.H.; Du, J. Epithelial-mesenchymal transition (EMT) induced by TNF-alpha requires AKT/GSK-3beta-mediated stabilization of snail in colorectal cancer. *PLoS ONE* **2013**, *8*, e56664.

41. Shao, Q.; Zhang, Z.; Cao, R.; Zang, H.; Pei, W.; Sun, T. CPA4 Promotes EMT in Pancreatic Cancer via Stimulating PI3K-AKT-mTOR Signaling. *Onco. Targets Ther.* **2020**, *13*, 8567–8580. [CrossRef] [PubMed]

42. Gasser, S.; Lim, L.H.K.; Cheung, F.S.G. The role of the tumour microenvironment in immunotherapy. *Endocr. Relat. Cancer* **2017**, *24*, T283–T295. [CrossRef] [PubMed]

43. Huang, T.X.; Fu, L. The immune landscape of esophageal cancer. *Cancer Commun.* **2019**, *39*, 79. [CrossRef]

44. Crispen, P.L.; Kusmartsev, S. Mechanisms of immune evasion in bladder cancer. *Cancer Immunol. Immunother.* **2020**, *69*, 3–14. [CrossRef] [PubMed]

45. Hatogai, K.; Sweis, R.F. The Tumor Microenvironment of Bladder Cancer. *Adv. Exp. Med. Biol.* **2020**, *1296*, 275–290.

46. Biswas, S.K.; Allavena, P.; Mantovani, A. Tumor-associated macrophages: Functional diversity, clinical significance, and open questions. *Semin. Immunopathol.* **2013**, *35*, 585–600. [CrossRef]

47. Yang, G.; Shen, W.; Zhang, Y.; Liu, M.; Zhang, L.; Liu, Q.; Lu, H.H.; Bo, J. Accumulation of myeloid-derived suppressor cells (MDSCs) induced by low levels of IL-6 correlates with poor prognosis in bladder cancer. *Oncotarget* **2017**, *8*, 38378–38388. [CrossRef]

48. Ibrahim, O.M.; Pandey, R.K.; Chatta, G.; Kalinski, P. Role of tumor microenvironment in the efficacy of BCG therapy. *Trends Res.* **2020**, *3*, 1–5. [CrossRef]

49. Stockis, J.; Roychoudhuri, R.; Halim, T.Y.F. Regulation of regulatory T cells in cancer. *Immunology* **2019**, *157*, 219–231. [CrossRef]

50. Ge, Z.; Ding, S. The Crosstalk Between Tumor-Associated Macrophages (TAMs) and Tumor Cells and the Corresponding Targeted Therapy. *Front. Oncol.* **2020**, *10*, 590941. [CrossRef]

51. Zhu, H.; Gu, Y.; Xue, Y.; Yuan, M.; Cao, X.; Liu, Q. CXCR2(+) MDSCs promote breast cancer progression by inducing EMT and activated T cell exhaustion. *Oncotarget* **2017**, *8*, 114554–114567. [CrossRef] [PubMed]

52. Singh, S.; Chakrabarti, R. Consequences of EMT-Driven Changes in the Immune Microenvironment of Breast Cancer and Therapeutic Response of Cancer Cells. *J. Clin. Med.* **2019**, *8*, 642. [CrossRef] [PubMed]

 *biology*

Article

Architectural Distortion-Based Digital Mammograms Classification Using Depth Wise Convolutional Neural Network

Khalil ur Rehman [1], Jianqiang Li [1,2], Yan Pei [3,*], Anaa Yasin [1], Saqib Ali [1] and Yousaf Saeed [1]

1 The School of Software Engineering, Beijing University of Technology, Beijing 100024, China; rehmankhalilur@emails.bjut.edu.cn (K.u.R.); lijianqiang@bjut.edu.cn (J.L.); yasinanaa@emails.bjut.edu.cn (A.Y.); alisaqib@emails.bjut.edu.cn (S.A.); b20196112w@emails.bjut.edu.cn (Y.S.)
2 Beijing Engineering Research Center for IoT Software and Systems, Beijing 100124, China
3 Computer Science Division, University of Aizu, Aizuwakamatsu 965-8580, Fukushima, Japan
* Correspondence: peiyan@u-aizu.ac.jp

Simple Summary: Breast cancer is leading cancer increases the death rate in women. Early diagnosis of breast cancer in women can save their lives. The current study proposed a novel scheme to detect architectural distortion from mammogram images to predict breast cancer using a deep learning approach. Results are evaluated on a public and a private dataset which may help to improve the diagnostic ability of breast cancer of radiologists and doctors in daily clinical routines. Furthermore, the proposed method achieved maximum accuracy as compared with previous approaches. This study can be interesting and valuable in the healthcare predictive modeling domain and will add a real contribution to society.

Citation: Rehman, K.u.; Li, J.; Pei, Y.; Yasin, A.; Ali, S.; Saeed, Y. Architectural Distortion-Based Digital Mammograms Classification Using Depth Wise Convolutional Neural Network. *Biology* 2022, 11, 15. https://doi.org/10.3390/biology11010015

Academic Editors: Shibiao Wan, Yiping Fan, Chunjie Jiang and Shengli Li

Received: 30 November 2021
Accepted: 17 December 2021
Published: 23 December 2021

Publisher's Note: MDPI stays neutral with regard to jurisdictional claims in published maps and institutional affiliations.

Abstract: Architectural distortion is the third most suspicious appearance on a mammogram representing abnormal regions. Architectural distortion (AD) detection from mammograms is challenging due to its subtle and varying asymmetry on breast mass and small size. Automatic detection of abnormal ADs regions in mammograms using computer algorithms at initial stages could help radiologists and doctors. The architectural distortion star shapes ROIs detection, noise removal, and object location, affecting the classification performance, reducing accuracy. The computer vision-based technique automatically removes the noise and detects the location of objects from varying patterns. The current study investigated the gap to detect architectural distortion ROIs (region of interest) from mammograms using computer vision techniques. Proposed an automated computer-aided diagnostic system based on architectural distortion using computer vision and deep learning to predict breast cancer from digital mammograms. The proposed mammogram classification framework pertains to four steps such as image preprocessing, augmentation and image pixel-wise segmentation. Architectural distortion ROI's detection, training deep learning, and machine learning networks to classify AD's ROIs into malignant and benign classes. The proposed method has been evaluated on three databases, the PINUM, the CBIS-DDSM, and the DDSM mammogram images, using computer vision and depth-wise 2D V-net 64 convolutional neural networks and achieved 0.95, 0.97, and 0.98 accuracies, respectively. Experimental results reveal that our proposed method outperforms as compared with the ShuffelNet, MobileNet, SVM, K-NN, RF, and previous studies.

Keywords: architectural distortion; image processing; depth-wise convolutional neural network; breast cancer; mammography

1. Introduction

Breast cancer is leading cancer worldwide in 2020, with 11.7% overall reported cases per world health organization [1] and one of the major causes of death in women. The mortality rate was increased from 6.6% to 6.9% this year due to breast cancer. Initially, these breast cancer tumors are screened on an X-ray machine for breast cancer diagnosis and manually interpreted by the radiologist to predict benign and malignant tumors. Screening methods such as ultrasound, and mammography are used to diagnose breast cancer, while the standard screening method is mammography at the early stage. Computer-aided diagnostic systems automatically detected abnormal regions in mammograms to help radiologists and doctors detect disease in less time to avoid unnecessary biopsies [2].

Breast composition containing attenuating tissue is an essential element for evaluating mammogram reports to predict malignant and benign cases. Architectural distortion (AD) is the third most suspicious appearance on a mammogram representing abnormal regions that can be found visible on mammography projection [3]. The main parameters such as global asymmetry, focal asymmetry, and developing asymmetry of tissue can be calculated using machine and deep learning algorithms to track AD in mammograms. Asymmetries are the isodense tissues obscured by adjacent fibro glandular mass, representing true malignancy in mammograms. Architectural distortion tracking from mammograms is very difficult due to its subtle and varying asymmetry on breast mass and small size. Therefore, the manual interpretation of architectural distortion is a challenging task for radiologists to figure out abnormalities during the examination of mammograms. The leading types of cancer that can present architectural distortion on mammography are invasive lobular carcinoma (ILC) and invasive ductal carcinoma (IDC). The ILC and IDC on mammography having a star-shaped pattern are likely to be malignant, while the complex and radial sclerosing lesions architectural distortion having larger than 1 cm is probably benign [4].

Several studies reported hand-crafted feature extraction techniques on mammogram images for AD ROI classification using machine learning and deep learning [5]. These methods successfully achieved remarkable accuracy in the diagnosis of breast cancer. However, many factors are involved in detecting architectural distortion, such as tinny size, subtle appearance inside mass, shape, noise, imaging artefact from digital mammograms. Due to a limited number of studies that reported AD ROI's classification in the literature, this primarily discusses the most relevant studies in the first phase. The second phase discusses deep learning, machine learning, and mass segmentation, to determine the limitations of predicting breast cancer. There are many limitations in these studies for detecting architectural distortion ROIs and classification. For example, Murali S. et al. [6] proposed a model-based approach to detect architectural distortion from mammograms and classify with a support vector machine to achieve 89.6 accuracy. A total of 150 ROI's were selected from the DDSM dataset to evaluate the performance. Banik et al. [7] employed the gobar filter and phase portrait analysis method to detect architectural distortion in prior mammograms by evaluating 4224 ROI's from a private dataset and achieved 90% sensitivity at 5.7 FP/image. J. et al. [8] presented a two-step method such as detecting ROIs with potential AD on analyzing the Gabor filter and recognizing AD's using a $2D$ Fourier transform. Experimental results were evaluated on 33 mammograms containing AD's from DDSM and obtained 83.50 accuracy. All three authors employed Gabor filter to the texture feature analysis of images while locating the boundary of ADs ROIs was still a limitation. As a result, these hand-crafted feature extraction methods decrease the computational time and affect the model's classification accuracy.

The classification of AD ROIs based on texture analysis model using support vector machine was implemented on mammogram images by Kamra A. et al. [9]. The texture analysis ROIs were selected from the digital database for screening mammography (DDSM) dataset to evaluate the model's performance and reported 92.94% accuracy. Liu et al. [10] employed a new method for architectural distortion ROIs recognition based on texture features from gray-level co-occurrence matrix (GLCM) matrix, spiculated and entropy features from mammogram images, and the sparse representation classifier was used for the classification of ROIs. The performance of the model was evaluated on the DDSM dataset by obtaining 91.79 accuracy. Ioana B. et al. [11] proposed radiomic analysis of contrast-enhanced spectral mammography approach for breast cancer prediction and classification using k-nearest neighbors (K-NN). Another radiomic feature reduction approach was proposed by Raffaella M. et al. [12] for mammogram classification to predict breast cancer. D. H. et al. [13] proposed a micro-pattern texture descriptor for the detection of architectural distortion from mammogram images using a local binary pattern, local map pattern, and haralick's descriptors. A total of 400 ROIs from the full-field digital mammography (FFDM) dataset were selected for the evaluation of the model and achieved 83% accuracy. Casti P. et al. [14] was introduced a new paradigm to detect AD track in digital breast tomosynthesis (DBT) exam by using a cross-cutting approach exploiting $3D$ imaging modality. The proposed approach achieves 0.9 sensitivity after evaluating the model on 37 sets of DBT from the FFDM dataset. Palma et al. [15] presented a fuzzy contrary-based approach for detecting masses and architectural distortion from digital breast tomosynthesis.

Another essential factor is noise removal from ADs ROIs which was still a limitation with these traditional methods. Moreover, all of the studies were employed traditional machine learning algorithms, which were limited to the lower classification accuracy. The architectural distortion star shapes heterogeneous pattern detection inside the denser mass using the texture analysis was still a limitation. Cai et al. [16] employed a method for identifying architectural distortion in mammogram images using a dense net deep neural network to train the image net model for breast mass dataset to classify the breast masses. Bahl et al. [17] was presented a retrospective review for the presence of architectural distortion on mammogram images and concluded that the presence of architectural diction on mammography has the chance of malignancy in approximately three fourth of the cases. Shu et al. [18] proposed a region-based pooling structure using a deep convolutional neural network to classify mammogram images. The whole region of images as an input to a deep neural network is limited to identifying the subtle location of ADs inside denser breast masses. Conventional deep neural networks only use a single channel for image feature maps which is not limited to neural networks but decreases the overall modal accuracy.

The current study investigated the gap to detect architectural distortion ROIs from mammograms using computer vision techniques. This study employed a depth-wise $2D$ V-net 64 convolutional neural network to classify these architectural distortion ROIs into benign and malignant ADs. With this approach, the above limitation is no longer. Computer vision is a powerful technology for removing the noise and detecting the object from hidden star-shape patterns. The Depth-wise neural network uses each input channel for creating a feature map that increases the modal efficiency and accuracy. Therefore, this study aim to develop a computer-aided diagnostic system using computer vision and a deep learning model to classify architectural distortions ROIs from digital mammograms at early stages.

The principal outcome of our study is reported as follows:

- Proposed an automated computer-aided diagnostic system based on architectural distortion using computer vision and depth-wise deep learning techniques to predict breast cancer from digital mammograms. Applied the image pixel-wise segmentation using a computer vision algorithm to extract architectural distortion ROIs from the digital mammogram image in the first phase.
- In the second phase, employed a depth-wise V-Net 64 convolutional neural network to extract automatic features from ADs ROIs and classify them into malignant and benign ROIs. Moreover, use machine learning and deep learning algorithms, such as shuffelnet, mobilenet, support vector machine, k-nearest neighbor, and random forest, to classify these ROIs.
- Proposed method obtained higher accuracy than machine learning and with the previous studies. Furthermore, evaluated proposed model with other metrics to enhance the diagnostic ability of the model.
- Evaluated the proposed method on three datasets, the local private PINUM and publicly available CBIS-DDSM and DDSM dataset that makes a fair comparison of the proposed model with others.

2. Related Works and Techniques

2.1. Conventional Deep Learning Mammogram Classification

The researchers presented several computer-aided diagnostic systems using deep convolutional neural networks to predict breast cancer from digital mammograms. Studied that reported deep learning algorithms for the classification of mammogram images herein briefly reported. Feature fusion bases-deep CNN was applied using extreme learning machines to predict breast cancer from mammograms by wang et al. [19]. An improved ResNet-based convolutional neural network was employed to the classification of mammogram images and significantly improve the area under the curve by Wu et al. [20]. Khan et al. [21] developed multi-view feature fusion-based CAD to detect abnormal and normal patterns from mammograms using a deep neural network to increase the accuracy in breast classification. On segmentation of the pectoral muscle-based approach using a deep convolutional neural network was developed by Soleiman et al. [22] to classify mammogram images. Hao et al. [23] presented an automated framework for identifying mislabeled data using cross-entropy and metric function, and the model was trained using a deep convolutional neural network to improve the classification performance. Sun et al. [24] was presented with an automated computer-aided diagnostic system based on a multimodal deep neural network for the integration of multi-dimensional data to prognosis prediction of breast cancer.

A region of interest-based approach was employed by Guan et al. [25] using u-net deep convolution neural network for locating asymmetric patterns to the diagnosis of breast cancer in digital mammograms. The generative adversarial neural network employed for tumor segmentation from digital mammogram by Singh et al. [26]. Song R. et al. [27] developed a combined feature-based model using a deep convolutional neural network for the classification of breast masses into normal, benign, and malignant classes. To overcome the drawbacks of pixel-wise segmentation of mammogram images, Shen et el. [28] was presented a hierarchical model using a deep convolutional neural network and fuzzy learning for breast cancer diagnosis. Guan et al. [29] applied a generative adversarial network for ROIs cropping from digital mammograms, and then the deep convolutional neural network was implemented for the classification of normal and abnormal ROIs. An improved dense net deep learning model was proposed by Li et al. [30] to classify benign and malignant mammograms. A whole image classification based-method was built using a deep neural network using by Iones et al. [31]. Falcon et al. [32] was employed transfer learning techniques to predict abnormalities in digital mammograms with a deep mobile net neural network.

Gnana S. et al. [33] developed a computer-aided diagnostic system using a deep convolutional neural network to classify malignant and benign masses. A deep active and self-paced learning-based framework was emphasized for detecting breast mass from digital mammograms by Shen et al. [34] to reduce the annotation effort for radiologists. Shen et al. [35] presented a method for lesion segmentation and disease classification using a mixed-supervision-guided residual u-net deep learning modal. Shayma A.H et al. [36] propose a novel method for cancer detection from breast mass using feature matching of different regions by applying maximally stable extremal regions. A hybrid deep learning-based framework was employed by Wang et al. [37] for the classification of breast mass for multi-view data. Wang et al. [38] employed a multi-level nested pyramid deep neural network to segment breast mass to classify malignant and benign classes using a public dataset. Birhanu et al. [39] proposed a breast density classification method to predict cancer from digital mammograms using a deep convolutional neural network. Rehman et al. Proposed a computer vision based deep learning method for the classification of microcalcification ROIs into malignant and benign classes.

2.2. Conventional Machine Learning Mammogram Classification

Machine learning modalities such as SVM, KNN, and random forest were adopted to classify digital mammograms to diagnose breast cancer. Machine learning-based classification CAD systems used hand-crafted feature extraction techniques, which are computationally slow and reduce the performance model. Fan et al. [40] proposed a novel method based on single-nucleotide polymorphism to predict breast cancer risk by extracting architectural distortion features from mammograms. Loizidou et al. [41] presented subtraction of temporally sequential mammogram technique to detect microcalcification clusters and classification performed using a support vector machine. The breast boundary is eliminated with the thresholding technique, and a machine learning-based hybrid model is proposed to classify breast mammograms into malignant and benign classes by Zebari et al. [42]. A computer-aided diagnostic system was built to generate an image feature map using fast Fourier transforms on digital mammograms by Heidar et al. [5]. Chakaraborty et al. [43] presented a machine learning-based hybrid approach for automatic detection of mammographic masses using low-to-high level intensity thresholding and performed classification using FLDA, Bayesian, and ANN. Beham et al. [44] applied wavelet transforms for feature extraction from the digital mammogram, and the K-nearest neighbor algorithm was employed for classification into benign and malignant classes. Liu et al. [45] was proposed a novel approach for breast cancer prediction, which employed information gain simulated annealing genetic algorithm for feature selection and const sensitive support vector machine for classification. Another support vector machine-based approach was employed by Yang et al. [46] to diagnose breast tumors using textual features from mammogram images. Obaidullah et al. [47] presented an image descriptor-based approach for mammogram mass classification using a random forest algorithm. Saqib et al. presented the comparison of machine learning techniques for the prediction of multi-organ cancers.

3. Materials

Databases

This study validated the proposed method on three databases, the PINUM (Punjab institute of nuclear medicine) [48], the CBIS-DDSM (curated breast imaging digital database for screening mammography) [49] and DDSM (digital database for screening mammography) [50]. The PINUM private dataset was collected from a local hospital in Pakistan with the approval of diagnostic imaging nuclear medicine and radiology. A total of 289 patient data in the form of DICOM (Digital Imaging and Communications in Medicine) images were collected ranging age between 32-73 with a mean age of 48.5 years. The dataset includes 577 original images containing 425 benign and 152 malignant images with MLO (mediolateral-oblique) and CC (craniocaudal) views at the resolution of 4096×2047 are shown in Figure 1. The proposed study is based on architectural distortion, so that the validation set of mammogram images is labeled by the radiologist for benign and malignant architectural distortion ROIs. A total of 150 AD ROIs are cropped from full mammograms for validating the training set with the proposed algorithm. The radiologist team consisted of two members, one being a senior radiologist and physicist holding a Ph.D. degree in nuclear medicine with 10 years of experience and the second being a junior radiologist with a Master's degree in radiology. The mammography exam of the PINUM dataset was acquired with Hologic 2D, 3D mammography. The PINUM dataset images have MLO and CC views. The size of the PINUM dataset was artificially inflated using augmentation techniques up to 3462 images.

The CBIS-DDSM (digital database for screening mammography) was a public dataset and enhanced version of the DDSM dataset provided by the University of Florida. The mammogram images are in DICOM files at the complete mammography and abnormality levels. Both MLO and CC views of the mammograms are included in the full mammography pictures. Abnormalities are represented as binary mask images that are the same size as the mammograms they are connected with. The ROI of each anomaly is defined by these mask images. Within each mammogram's abnormality mask, users may make an element-by-element selection of pixels. Due to the unavailability of AD ROIs in the CBIS-DDSM dataset, our radiologist team labeled ADs ROIs manually on full mammogram images. A total of 200 AD ROIs are cropped from full mammograms for validation. We included 3568 mammogram images, including 1740 benign and 1828 malignant images with MLO and CC views, as shown in Figure 2. The DDSM is a public dataset provided by Massachusetts General Hospital, Wake Forest University School of Medicine, and Sacred Heart Hospital and maintained by the University of Florida. The DDSM datasets contain 2500 studies including normal, benign, and malignant cases. Each study comprises two images of the breast as well as some patient data such as age at the time of the study, ACR breast density rating, and subtlety rating for abnormalities. Suspicious lesions in images are correlated with pixel-level ground truth information about their positions and kinds. The DDSM datasets contain 200 ADs ROIs of benign and malignant images. In this study, the predefined ADs are considered validation test datasets. A total of 5500 images (2500 benign, 3000 malignant) were included for training and testing the neural networks from the DDSM dataset. Figure 3 shows benign and malignant mammogram images from DDSM dataset. A detailed description of the datasets is in Table 1.

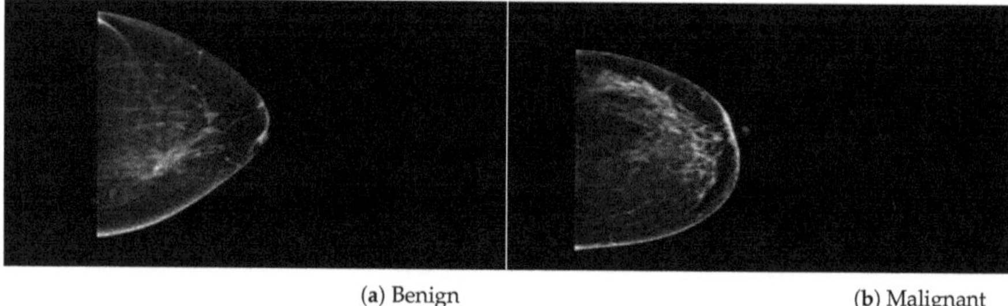

(a) Benign   (b) Malignant

Figure 1. An example of breast mammogram images from PINUM dataset. (**a**) The Benign image (**b**) The Malignant image verified by the Expert radiologist.

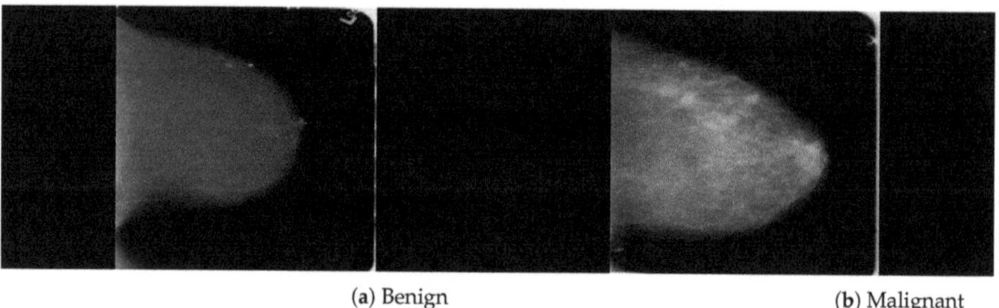

(a) Benign   (b) Malignant

Figure 2. An example of breast mammogram images from CBIS-DDSM dataset. (**a**) The Benign image (**b**) The Malignant image with verified pathology information.

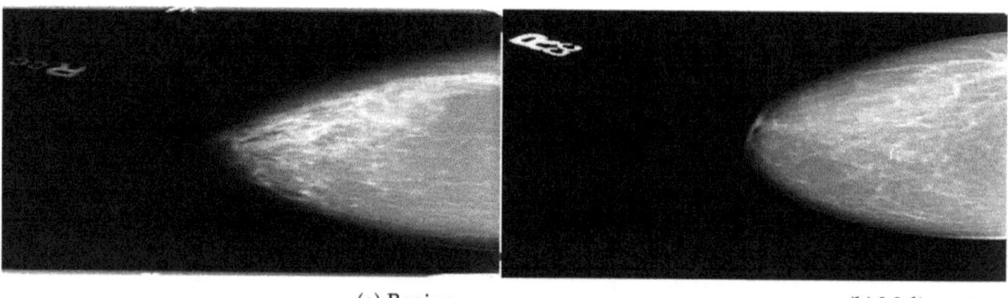

(a) Benign   (b) Malignant

Figure 3. An example of breast mammogram images from DDSM dataset. (**a**) The Benign image (**b**) The Malignant image with verified ground truth information.

Table 1. Data Set Description and Detail.

Mammogram Label	Category	Images	Dataset
Benign (0)	Original	425	PINUM
Malignant (1)	Original	152	PINUM
Benign (0)	Augmented	2550	PINUM
Malignant (1)	Augmented	912	PINUM
Benign (0)	AD ROIs	75	PINUM
Malignant (1)	AD ROIs	75	PINUM
Benign (0)	Original	1740	CBIS-DDSM
Malignant (1)	Original	1828	CBIS-DDSM
Benign (0)	AD ROIs	100	CBIS-DDSM
Malignant (1)	AD ROIs	100	CBIS-DDSM
Benign (0)	Original	2500	DDSM
Malignant (1)	Original	3000	DDSM
Benign (0)	AD ROIs	100	DDSM
Malignant (1)	AD ROIs	100	DDSM

4. Methods

4.1. Proposed Method

In this study, proposed a novel approach for the classification of architectural distortion using a depth-wise $2D$ V-net 64 convolutional neural network. The proposed method pertains to two steps: in the first step, a computer vision algorithm is used for AD ROIs extraction from digital mammogram images. In the second step, the extracted AD ROIs are classified using a depth-wise convolutional neural network. The proposed method can achieve higher accuracy than the deep machine learning methods such as shuffelnet, mobilenet, support vector machine, k-nearest neighbor, and random forest and previous studies. Furthermore, evaluate the performance of the proposed method with other evaluation metrics such as f1_score, precision, recall, sensitivity, specificity, and area under the curve (AUC). The proposed framework of proposed method for mammogram classification based on architectural distortion is presented in Figure 4. The details about the proposed methodology are determined in subsequent sections.

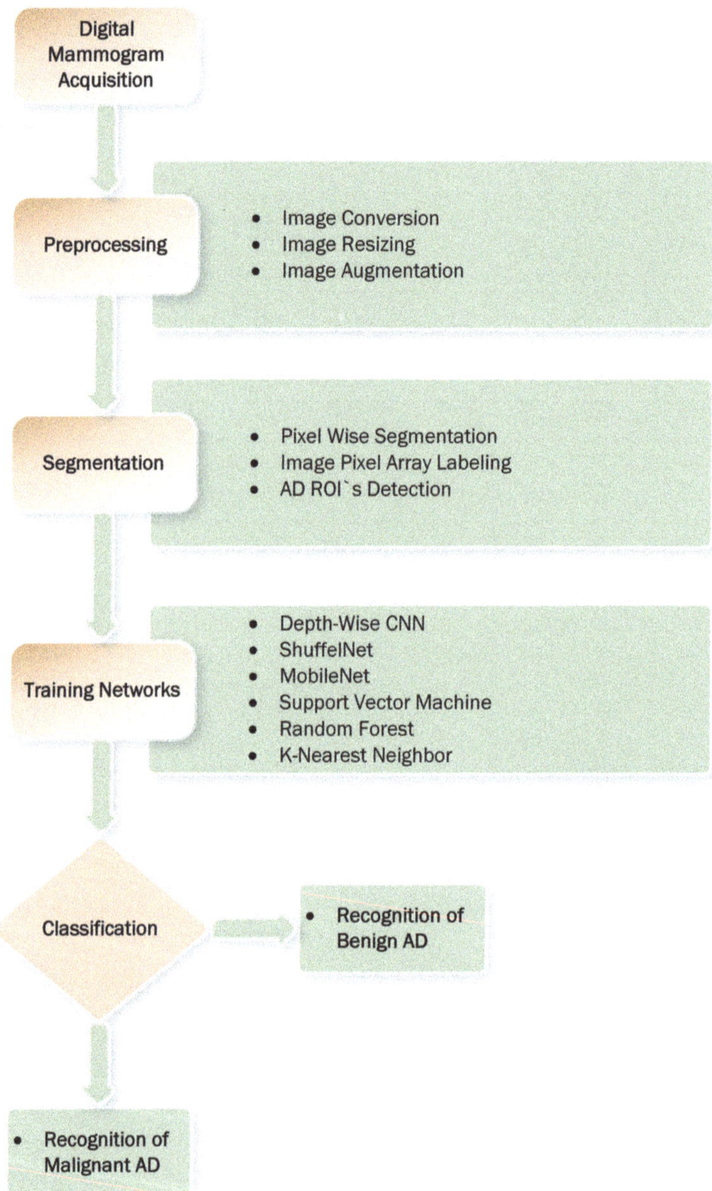

Figure 4. The proposed mammogram classification framework pertains to four steps: image preprocessing and augmentation, pixel wise segmentation and image pixel array labeling, architectural distortion ROI's detection, training deep learning, and machine learning networks to classify AD's ROIs into malignant and benign classes.

4.2. Image Preprocessing

Image conversion and resizing are employed in the preprocessing step to remove noise, artifacts, and irrelevant information. The original mammograms were acquired from three databases such as the PINUM [48] local database and the public database CBIS-DDSM [49], and DDSM [50]. The original databases PINUM and CBIS-DDSM were in the DICOM (digital imaging and communications in medicine) format containing images and patient data. In the first step, the DICOM images are converted into PNG format using an automated OpenCV conversion method, and the patient data is stored in a CSV file. The image preprocessing Algorithm 1 is reported below the complete steps. The converted PNG breast mammogram images are very high-resolution images with a 4096×2047 width and height. We employed the automatic image resizing method with a two-integer argument width and height by downsizing resolution up to 320×240 pixels to make fixed-size images before training a deep convolutional neural network. The DDSM database images are in gif format and converted into PNG format using the automated conversion method.

Algorithm 1 Image preprocessing algorithm 1.

Step 1: Select the DICOM file using read method.;

Step 2: Read DICOM Description values.;

Step 3: Create input vector of DICOM file;

Step 4: Write image description;

Step 5: Read patient data;

Step 6: Read image pixel values;

Step 7: Apply image function zoom in/out;

Step 8: Apply Linear Interpolation function;

Step 9: Create new input vector for new format;

Step 10: Replace Pixels DICOM format to PNG;

Step 11: Write patient data;

Step 12: Save converted image and patient data;

Step 13: Display PNG Image;

4.3. Image Augmentation

Deep learning is a data-driven method so that the small size of data and non-standardization are the main challenges for the generalization of the model. However, to handle the generalization, overfitting, and improving the robustness of the deep learning model, we artificially inflate the PINUM database five times from the original images to increase the dataset size. The data augmentation techniques such as rotating, flipping, sharpening, d-skew, brightness, and contrast are employed to increase the dataset's size, as shown in Table 2. In addition, the overfitting and generalization of the deep learning model can be improved by applying augmentation [51]. The mammogram images are rotated at 45, 90, 135, 180, and 360 degrees and return a new object of the rotated images within a described resolution to increase dataset size up to 3462. Moreover, we rotated a single mammogram at five angles that produce five rotated images and one original image and employed augmentation methods, as shown in Figure 5. The volume of the CBIS-DDSM and the DDSM dataset is 3568, 5500 images; therefore, the data augmentation was not employed on both datasets as the modal overfitting and generalization was not a challenging issue.

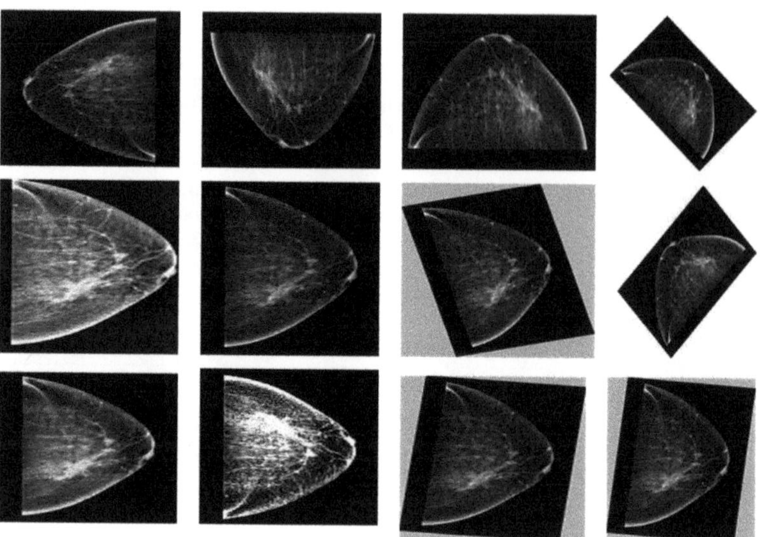

Figure 5. The augmented images of PINUM dataset starting from original to augmented images.

Table 2. Data augmentation techniques with performance value.

Sr	Augmentation Techniques	Performance Values
1	Rotation	$45°, 90°, 135°, 180°, 360°$
2	Flipping	Left, Right, Top, Bottom
3	Sharpen (lightness value)	0.5–1.5
4	D-skew (angle)	$15°, 40°$
5	Contrast (intensity value)	20–60%
6	Brightness (darkness values)	15–55%

4.4. Pixel Wise Segmentation

The image pixel-wise segmentation method maps each pixel of the image that belongs to the image's object or shape and gives a label. M. Wang et al. [52] employed image path-based pixel segmentation using a label fusion algorithm. The image pixel-wise segmentation method maps each pixel of the image that belongs to the image's object or shape and gives a label. Pixels have the same attribute locating an object of the image. Computer vision is a powerful technology for detecting objects as compared with other object detection techniques. Employed a computer vision-based object detection technique and create an image pixel array. Each pixel array has labeled with a class *label*0 and *label*1. The detailed process is as follow:

1. The image is to be segmented as a targeted image $P = (x, y, N)^{w \times h}$, where P representing a pixel array vector having N elements that has belongs to the specific category as:

$$\sum_p P \in (x, y)^{w \times h} = L \in [0, 1] \tag{1}$$

2. The pixel $x \in (x1, x2, \dots w)$ and $y \in (y1, y2, \dots h)$ represents the vertical w and horizontal h pixels, where $x1$ and $y1$ are the elements of pixel vector. The dot product has performed as:

$$P(x, y) = P(x, y).L \tag{2}$$

3. $L \in [0,1]$ represents each object in a pixel array belonging to classes 0 and 1. The pixel-wise prediction can be improved on which we can generate the segmentation results.

4.5. Architectural Distortion ROI's Detection

Architectural distortion is the third most suspicious appearance on a mammogram that represents abnormal regions. Architectural distortion tracking from mammograms is challenging due to its subtle and varying asymmetry on breast mass and small size. The architectural distortion associated with ILC or IDC on mammography represents the abnormality, and having a star-shaped pattern is likely to be malignant, while the complex and radial sclerosing lesions architectural distortion having larger than 1 cm is probably benign [4]. Employed computer vision-based pixel-wise segmentation for the detection of AD ROIs from digital mammograms. In the first step, the computer vision object detection algorithm was applied to create a segmented pixel array. In the second step, the area having a star shape pattern and larger radios than 1 cm was considered as ADs ROIs. The segmented architectural distortion ROIs input to the dept-wise convolutional neural network for classification. Figures 6–8 presented segmented benign and malignant ROIs from the PINUM, CBIS-DDSM datasets and DDSM. Moreover, we pertain to the same procedure for the segmentation of AD ROIs from the CBIS-DDSM dataset. The automated segmented ROIs are validated with manually marked ADs ROIs by the radiologist team. The DDSM dataset has predefined ground truth ADs ROIs and is included in the validation dataset. Samreen et al. [53] presented an imaging evaluation management algorithm on architectural distortion detection from digital breast tomosynthesis.

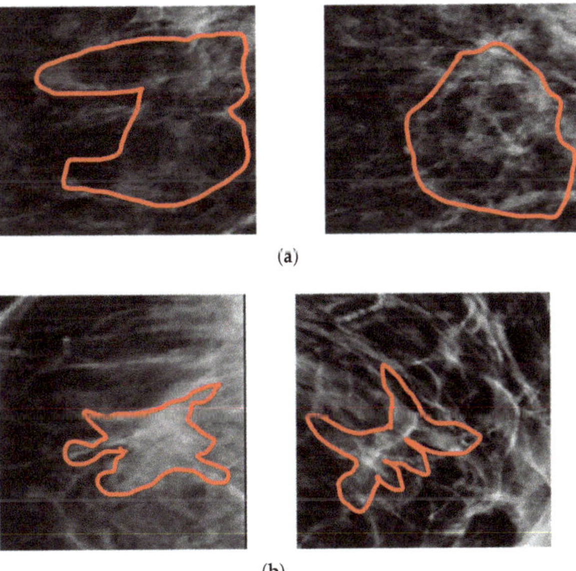

Figure 6. An example of the architectural distortion ROI's from PINUM dataset by the experts team of radiologists. (**a**) Radial shape (**b**) Star shape.

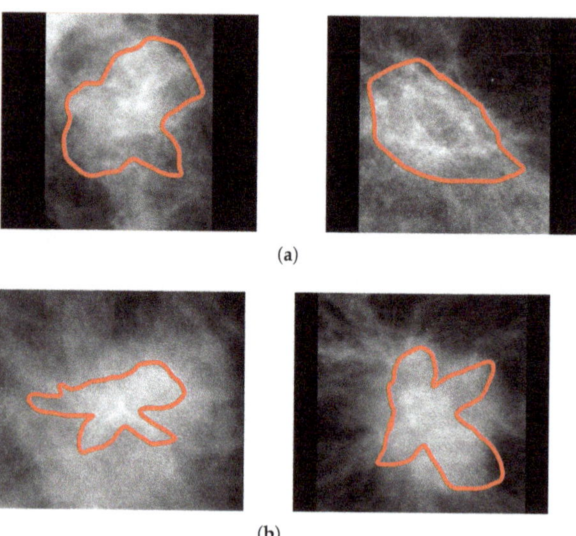

(a)

(b)

Figure 7. An example of the architectural distortion ROI's segmentation of CBIS-DDSM dataset by the radiologists. (**a**) Radial shape (**b**) Star shape.

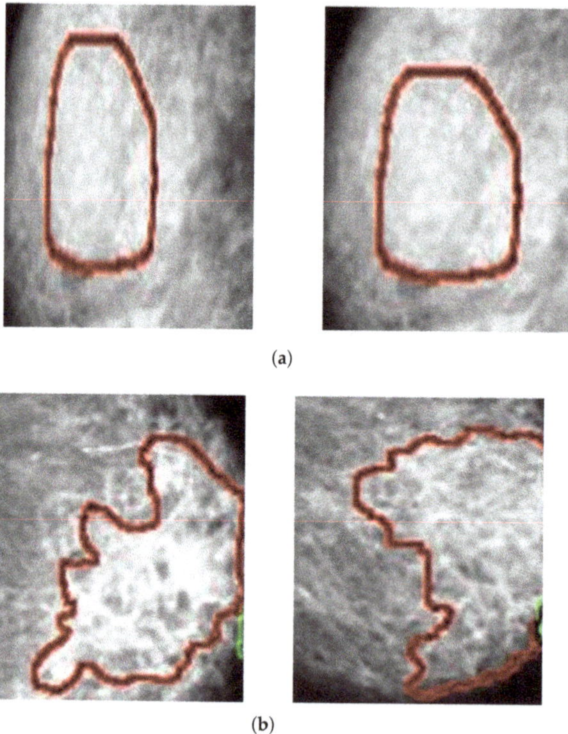

(a)

(b)

Figure 8. An example of the architectural distortion ROI's segmentation of DDSM dataset. (**a**) Radial shape (**b**) Star shape.

4.6. Depth-Wise-CNN Architecture

A deep convolutional neural network using a computer vision-based method has improved pattern recognition and architectural distortion classification. The standard convolutional neural network uses input and output with only width and height parameters. For input with only width and height, the neural network increases the parameters and can be overfitting. Employed a depth-wise 2D convolutional neural network using V-net 64 architecture with three convolutional layers, three max-pooling layers, one fully connected flatten layer, and one dense layer followed by the sigmoid classifier. The depth-wise convolution only uses one input channel for each depth level of input and then performs convolution. The depth-wise convolutional neural network architecture is presented in Figure 9. In the convolutional layer, use a 3×3 kernel using the Relu activation function and the input vector mapping the features to the convolutional layer as $dim(image) = (n_h, n_w, n_c)$ Where n_h is the size of height, n_w size of width and n_c is the number of channels. The input image of the lth layer we use $a^{[l-1]}$ filters with the size of $(n_h^{[l-1]}, n_w^{[l-1]}, n_c^{[l-1]}), a^{[0]}$. The stride parameter is: $s^{[l]}$ and the number of filters denoted as $n_c^{[l]}$ where for each K^n is size of $(f^{[l]}, f^{[l]}, n_c^{[l-1]})$. The activation function ReLu is: $\varphi^{[l]}$ and the output image is $a^{[l]}$ with the size of $(n_h^{[l]}, n_w^{[l]}, n_c^{[l]})$. Equations (3) and (4) shows the input and output of convolutional layer. For all n belongs to $[1, 2, ..., n_c^{[l]}]$.

$$Conv(a^{[l-1]}, K^n)_{x,y} = \varphi^{[l]} \left(\sum_{i=1}^{n_h^{[l-1]}} \sum_{j=1}^{n_w^{[l-1]}} \sum_{k=1}^{n_c^{[l-1]}} \right.$$
$$\left. K_{i,j,k}^n a_{x+i-1,y+j-1,k}^{l-1} + b_n^l \right) \tag{3}$$
$$dim(conv(a^{[l-1]}, K^n)) = (n_h^{[l]}, n_w^{[l]})$$

$$[\varphi^{[l]}(Conv(a^{[l-1]}, K^1)), \varphi^{[l]}(Conv(a^{[l-1]}, K^2)), ...$$
$$\varphi^{[l]}(Conv(a^{[l-1]}, K^{(n_c^{[l]})}))$$
$$dim(a^{[l]} = (n_h^{[l]}, n_w^{[l]}, n_c^{[l]}) \tag{4}$$
$$n_c^{[l]} = number of filters$$

where f is activation, x and y the actual pixels location on height and width dimension of input image. The learning parameters of convolutional layer at l^{th} layers are $(f^{[l]} \times f^{[l]} \times f_c^{[l-1]}) \times n_c^{[l]}$ filters. In the max-pooling layer, uses a 2×2 kernel size to down-sampling the features and the input size is $a^{[l-1]}$ with the size of $(n_h^{[l-1]}, n_w^{[l-1]}, n_c^{[l-1]}), a^{[0]}$. The filter size of pooling layer is denoted as $f^{[l]}$ and the pooling function $\phi^{[l]}$. The Equations (5) and (6) performs the pooling function.

$$a_{x,y,z}^{[l]} = pool(a^{[l-1]})_{x,y,z} = \phi^{[l]}$$
$$((a_{x+i-1,y+j-1,z}^{[l-1]})_{(i,j)\in[1,2,...f^{[l]}]^2})$$
$$dim(a^{[l]}) = (n_h^{[l]}, n_w^{[l]}, n_c^{[l]}) \tag{5}$$
$$n_c^{[l]} = n_c^{[l-1]}$$

where (i, j) belongs to $[1, 2, ..., \phi^{[l]}]$, x, y are the pixels location and z is the input channel. The last fully-connected layer a fine number of neurons as input vector considering the jth nodes of the ith layer can be calculated with Equation (6).

$$z_j^{[j]} = \sum_{l=1}^{n_{i-1}} w_{j,l}^{[i]} a_l^{[i-1]} + b_j^{[i]}$$

$$\rightarrow a_j^{[i]} = \varphi^{[i]}(z_j^{[i]})$$

(6)

The input $a^{[i-1]}$ the result of the convolutional and pooling layer with the dimensions $(n_h^{[i-1]}, n_w^{[i-1]}, n_c^{[i-1]})$. Finally the $1D$ flatten layer has the dimensions $(n_h^{[i-1]} \times n_w^{[i-1]} \times n_c^{[i-1]}, 1)$. and the nodes are:

$$n_{i-1} = n_h^{[i-1]} \times n_w^{[i-1]} \times n_c^{[i-1]}$$

where $w_{j,l}$ are weights with learned parameters $n_{[l-1]} \times n_l$ parameters at $l^t h$ layer. The proposed depth-wise convolutional neural network significantly outperformed without overfitting and achieved the highest accuracy.

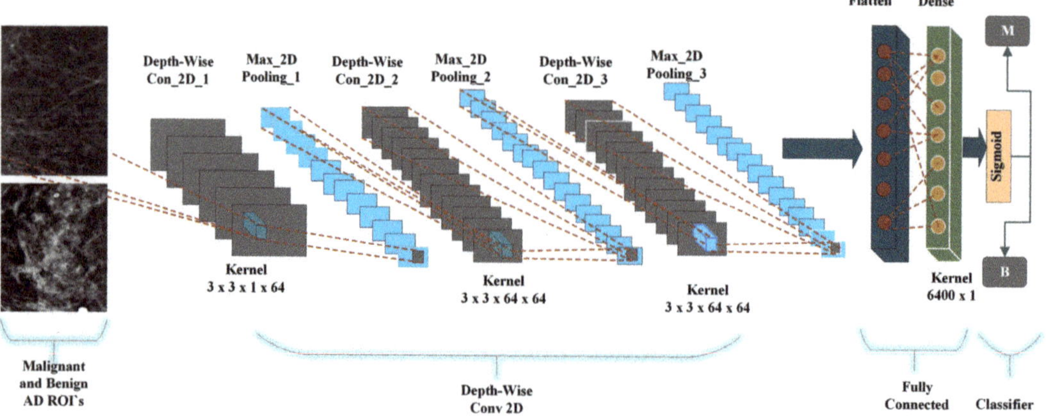

Figure 9. The proposed depth-wise CNN architecture for the classification of benign and malignant architectural distortion ROIs.

4.7. Depth-Wise-V-Net64 Training

The depth-wise 2D convolutional neural network is evaluated on three databases, the local PINUM, the public CBIS-DDSM, and the DDSM dataset. Split the data into the training, testing, and validation data for the proposed deep neural modal. The dataset was randomly divided into 60% for training, 20% for testing, and 20% for cross-validation. For the deep learning model's regularization and adequate robustness, the data augmentation object is used in our deep learning network for both datasets. Build a depth-wise 2D V-net 64 architecture with three convolutions, three max-pooling, and two fully connected layers for the training of our dataset. The sigmoid classifier has pertained to the classification of malignant and benign ADs ROIs. The epochs size was set 20 to reduce the learning rate by 0.1 factor after every 2.5 epochs, the batch size was 16, and the class weight and *"binary_crossentropyloss"* function were used to deal with training data imbalance. The proposed deep learning models learning ability was increased as the training ephods increases. Figures 10–12 shows that the noise around the data is higher at first layer of network. As well as the modal learns more the noise around the data decreases till the last layer. The training loss continues decreases after the 10th epochs and training accuracy increases and reached up to 100. The training graphs shows that modals learning ability is better and well regularized. The network structure considered in experiments is summarized in Table 3.

Table 3. The proposed network layers architecture.

Network Layers	Filters	Filter Size	Padding	Stride	Output Shape
Input Image	-	-	-	-	$240 \times 320 \times 3$
DW_Conv2D	64	$3 \times 3 \times 64$	same	1×1	$100 \times 100 \times 64$
Activataion_Relu	-	-	-	-	$98 \times 98 \times 64$
Max_Pooling	1	2×2	-	0	$49 \times 49 \times 64$
DW_Conv2D	64	$3 \times 3 \times 64$	same	1×1	$47 \times 47 \times 64$
Activataion_Relu	-	-	-	-	$47 \times 47 \times 64$
Max_Pooling	1	2×2	-	0	$23 \times 23 \times 64$
DW_Conv2D	64	$3 \times 3 \times 64$	same	1×1	$21 \times 21 \times 64$
Activataion_Relu	-	-	-	-	$21 \times 21 \times 64$
Max_Pooling	1	2×2	-	0	$10 \times 10 \times 64$
Dropout (0.5)	-	-	-	-	$10 \times 10 \times 64$
FC1_Flatten_4	-	-	-	-	(6400)
FC2_Dense_5	64	-	-	-	(6400)
Sigmoid	-	-	-	-	[0/1]

4.8. Standard Classifiers

ShuffleNet, developed by Magvi Inc, is a highly efficient convolutional neural network architecture optimized for mobile devices with low processing capacity. The new design makes use of two procedures to decrease computing costs while maintaining or improving accuracy and perform groups convolutions pointwise and the Channel Shuffle. The Channel Shuffle is a novel procedure performed to create additional feature map channels, which aids in the encoding of more information and improves the robustness of feature recognition. Group Convolution, introduced in AlexNet, is a form of convolution in which the channels are divided into groups and then the kernel is convolved individually on each group and then re concatenated. This procedure contributes to the retention of existing connections and reduces the connection count

MonileNet is a deep convolutional neural network that uses a depth-wise separable convolutional neural network. Compared to a network with normal convolutions of the same depth in the nets, it substantially reduces the number of parameters. MobileNet is an open-source neural network provided by Google. The actual difference between the MobileNet design and a conventional CNN is that instead of a single 3×3 convolutional layer followed by the batch norm and ReLU, the MobileNet architecture uses several 3×3 convolutional layers. The mobile nets divide the convolution into a 3×3 depth-wise convolution and a 1×1 point-wise convolution.

Loi et al. [41] presented subtraction of temporally sequential mammogram technique to predict breast cancer using a support vector machine algorithm. To validate the proposed method, perform a classification task using a support vector machine algorithm. A computer vision-based object detection method was employed for architectural distortion ROIs detection in the preprocessing phase. we extracted pixel-wise features using a computer-vision algorithm for creating input to SVM and for other machine learning algorithms. We use the non-linear kernel function in the support vector machine algorithm to classify ADs ROIs. It has been observed that the support vector machine algorithm provides more general results where the number of samples is relatively low [54]. In our SVM model, we employed a 5-fold cross-validation function for the validation of SVM.

K-NN is a supervised machine learning algorithm for binary class, multiclass, and regression problems. Beham et al. [44] applied wavelet transforms for feature extraction from the digital mammogram, and the K-nearest neighbor algorithm was employed for classification into benign and malignant classes. We employed K-NN for binary classification to evaluate and compare the performance of our deep neural network. The image segmentation and ROIs detection method were the same as we use for the SVM algorithm. We set the maximum value for K as 40 and the optimal error rate is 0.17 which shows the K-NN classifier was not overfitted.

Random forest is a supervised machine learning algorithm that ensembles a tree. Obaidullah et al. [47] presented an image descriptor-based approach for mammogram mass classification using a random forest algorithm. In each node of a tree gets a vote for predicting the output. We use a computer vision-based feature selection method for a random forest classifier. We trained a multiple-time random forest classifier to classify ADs' ROIs, compare it with our proposed method, and observe that random forest performance was low.

4.9. Evaluation Metrics

The proposed method was able to classify detected architectural distortion ROIs into malignant and benign classes and significantly improve model accuracy. The performance of the proposed method is evaluated on the local PINUM, the public CBIS-DDSM, and the DDSM database. The evaluation metrics such as accuracy, sensitivity, f1-score, precision, recall, and area under the curve (AUC) are used to assess the performance of the proposed method. The following equations are employed to calculate the accuracy, sensitivity, f1-score, and area under the curve. Accuracy measures the corrected classified sample of the binary class. Sensitivity measures the corrected true-positive cases from false-positive. The area under the curve calculates the ratio between true-positive and false-positive. F1-score can be calculated to compute precision and recall.

$$Accuracy = \frac{TP + TN}{FP + FN + TP + TN} \tag{7}$$

$$Sensitivity = \frac{TP}{TP + FN} \tag{8}$$

$$F1 - Score = 2 * \frac{\left(\frac{TP}{TP+FP}\right) * \left(\frac{TP}{TP+FN}\right)}{\left(\frac{TP}{TP+FP}\right) + \left(\frac{TP}{TP+FN}\right)} \tag{9}$$

$$AUC = \frac{1}{2} * \left(\frac{TP}{TP + FN} + \frac{TN}{TN + FP}\right) \tag{10}$$

where TP: true positive, TN: True negative, FP: False positive, FN: False Negative.

5. Results Analysis

The proposed method was designed on scientific fundamentals to predict breast cancer from digital mammograms. The computer vision-based image preprocessing method has pertained to detecting the architectural distortion ROIs from digital mammograms for all models. The experiments were carried out on six pre-trained models (Proposed-CNN, ShuffelNEt, MobileNet, SVM, K-NN, RF) to evaluate the two databases. The experimental results reveal that our proposed method outperforms as compared with other and previous studies.

5.1. Experimental Configuration

In the current study, experimental work was performed on google collab GPU, 12 GB RAM, and Windows 10 operating system. All experimental algorithms are implemented in python 3.6 using TensorFlow/Keras library. The computation time was 30 min for training and testing on PINUM datasets, 40 min on the CBIS-DDSM dataset, and 50 min on the DDSM for all neural networks. Furthermore, image preprocessing and augmentation are performed in Python. Pertained to the best hyperparameters, such as batch size, loss function, learning rate, target size, and optimization function, as presented in Table 4.

Table 4. Hyper parameter configuration detail.

Configuration	Values
Batch Size	16
Learning Rate	0.001
Epochs	20
Optimization function	Adam
Loss Function	binary_crossentropy
Target Size	[320, 240]
histogram_freq	1
Tarin Split	0.6
Validation Split	0.2

5.2. Comparison between Proposed Method, ShuffelNet, MobileNet and SVM, KNN, RF

The results of the proposed method were compared with well-known three machine learning and two deep learning algorithms. It could be observed that in Tables 5–7 the performance of the proposed method was much better than the ShuffelNet, MobileNet, SVM, K-NN, and random forest. The performance of experimental results was evaluated using a five-fold cross-validation test on the PINUM, the CBIS-DDSM, and DDSM datasets. The deep learning models training accuracy and training loss for all datasets has shown in Figures 10–12. In Figure 10, after the 7th epochs, the training loss continuously decreases while the training accuracy remains constant over the iterations, while the loss and accuracy of shuffelnet and mobilenet are lower which shows our model perfectly fitted on the PINUM dataset. Figures 11 and 12 for the CBIS-DDSM and DDSM datasets after the 10th epoch, the training loss steadily decreases while the training accuracy remains higher until the last epochs as compared to shuffelnet and mobilenet. The training accuracy on all datasets reaches 99% after the 17th epochs, which indicates that our model was regularized and perfectly fitted.

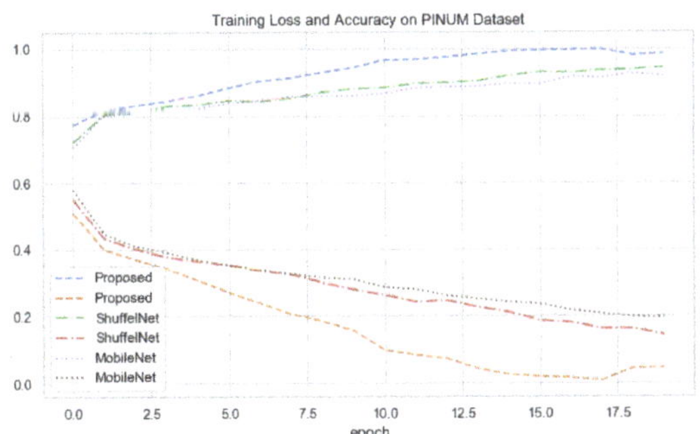

Figure 10. All Deep Networks Training Loss and Accuracy on PINUM Dataset.

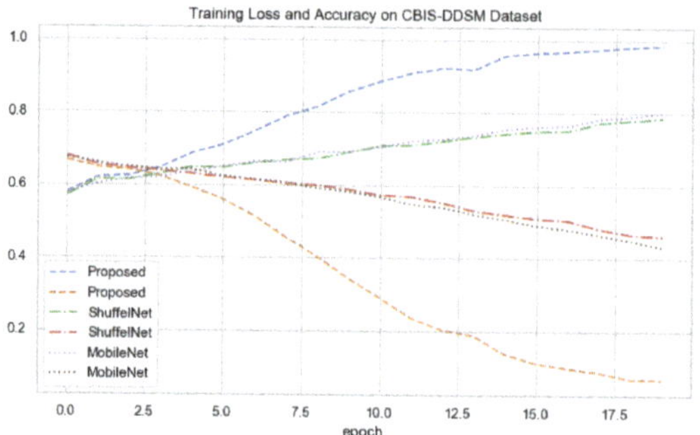

Figure 11. All Deep Networks Training Loss and Accuracy on CBIS-DDSM Dataset.

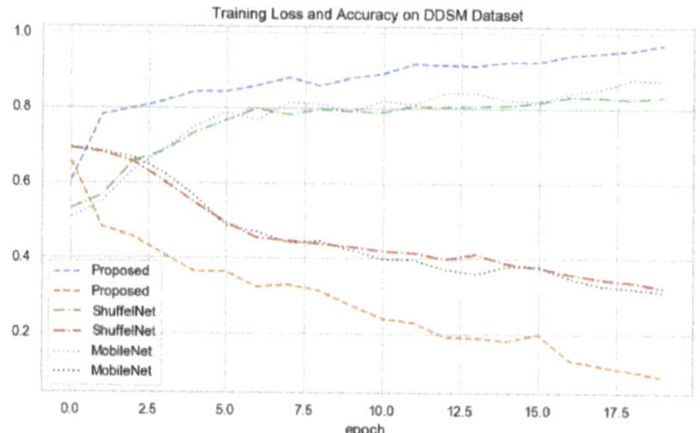

Figure 12. All Deep Networks Training Loss and Accuracy on DDSM Dataset.

Figures 13–15 show that the proposed method yielded the best performance and achieved 0.95, 0.97 and 0.98 accuracies on the PINUM, CBIS-DDSM and DDSM datasets, respectively. Shuffelnet,MobileNet, SVM, K-NN, and RF accuracies were 0.91, 0.89, 0.87, 0.83, and 0.90 on the PINUM dataset, 0.93, 0.90, 0.73, 0.80, and 0.95 on the CBIS-DDSM dataset and 0.87, 0.90, 0.80, 0.81 and 0.91 on DDSM dataset. The proposed method achieves 4%, 6%, 8%, 12%, and 5% higher accuracy than ShuffelNet, MobileNet, SVM, K-NN, and RF on the PINUM dataset, 4%, 7%, 24%, 17%, and 2% on the CBIS-DDSM dataset and 11%, 8%, 18%, 17% and 7% on DDSM dataset.

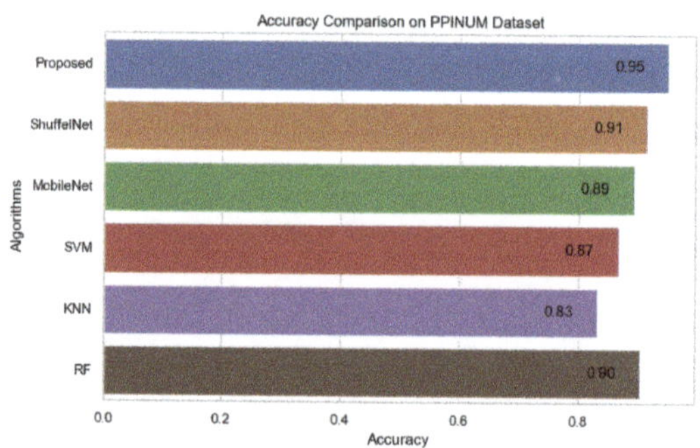

Figure 13. Accuracy Comparison on PINUM Dataset.

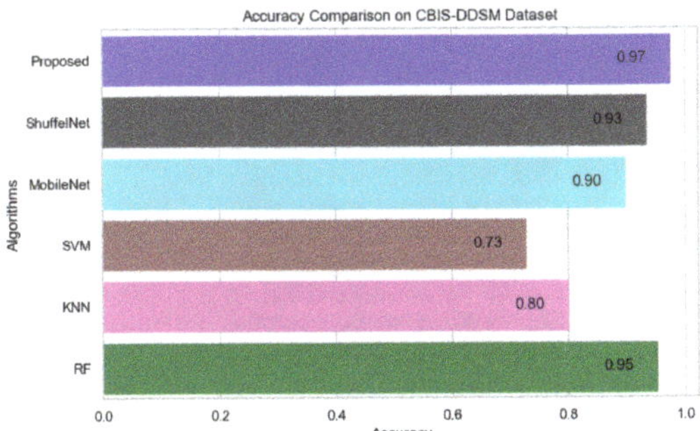

Figure 14. Accuracy Comparison on CBIS-DDSM Dataset.

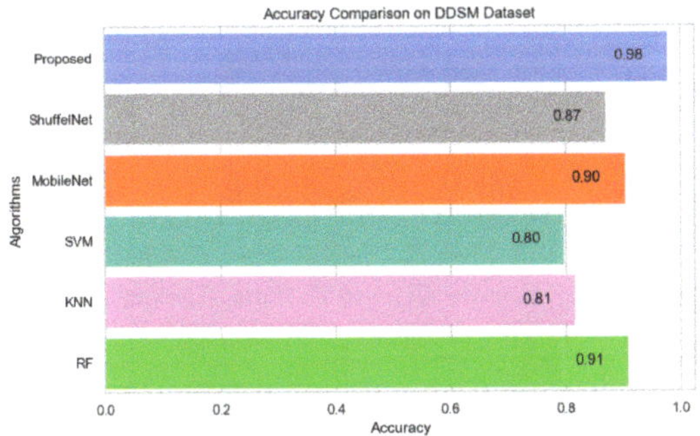

Figure 15. Accuracy Comparison on DDSM Dataset.

Figures 16–18 reveals that the proposed method was achieved 0.87, 0.90, 0.89 f1-score, precision, and recall on the PINUM dataset, 0.96, 0.94, and 0.98 on the CBIS-DDSM dataset, and 0.90, 0.96 and 0.86 on DDSM dataset which was higher as compared with ShuffelNet,MobileNEt, SVM, K-NN, and RF, respectively. In addition, the performance of the f1-score of the proposed method was 6%, 10%, 15%, 24%, and 6% higher than ShuffelNet, MobileNet, SVM, K-NN, and RF on the PINUM dataset. Furthermore, f1-score was 27%, 3%, 27%, 18%, and 1% higher than ShuffelNet, MobileNet, SVM, K-NN, and RF on the CBIS-DDSM dataset and 16%, 6%, 14% 12% and 2% on the DDSM dataset. Moreover, the precision and recall of the PINUM dataset of the proposed model was 4%, 29%5, 2%, 6%, 1%, and 13%, 16%, 28%, 38%, 14%, respectively, higher than the ShuffelNet, MobileNet, SVM, K-NN, and random forest. For the CBIS-DDSM and DDSM data set, the proposed method precision and recall performance was 19%, 12%, 21%, 15%, 1% and 25%, 15%, 32%, 20%, 1% and 13%, 11%, 24%, 21%, 1% and 10%, 2%, 9%, 4%, 4% better than the ShuffelNet, MobileNet, SVM, K-NN, and RF.

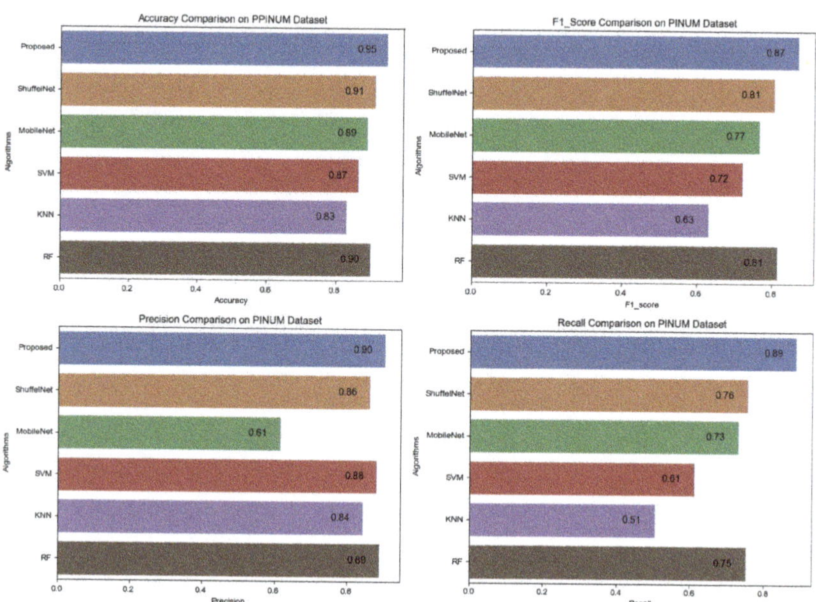

Figure 16. Comparison of Accuracy, F1-Score, Precision and Recall on PINUM Dataset.

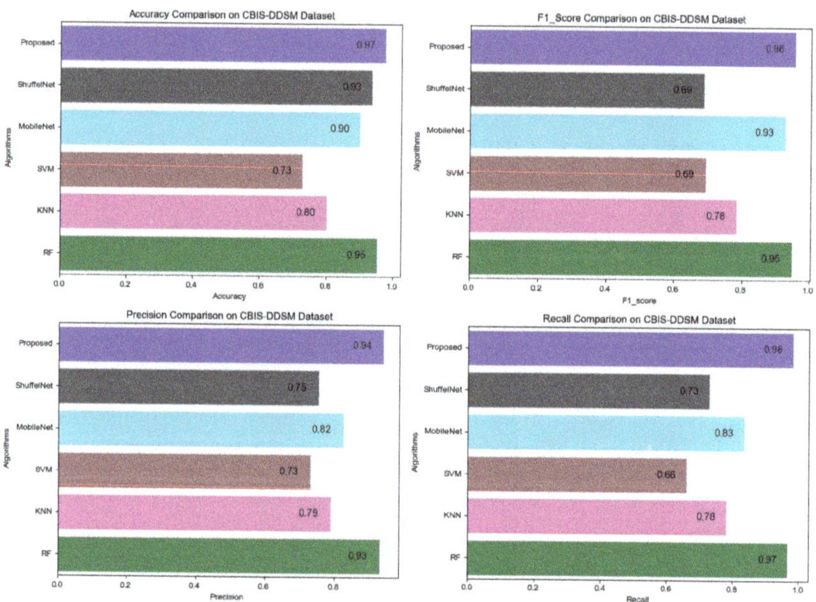

Figure 17. Comparison of Accuracy, F1-Score, Precision and Recall on CBIS-DDSM Dataset.

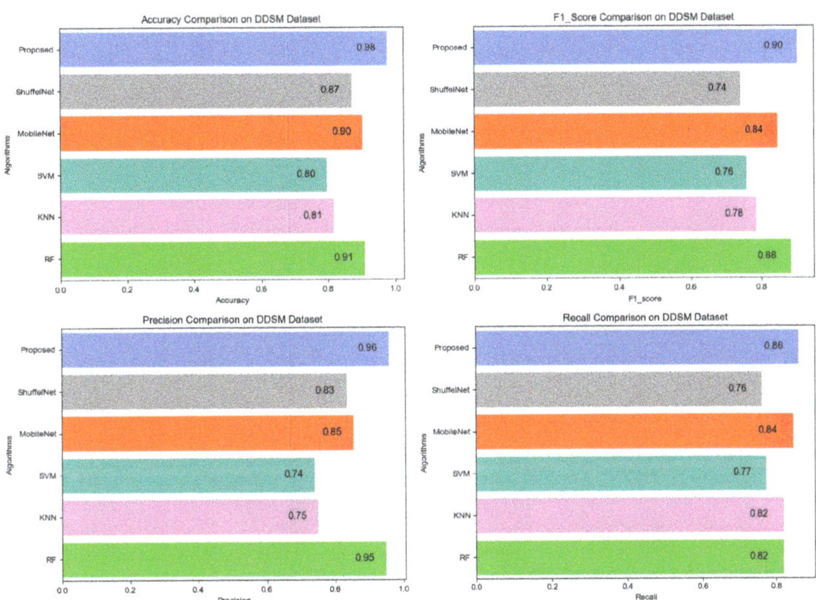

Figure 18. Comparison of Accuracy, F1-Score, Precision and Recall on DDSM Dataset.

On the other hand, when comparing the sensitivity of the proposed model with ShuffelNet, MobileNet, SVM, K-NN, and RF on the PINUM and CBIS-DDSM is 3% 13%, 11%, 8%, 2%, 1%, 1%, and 16%, 13%, 1% higher, respectively as shown in Figures 19 and 20. Figure 21 reveals that the sensitivity of the proposed method on the DDSM dataset was 7%, 8%, 15%, 15%, and 6% higher than ShuffelNet, MobileNet, SVM, K-NN, and RF. The area under the curve (AUC) was calculated of our proposed model, as shown in Figures 22–24. The AUC curve of our model was higher than the ShuffelNet, MobileNet, SVM, K-NN, and random forest. The above aforementioned deep analysis of all datasets stated that the proposed method significantly outperforms rather than the ShuffelNet, MobileNet, SVM, K-NN, and RF. The experimental results demonstrated the effectiveness of a deep convolutional neural network to classify architectural distortion ROIs that can help doctors and radiologists to predict breast cancer at initial stages.

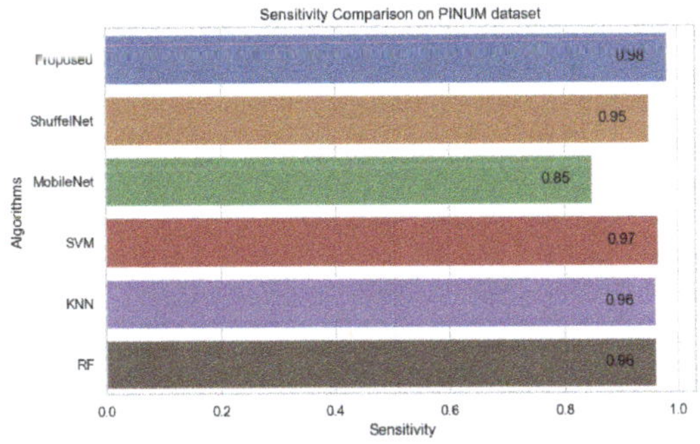

Figure 19. Sensitivity Comparison on PINUM Dataset.

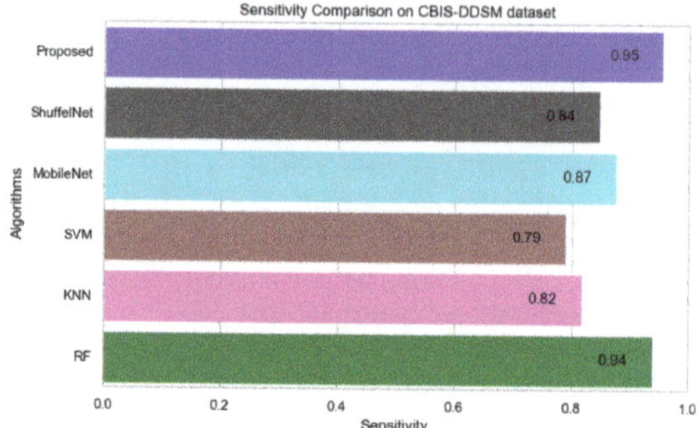

Figure 20. Sensitivity Comparison on CBIS-DDSM Dataset.

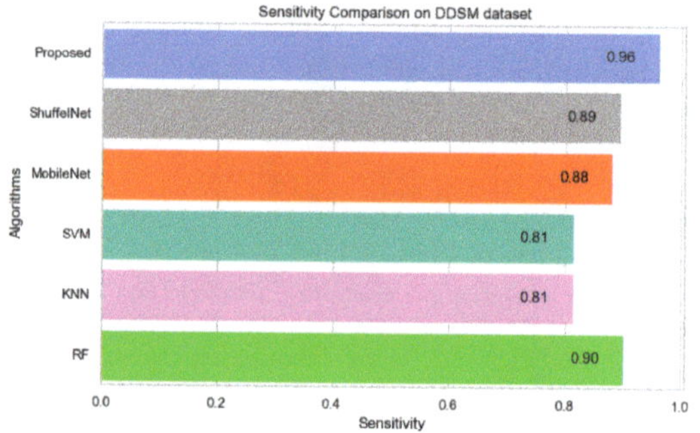

Figure 21. Sensitivity Comparison on DDSM Dataset.

Table 5. Performance Evaluation compression of proposed method and with ShuffelNet, MobileNet, SVM, K-NN and RF on PINUM dataset.

Algorithms	Accuracy	F1-Score	Precision	Recall	Sensitivity	AUC
Proposed	0.95	0.87	0.90	0.89	0.99	0.91
ShuffelNet	0.91	0.81	0.86	0.76	0.95	0.79
MobileNet	0.89	0.77	0.61	0.73	0.85	0.79
SVM	0.87	0.72	0.88	0.61	0.97	0.69
KNN	0.83	0.63	0.84	0.51	0.96	0.59
RF	0.90	0.81	0.89	0.75	0.96	0.75

Table 6. Performance Evaluation compression of proposed method and with ShuffelNet, MobileNet, SVM, K-NN and RF on CBIS-DDSM dataset.

Algorithms	Accuracy	F1-Score	Precision	Recall	Sensitivity	AUC
Proposed	0.97	0.96	0.94	0.98	0.95	0.98
ShuffelNet	0.93	0.69	0.75	0.73	0.84	0.69
MobileNet	0.90	0.93	0.82	0.83	0.87	0.61
SVM	0.73	0.69	0.73	0.66	0.79	0.67

Table 6. *Cont.*

Algorithms	Accuracy	F1-Score	Precision	Recall	Sensitivity	AUC
KNN	0.80	0.78	0.79	0.78	0.82	0.81
RF	0.95	0.95	0.93	0.97	0.94	0.89

Table 7. Performance Evaluation compression of proposed method and with ShuffelNet, MobileNet, SVM, K-NN and RF on DDSM dataset.

Algorithms	Accuracy	F1-Score	Precision	Recall	Sensitivity	AUC
Proposed	0.98	0.90	0.96	0.86	0.96	0.85
ShuffelNet	0.87	0.74	0.83	0.76	0.89	0.69
MobileNet	0.90	0.84	0.85	0.84	0.88	0.81
SVM	0.80	0.76	0.74	0.77	0.81	0.79
KNN	0.81	0.78	0.75	0.82	0.81	0.81
RF	0.91	0.88	0.95	0.82	0.90	0.78

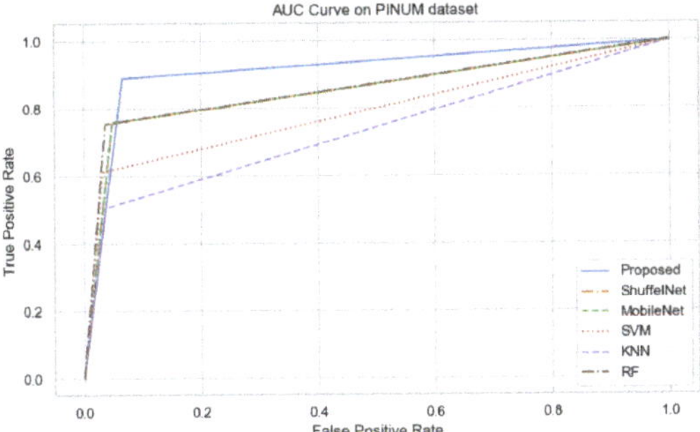

Figure 22. The AUC curves of algorithms on PINUM Dataset.

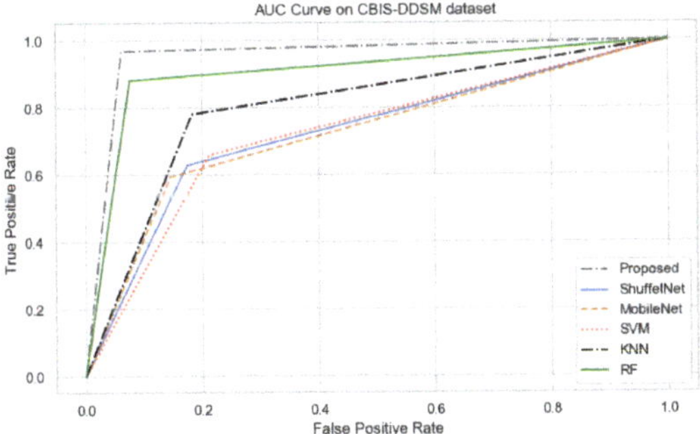

Figure 23. The AUC curves of algorithms on CBIS-DDSM Dataset.

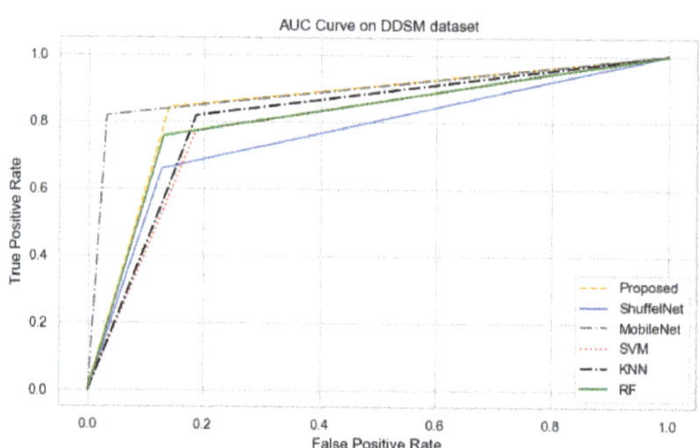

Figure 24. The AUC curves of algorithms on DDSM Dataset.

5.3. Results Comparison between Proposed Method and Previous Studies

The proposed method is validated by comparing it with previous studies using the same dataset and the private dataset. The experimental results reveal that the performance of the proposed method was much better than the previous studies. Table 8 summarized that the proposed method was achieved 0.95, 0.97, and 0.98 accuracies on the PINUM, CBIS-DDSM, and DDSM datasets, respectively, which were higher comparatively from previous studies. Murali. et al. [6] pertain SVM and MLP for classifying architectural distortion ROIs and achieved 89.6% accuracy on the DDSM dataset. [7] implemented the Gober filter-based method to detect architectural distortion and achieve 90% sensitivity. The authors [8–10] employed a machine learning-based classification algorithm to detect architectural distortion from the DDSM data set and reporting 83.50%, 92.94%, and 91.79% accuracies, respectively. Another study by [13] applied a multilayer-perception network to detect architectural distortion evaluating 300 images and reported 83% accuracy. The authors [14] used the LDA classifier to detect architectural distortion tracking from digital breast tomosynthesis and achieved 0.90 sensitivity.

Table 8. Comparison of results with previous studies and proposed method.

Authors	Problem	Method	Database	Images	Accuracy
[6]	Architectural Distortion Detection	SVM, MLP	DDSM	190	0.89
[7]	Architectural Distortion Detection	Bayesian, SELF ANN	Private	1745	N/A
[8]	Architectural Distortion Detection	Differential direction method	DDSM	33	0.83
[9]	Architectural Distortion Detection	SVM	DDSM	147	0.92
[10]	Architectural Distortion Detection	Sparse classifier	DDSM	69	0.91
[13]	Architectural Distortion Detection	MLP	FFDM	300	0.83
[14]	Architectural Distortion tracking	LDA	FFDM	37	N/A
[55]	Architectural Distortion tracking	CNN	CBIS-DDSM	334	0.92
Proposed	Architectural Distortion Detection	Depth-wise 2DCNN	Private (PINUM)	3462	0.95
Proposed	Architectural Distortion Detection	Depth-wise 2DCNN	CBIS-DDSM	3568	0.97
Proposed	Architectural Distortion Detection	Depth-wise 2DCNN	DDSM	5500	0.98

The proposed method depth-wise 2D convolutional neural network achieved 0.95, 0.97, and 0.98 accuracies on the 3264 PINUM, 3568 CBIS-DDSM, and 5500 DDSM datasets images, respectively, which were better than previous studies. The proposed model has achieved 0.98 accuracy which was 6% and 15% higher than the previous studies on the DDSM dataset which indicates that the performance of the proposed modal was much better. The performance of the proposed method on a private dataset was also better than the previous studies.

6. Discussion

In the current study, proposed a state-of-the-art computer-aided diagnostic system using a computer vision and depth-wise 2D convolutional neural network to detect and classify architectural distortion ROIs from digital mammograms. The proposed mammogram classification framework pertains to four steps: image preprocessing and augmentation, image pixel-wise segmentation, architectural distortion ROI's detection, training deep learning, and machine learning networks to classify AD's ROIs into malignant and benign classes. Image classification using the deep convolutional neural network, a minimum number of images is approximately 1000 required, and it can be increased for pre-trained models to regularize the neural network. [56]. Deep learning is a data-driven method so that the small size of data and non-standardization are the main challenges for the generalization of the model. However, to handle the generalization, overfitting, and improving the robustness of the deep learning model, we artificially inflate the PINUM database up to 3462 using data augmentation techniques as discussed above. The CBIS-DDSM dataset consists of 3568 mammogram images, including 1740 benign and 1828 malignant images with MLO and CC views. The 5500 images were included from the DDSM dataset. Split the data into the training, testing, and validation data for the proposed deep neural modal. The dataset was randomly divided into 60% for training, 20% for testing, and 20% for cross-validation.

In the context of comparing results with the ShuffleNet, MobileNet, SVM, K-NN, and RF the obtained results of the proposed method are comparable, encouraging, and better in many aspects. The proposed method yielded better accuracy, f1-score, precision, recall, sensitivity, and area under the curve. When we are seeing the training accuracy of the proposed method on both datasets it reaches 100% as compared with the ShuffelNet and MobileNet. On the other hand, the training loss of our proposed method is consistently decreasing after the 7th epochs which shows the noise around the proposed method is much lower than the ShuffleNet and MobileNet on the PINUM, CBIS-DDSM, and DDSM datasets. In comparison to the findings of previous research on architectural distortion, the current study's findings for malignant and benign ADs are promising, better, and outperforms. The authors [6,8–10] achieved 89.6%, 83.50%, 92.94%, and 91.79% accuracies, respectively. The experimental results demonstrated that the proposed approach significantly outperforms the ShuffelNet, MobileNet, SVM, K-NN, RF, and previous studies. The proposed approach achieved 0.95%, 0.97%, 0.98% accuracies on the PINUM, CBIS-DDSM, and DDSM dataset, while the maximum accuracy in previous studies was 92.94% [9] on the DDSM dataset, which healed our model. On the other hand, the highest accuracy was achieved by the random forest algorithm are 0.90, 0.95 on the PINUM and CBIS-DDSAM dataset, which is still lower than our proposed model. Furthermore, to enhance the effectiveness of the proposed model, compared it with other evaluation metrics such as f1-score, precision, recall, and sensitivity; the model achieved better results, as seen in Tables 5–7.

Fully automatic identification of architectural distortion in mammograms of interval-cancer cases is more challenging because extensive comparative analysis, which was not investigated in our study, is still a limitation. The diagnostic mammograms were not accessible in the current investigation on interval-cancer patients, including benign control cases, because of localizing the areas of architectural distortion on mammograms.

The current study observed that the classification approach using depth-wise 2D convolutional neural networks was much better than the machine learning algorithms such as SguffelNet, MobileNet, SVM, K-NN, and RF. Moreover, computer-vision technology is more potent for image segmentation and ROIs detection than the traditional and hand-crafted approaches. The proposed fully automated CAD system could predict breast cancer more accurately than the older one and help the clinical staff with disease diagnostic. To enhance the validity of the model, employed it on the three databases, the public and the private. The proposed approach with a computer vision and depth-wise 2D convolutional neural network is a novel approach for architectural distortion ROIs detection and classification into benign and malignant ROIs.

7. Conclusions

Mammogram screening is an effective and initial screening method for the diagnosis of breast cancer in women. Architectural distortion is the third most suspicious appearance on a mammogram that represents abnormal regions. Architectural distortion detection from mammograms is challenging due to its subtle and varying asymmetry on breast mass and small size. Therefore, the manual interpretation of Architectural Distortion is a challenging task for radiologists to figure out abnormalities during the examination of mammograms due to its subtle appearance on fatty denser mass. In the current study, proposed an automated computer-aided diagnostic system based on computer vision and deep learning to predict breast cancer from the digital mammogram. Proposed a state-of-the-art- method for breast cancer detection from architectural distortion ROIs. The proposed method consists of two major phases, in the first phases the architectural distortion ROIs are extracted using a computer vision algorithm and verified by the expert radiologists, in the 2nd phase these ROIs are classified with the proposed deep learning method to classify into malignant and benign ROIs. Experimental results reveal that our proposed method outperforms as compared with the ShuffelNet, MobileNet, SVM, K-NN, RF, and previous studies. Although the results are very promising and better, further investigate new techniques for localizing the patterns for detecting architectural distortion ROIs that are not limited to spiculated patterns. Furthermore, will investigate other deep learning models to detect architectural distortion from other public and larger private datasets. In addition, we will also analyze our modal to improve the true-positive rate and detect ADs tracks from DBT slices. Another, limitation to this study is the use of transfer learning for handling small label datasets which will be further considered in future studies.

Author Contributions: K.u.R. conceived this study. Y.P. and A.Y. contribute to the design of this study. J.L. reviewed, drafted, and revise the study. S.A. and Y.S. have done proofreading of this study. All authors have read and agreed to the published version of the manuscript.

Funding: This study is supported by the National Key R&D Program of China with project no. 2020YFB2104402.

Institutional Review Board Statement: Ethical review and approval were waived from Local hospital and from the university for private dataset while there are no ethical implications on public dataset.

Informed Consent Statement: Patient consent was waived for local private dataset and included after the approval. There is no ethical implications on public dataset.

Data Availability Statement: The CBIS-DDSM [49] and DDSM [50] dataset is publicly available and the Private PINUM [48] data set is collected from local hospital.

Acknowledgments: Authors would like to thank the National Key R&D Program of China for providing experimental facilities to conduct these experimentations. Furthermore, all contributing authors declare no conflict of interest.

Conflicts of Interest: The authors declare that they have no conflict of interest.

References

1. WHO. Fact Sheet World Health Organization. WHO. 2019. Available online: https://www.who.int/news-room/fact-sheets/detail/cancer (accessed on 16 December 2021).
2. Ribli, D.; Horváth, A.; Unger, Z.; Pollner, P.; Csabai, I. Detecting and classifying lesions in mammograms with deep learning. *Sci. Rep.* **2018**, *8*, 1–7.
3. Radiology, A.C. Mammography and Breast Imaging Resoruces. Available online: https://www.acr.org/Clinical-Resources/Breast-Imaging-Resources (accessed on 16 December 2021).
4. Gaur, S.; Dialani, V.; Slanetz, P.J.; Eisenberg, R.L. Architectural distortion of the breast. *Am. J. Roentgenol.* **2013**, *201*, 662–670. [CrossRef]

5. Heidari, M.; Mirniaharikandehei, S.; Liu, W.; Hollingsworth, A.B.; Liu, H.; Zheng, B. Development and assessment of a new global mammographic image feature analysis scheme to predict likelihood of malignant cases. *IEEE Trans. Med. Imaging* **2019**, *39*, 1235–1244. [CrossRef]

6. Murali, S.; Dinesh, M. Model based approach for detection of architectural distortions and spiculated masses in mammograms. *Int. J. Comput. Sci. Eng.* **2011**, *3*, 3534.

7. Banik, S.; Rangayyan, R.M.; Desautels, J.L. Detection of architectural distortion in prior mammograms. *IEEE Trans. Med. Imaging* **2010**, *30*, 279–294. [CrossRef] [PubMed]

8. Jasionowska, M.; Przelaskowski, A.; Rutczynska, A.; Wroblewska, A. A two-step method for detection of architectural distortions in mammograms. In *Information Technologies in Biomedicine*; Springer: Berlin, Germany, 2010; pp. 73–84.

9. Kamra, A.; Jain, V.; Singh, S.; Mittal, S. Characterization of architectural distortion in mammograms based on texture analysis using support vector machine classifier with clinical evaluation. *J. Digit. Imaging* **2016**, *29*, 104–114. [CrossRef]

10. Liu, X.; Zhai, L.; Zhu, T.; Yang, Z. Architectural distortion recognition based on a subclass technique and the sparse representation classifier. In Proceedings of the 2016 9th International Congress on Image and Signal Processing, BioMedical Engineering and Informatics (CISP-BMEI), Datong, China, 15-17, Octobar, 2016; pp. 422–426.

11. Ciurea, A.I.; Ciortea, C.A.; Ștefan, P.A.; Lisencu, L.A.; Dudea, S.M. Differentiating Breast Tumors from Background Parenchymal Enhancement at Contrast-Enhanced Mammography: The Role of Radiomics A Pilot Reader Study. *Diagnostics* **2021**, *11*, 1248–1265.

12. Massafra, R.; Bove, S.; Lorusso, V.; Biafora, A.; Comes, M.C.; Didonna, V.; Diotaiuti, S.; Fanizzi, A.; Nardone, A.; Nolasco, A.; et al. Radiomic Feature Reduction Approach to Predict Breast Cancer by Contrast-Enhanced Spectral Mammography Images. *Diagnostics* **2021**, *11*, 684–699. [CrossRef]

13. de Oliveira, H.C.; Moraes, D.R.; Reche, G.A.; Borges, L.R.; Catani, J.H.; de Barros, N.; Melo, C.F.; Gonzaga, A.; Vieira, M.A. A new texture descriptor based on local micro-pattern for detection of architectural distortion in mammographic images. In Proceedings of the Medical Imaging 2017: Computer-Aided Diagnosis. International Society for Optics and Photonics, SPIE Medical Imaging, Orlando, FL, USA, 3 March 2017; Volume 10134, pp. 101342–101357.

14. de Oliveira, H.C.; Mencattini, A.; Casti, P.; Catani, J.H.; de Barros, N.; Gonzaga, A.; Martinelli, E.; da Costa Vieira, M.A. A cross-cutting approach for tracking architectural distortion locii on digital breast tomosynthesis slices. *Biomed. Signal Process. Control* **2019**, *50*, 92–102. [CrossRef]

15. Palma, G.; Bloch, I.; Muller, S. Detection of masses and architectural distortions in digital breast tomosynthesis images using fuzzy and a contrario approaches. *Pattern Recognit.* **2014**, *47*, 2467–2480. [CrossRef]

16. Cai, Q.; Liu, X.; Guo, Z. identifying architectural distortion in mammogram images via a SE-DenseNet model and twice transfer learning. In Proceedings of the 2018 11th International Congress on Image and Signal Processing, BioMedical Engineering and Informatics (CISP-BMEI), Beijing, China, 13–15 October 2018; pp. 1–6.

17. Bahl, M.; Baker, J.A.; Kinsey, E.N.; Ghate, S.V. Architectural distortion on mammography: correlation with pathologic outcomes and predictors of malignancy. *Am. J. Roentgenol.* **2015**, *205*, 1339–1345. [CrossRef] [PubMed]

18. Shu, X.; Zhang, L.; Wang, Z.; Lv, Q.; Yi, Z. Deep neural networks with region-based pooling structures for mammographic image classification. *IEEE Trans. Med. Imaging* **2020**, *39*, 2246–2255. [CrossRef] [PubMed]

19. Wang, Z.; Li, M.; Wang, H.; Jiang, H.; Yao, Y.; Zhang, H.; Xin, J. Breast cancer detection using extreme learning machine based on feature fusion with CNN deep features. *IEEE Access* **2019**, *7*, 105146–105158. [CrossRef]

20. Wu, N.; Phang, J.; Park, J.; Shen, Y.; Huang, Z.; Zorin, M.; Jastrzebski, S.; Fevry, T.; Katsnelson, J.; Kim, E.; et al. Deep neural networks improve radiologists performance in breast cancer screening. *IEEE Trans. Med. Imaging* **2019**, *39*, 1184–1194. [CrossRef]

21. Khan, H.N.; Shahid, A.R.; Raza, B.; Dar, A.H.; Alquhayz, H. Multi-view feature fusion based four views model for mammogram classification using convolutional neural network. *IEEE Access* **2019**, *7*, 165724–165733. [CrossRef]

22. Soleimani, H.; Michailovich, O.V. On Segmentation of Pectoral Muscle in Digital Mammograms by Means of Deep Learning. *IEEE Access* **2020**, *8*, 204173–204182. [CrossRef]

23. Hao, D.; Zhang, L.; Sumkin, J.; Mohamed, A.; Wu, S. Inaccurate Labels in Weakly-Supervised Deep Learning: Automatic Identification and Correction and Their Impact on Classification Performance. *IEEE J. Biomed. Health Inform.* **2020**, *24*, 2701–2710. [CrossRef]

24. Sun, D.; Wang, M.; Li, A. A multimodal deep neural network for human breast cancer prognosis prediction by integrating multi-dimensional data. *IEEE/ACM Trans. Comput. Biol. Bioinform.* **2018**, *16*, 841–850. [CrossRef]

25. Guan, Y.; Wang, X.; Li, H.; Zhang, Z.; Chen, X.; Siddiqui, O.; Nehring, S.; Huang, X. Detecting Asymmetric Patterns and Localizing Cancers on Mammograms. *Patterns* **2020**, *1*, 100106–100120. [CrossRef]

26. Singh, V.K.; Rashwan, H.A.; Romani, S.; Akram, F.; Pandey, N.; Sarker, M.M.K.; Saleh, A.; Arenas, M.; Arquez, M.; Puig, D.; et al. Breast tumor segmentation and shape classification in mammograms using generative adversarial and convolutional neural network. *Expert Syst. Appl.* **2020**, *139*, 112855–112870. [CrossRef]

27. Song, R.; Li, T.; Wang, Y. Mammographic Classification Based on XGBoost and DCNN With Multi Features. *IEEE Access* **2020**, *8*, 75011–75021. [CrossRef]

28. Shen, T.; Wang, J.; Gou, C.; Wang, F.Y. Hierarchical Fused Model With Deep Learning and Type-2 Fuzzy Learning for Breast Cancer Diagnosis. *IEEE Trans. Fuzzy Syst.* **2020**, *28*, 3204–3218. [CrossRef]

29. Guan, S.; Loew, M. Breast cancer detection using synthetic mammograms from generative adversarial networks in convolutional neural networks. *J. Med. Imaging* **2019**, *6*, 031411–031432. [CrossRef] [PubMed]

30. Li, H.; Zhuang, S.; Li, D.a.; Zhao, J.; Ma, Y. Benign and malignant classification of mammogram images based on deep learning. *Biomed. Signal Process. Control* **2019**, *51*, 347–354. [CrossRef]

31. Ionescu, G.V.; Fergie, M.; Berks, M.; Harkness, E.F.; Hulleman, J.; Brentnall, A.R.; Cuzick, J.; Evans, D.G.; Astley, S.M. Prediction of reader estimates of mammographic density using convolutional neural networks. *J. Med. Imaging* **2019**, *6*, 031405–031425. [CrossRef]

32. Falconí, L.G.; Pérez, M.; Aguilar, W.G. Transfer learning in breast mammogram abnormalities classification with mobilenet and nasnet. In Proceedings of the 2019 International Conference on Systems, Signals and Image Processing (IWSSIP), Osijek, Croatia, 5–7 June 2019; pp. 109–114.

33. Gnanasekaran, V.S.; Joypaul, S.; Sundaram, P.M.; Chairman, D.D. Deep learning algorithm for breast masses classification in mammograms. *IET Image Process.* **2020**, *14*, 2860–2868. [CrossRef]

34. Shen, R.; Yan, K.; Tian, K.; Jiang, C.; Zhou, K. Breast mass detection from the digitized X-ray mammograms based on the combination of deep active learning and self-paced learning. *Future Gener. Comput. Syst.* **2019**, *101*, 668–679. [CrossRef]

35. Shen, T.; Gou, C.; Wang, J.; Wang, F.Y. Simultaneous segmentation and classification of mass region from mammograms using a mixed-supervision guided deep model. *IEEE Signal Process. Lett.* **2019**, *27*, 196–200. [CrossRef]

36. Shaymaa, A.H.; Sayed, M.S.; Abdalla, M.I.; Rashwan, M.A. Detection of breast cancer mass using MSER detector and features matching. *Multimed. Tools Appl.* **2019**, *78*, 20239–20262.

37. Wang, H.; Feng, J.; Zhang, Z.; Su, H.; Cui, L.; He, H.; Liu, L. Breast mass classification via deeply integrating the contextual information from multi-view data. *Pattern Recognit.* **2018**, *80*, 42–52. [CrossRef]

38. Wang, R.; Ma, Y.; Sun, W.; Guo, Y.; Wang, W.; Qi, Y.; Gong, X. Multi-level nested pyramid network for mass segmentation in mammograms. *Neurocomputing* **2019**, *363*, 313–320. [CrossRef]

39. Birhanu, M.A.; Karssemeijer, N.; Gubern-Merida, A.; Kallenberg, M.; et al. A deep learning method for volumetric breast density estimation from processed full field digital mammograms. In Proceedings of the Medical Imaging 2019: Computer-Aided Diagnosis. International Society for Optics and Photonics, SPIE Medical Imaging, 2019, San Diego, CA, USA, 13 March 2019; Volume 10950, pp. 109500–109525.

40. Fan, J.; Wu, Y.; Yuan, M.; Page, D.; Liu, J.; Ong, I.M.; Peissig, P.; Burnside, E. Structure-leveraged methods in breast cancer risk prediction. *J. Mach. Learn. Res.* **2016**, *17*, 2956–2970.

41. Loizidou, K.; Skouroumouni, G.; Nikolaou, C.; Pitris, C. An automated breast micro-calcification detection and classification technique using temporal subtraction of mammograms. *IEEE Access* **2020**, *8*, 52785–52795. [CrossRef]

42. Zebari, D.A.; Zeebaree, D.Q.; Abdulazeez, A.M.; Haron, H.; Hamed, H.N.A. Improved Threshold Based and Trainable Fully Automated Segmentation for Breast Cancer Boundary and Pectoral Muscle in Mammogram Images. *IEEE Access* **2020**, *8*, 203097–203116. [CrossRef]

43. Chakraborty, J.; Midya, A.; Mukhopadhyay, S.; Rangayyan, R.M.; Sadhu, A.; Singla, V.; Khandelwal, N. Computer-aided detection of mammographic masses using hybrid region growing controlled by multilevel thresholding. *J. Med Biol. Eng.* **2019**, *39*, 352–366. [CrossRef]

44. Beham, M.P.; Tamilselvi, R.; Roomi, S.M.; Nagaraj, A. Accurate Classification of Cancer in Mammogram Images. In *Innovations in Electronics and Communication Engineering*; Springer: Berlin, Germany, 2019; pp. 71–77.

45. Liu, N.; Qi, E.S.; Xu, M.; Gao, B.; Liu, G.Q. A novel intelligent classification model for breast cancer diagnosis. *Inf. Process. Manag.* **2019**, *56*, 609–623. [CrossRef]

46. Yang, L.; Xu, Z. Feature extraction by PCA and diagnosis of breast tumors using SVM with DE-based parameter tuning. *Int. J. Mach. Learn. Cybern.* **2019**, *10*, 591–601. [CrossRef]

47. Obaidullah, S.M.; Ahmed, S.; Gonçalves, T.; Rato, L. RMID: A novel and efficient image descriptor for mammogram mass classification. In Proceedings of the Conference on Information Technology, Systems Research and Computational Physics, Cham, Switzerland, 11–13 January 2020; pp. 229–240.

48. PAEC. Pakistan Atomic Energy Commission Punjab Institue of Nuclear Medicine (PINUM Faisalabad). 2020. Available online: http://www.paec.gov.pk/Medical/Centres/ (accessed on 16 December 2021).

49. Archive, C.I. Curated Breast Imaging Digital Database for Screening Mammography(DDSM). 2021 Available online: https://wiki.cancerimagingarchive.net/display/Public/CBIS-DDSM (accessed on 16 December 2021).

50. of South Florida, U. Digital Database for Screening Mammography (DDSM) 2021. Available online: http://www.eng.usf.edu/cvprg/Mammography/Database.html (accessed on 16 December 2021).

51. Ting, F.F.; Tan, Y.J.; Sim, K.S. Convolutional neural network improvement for breast cancer classification. *Expert Syst. Appl.* **2019**, *120*, 103–115. [CrossRef]

52. Wang, M.; Li, P. Label fusion method combining pixel greyscale probability for brain MR segmentation. *Sci. Rep.* **2019**, *9*, 1–10. [CrossRef]

53. Samreen, N.; Moy, L.; Lee, C.S. Architectural Distortion on Digital Breast Tomosynthesis: Management Algorithm and Pathological Outcome. *J. Breast Imaging* **2020**, *2*, 424–435. [CrossRef]

54. Guo, Q.; Shao, J.; Ruiz, V. *Investigation of Support Vector Machine for the Detection of Architectural Distortion in Mammographic Images*; Journal of Physics: Conference Series; Institute of Physics and Engineering in Medicine (IPEM) and IOP: London, UK, 2005; Volume 15, pp. 15–35.

55. Vedalankar, A.V.; Gupta, S.S.; Manthalkar, R.R. Addressing architectural distortion in mammogram using AlexNet and support vector machine. *Inform. Med. Unlocked* **2021**, *23*, 100551. [CrossRef]
56. Pete, W.B. How Many Images Do You Need to Train A Neural Net-Work? 2017. Available online: https://petewarden.com/2017/12/14/how-many-images-do-you-need-to-train-a-neural-network (accessed on 16 December 2021).

Article

Comprehensive Analysis of CPA4 as a Poor Prognostic Biomarker Correlated with Immune Cells Infiltration in Bladder Cancer

Chengcheng Wei [1,†], Yuancheng Zhou [1,†], Qi Xiong [2,†], Ming Xiong [1], Yaxin Hou [1], Xiong Yang [1,*] and Zhaohui Chen [1,*]

1   Department of Urology, Union Hospital, Tongji Medical College, Huazhong University of Science and Technology, Wuhan 430074, China; chengchengwei@hust.edu.cn (C.W.); m202175943@hust.edu.cn (Y.Z.); xiong_ming@hust.edu.cn (M.X.); m201975738@hust.edu.cn (Y.H.)
2   Chongqing Key Laboratory of Molecular Oncology and Epigenetics, Chongqing Medical University, Chongqing 400000, China; xiongqi@stu.cqmu.edu.cn
*   Correspondence: yangxiong1368@hust.edu.cn (X.Y.); zhaohuichen@hust.edu.cn (Z.C.)
†   These authors contributed equally to this work.

Simple Summary: The overexpression of Carboxypeptidase A4 (CPA4) has been observed in plenty of types of cancer and has been elucidated to promote tumor growth and invasion; however, its role in bladder urothelial carcinoma (BLCA) is still unclear. Therefore, we aimed to show the prognostic role of CPA4 and its relationship with immune infiltrates in BLCA. We confirmed that the overexpression of CPA4 is associated with shorter overall survival, disease-specific survival, progress-free intervals, and higher dead events. Moreover, we found that several infiltrating immune cells (Th1cell, Th2 cell, T cell exhaustion, and Tumor-associated macrophage) were correlated with the expression of CPA4 in bladder cancer using TIMER2 and GEPIA2. In conclusion, CPA4 may be a novel and great prognostic biomarker based on bioinformation analysis in BLCA.

Abstract: Carboxypeptidase A4 (CPA4) has shown the potential to be a biomarker in the early diagnosis of certain cancers. However, no previous research has linked CPA4 to therapeutic or prognostic significance in bladder cancer. Using data from The Cancer Genome Atlas (TCGA) database, we set out to determine the full extent of the link between CPA4 and BLCA. We further analyzed the interacting proteins of CPA4 and infiltrated immune cells via the TIMER2, STRING, and GEPIA2 databases. The expression of CPA4 in tumor and normal tissues was compared using the TCGA + GETx database. The connection between CPA4 expression and clinicopathologic characteristics and overall survival (OS) was investigated using multivariate methods and Kaplan–Meier survival curves. The potential functions and pathways were investigated via gene set enrichment analysis. Furthermore, we analyze the associations between CPA4 expression and infiltrated immune cells with their respective gene marker sets using the ssGSEA, TIMER2, and GEPIA2 databases. Compared with matching normal tissues, human CPA4 was found to be substantially expressed. We confirmed that the overexpression of CPA4 is linked with shorter OS, DSF(Disease-specific survival), PFI(Progression-free interval), and increased diagnostic potential using Kaplan–Meier and ROC analysis. The expression of CPA4 is related to T-bet, IL12RB2, CTLA4, and LAG3, among which T-bet and IL12RB2 are Th1 marker genes while CTLA4 and LAG3 are related to T cell exhaustion, which may be used to guide the application of checkpoint blockade and the adoption of T cell transfer therapy.

Keywords: CPA4; bladder urothelial carcinoma; immune cells; T cell exhaustion; checkpoint

Citation: Wei, C.; Zhou, Y.; Xiong, Q.; Xiong, M.; Hou, Y.; Yang, X.; Chen, Z. Comprehensive Analysis of CPA4 as a Poor Prognostic Biomarker Correlated with Immune Cells Infiltration in Bladder Cancer. *Biology* **2021**, *10*, 1143. https://doi.org/10.3390/biology10111143

Academic Editors: Shibiao Wan, Yiping Fan, Chunjie Jiang and Shengli Li

Received: 25 September 2021
Accepted: 4 November 2021
Published: 6 November 2021

Publisher's Note: MDPI stays neutral with regard to jurisdictional claims in published maps and institutional affiliations.

1. Introduction

Bladder Urothelial Carcinoma (BLCA) is the eighth most prevalent cancer worldwide, with 549,393 new cases reported worldwide in 2018 [1]. Additionally, in the USA alone, there are estimated to be more than 80,000 new cases and 17,000 deaths each year [2].

This disease is particularly heterogeneous [3]. They are classified as high-grade and low-grade diseases based on standardized histomorphological features, as described by the World Health Organization. The depth of an invasion in the bladder wall determines the tumor stage. Approximately 80% of BLCA patients present non-muscle-invasive bladder cancer (NMIBC) at the time of diagnosis, while the remainder present muscle-invasive bladder cancer (MIBC) or even distant metastases [4]. NMIBCs do not normally pose a threat to patient survival and have a much better prognosis due to effective therapeutic options [5]. However, they almost always relapse, and patients need to repeat intravesical treatments, endoscopic evaluations, and biopsies, which may take an extended period of time, resulting in expensive surgical and surveillance management [6–8]. MIBCs, on the other hand, are clinically aggressive and can progress rapidly to lymph nodes, brain, lungs, liver, and bone metastases, which are often fatal [3]. However, over the past three decades, clinical management and five-year survival rates have seen few substantial advances [9]. Therefore, it is significant to identify novel biomarkers and molecular targets for advancing the prognosis of BLCA.

Carboxypeptidase A4 (CPA4) is a member of the zinc-containing metallocarboxypeptidase family [10], which could specifically catalyze the peptide bonds released from carboxy-terminal amino acids [11,12]. CPA4 was first discovered when screening for upregulated mRNA during cancer cell differentiation induced by sodium butyrate [13]. From the cellular and biochemical characteristics, CPA4 is secreted from cells in the form of soluble proenzyme (pro-CPA4), which might play a role in creating a tumor microenvironment [10]. Previous studies have demonstrated that CPA4 is closely associated with the aggressiveness, growth, and differentiation in cancer cells [14,15]. However, the underlying mechanism of CPA4 in BLCA remains unclear.

Recently, CPA4 has shown the potential to be a biomarker in the early diagnosis for certain cancers. Sun et al. have reported that the higher expression level of CPA4 in pancreatic cancer tissues and serum is related to poor prognosis and higher aggressiveness [13]. Previously studied showed that upregulated mRNA levels of CPA4 in androgen-independent prostate cancer cells is associated with the Histone Hyperacetylation signaling pathway [16]. In liver cancer and lung cancer, studies have also shown that the higher expression of CPA4 was closely associated with early diagnosis and poor prognosis [13,17]. Despite the potential significance of CPA4 expression in plenty types of cancer, no previous studies have ever shown the expression levels of CPA4 in bladder cancer, especially with regard to its potential therapeutic and prognostic values. Additionally, the correlation with immune infiltrates of CPA4 in BLCA remains to be investigated. Shao et al. demonstrated that CPA4 overexpression promotes the progression of aggressive clinical stage in pancreatic cancer and that the downregulation of CPA4 inhibits non-small-cell lung cancer growth [15,18]. Therefore, we hypothesized that the level of CPA4 is associated with the prognosis and immune cell infiltration in BLCA.

To test this hypothesis, our study evaluated the role of CPA4 on tumorigenesis and clinical significance based on The Cancer Genome Atlas (TCGA). We compared the different expression level of BLCA in age; gender; pathologic T, N, and M stage; pathology; subtype; and OS. In this study, we found that CPA4 is upregulated in BLCA. Significantly, the risk factors of CPA4 upregulation are correlated with poor prognosis. Additionally, the correlation with immune infiltrates of CPA4 for BLCA is also evaluated. Eventually, we link high CPA4 levels and poor prognosis in BLCA.

2. Materials and Methods

2.1. Data Source

The Cancer Genome Atlas (TCGA) (https://portal.gdc.cancer.gov/, accessed on 7 September 2021) provides 33 types of clinical and pathological information on cancer for scholars and researchers for free [19]. The expression profiles of CPA4 and clinical information of TCGA cancer data were downloaded from the UCSC Xena (https://xenabrowser.net/datapages/, accessed on 7 September 2021) database. The

TCGA database is available publicly in open access format and is available where ethical approval and informed consent of the patients were not necessary [20].

2.2. CPA4 Methylation Level Analysis

UALCAN (http://ualcan.path.uab.edu/, accessed on 6 September 2021) is a comprehensive, user-friendly, and interactive web resource for analyzing cancer OMICS data and provides graphs and plots depicting expression profiles and patient survival information for protein-coding, miRNA-coding, and lincRNA-coding genes [21]. The UALCAN online tool was utilized to analyze the CPA4 methylation level in BLCA (TCGA data).

2.3. Analysis of Differentially Expressed Genes (DEGs)

Through the limma Package by R, patients with different CPA4 expression profiles in the high and low expression groups (HTSeq-TPM) were compared using unpaired Student's t-test to identify the DEGs [22]. A |log2Fold Change| > 2 and BH-adjusted p-values < 0.05 were considered the threshold for the DEGs in a Gene Ontology (GO) Enrichment Analysis. Metascape (https://metascape.org, accessed on 7 September 2021) is a tool used for gene annotation and pathway analysis [23]. In this study, Metascape was utilized to analyze the enrichment of CPA4-related DEGs in processes and pathways. A p-value < 0.01, a minimum count of 3, and an enrichment factor of > 1.5 were regarded as significant [24].

2.4. Gene Set Enrichment Analysis (GSEA)

GSEA was used as a statistical method in order to seek out whether gene exhibits are statistically significant and concordant between two biological states [25]. We used the R package Cluster Profiler to evaluate excessive function and pathway differences between groups with different expression of CPA4 expression [26]. Each analysis of the processes was repeated 1000 times. Adjusted p-value < 0.05 and false discovery rate (FDR) < 0.25 were considered statistically significant enrichments [27]. We chose the potential pathway in which FDR < 0.05 with higher NES after analysis.

2.5. Comprehensive Analysis of Protein–Protein Interaction

The Search Tool for the Retrieval of Interacting Genes/Proteins (STRING) website (https://string-db.org/, accessed on 7 September 2021) is a database of known and predicted protein–protein interactions that hosts a collection of integrated and consolidated protein–protein interaction data including direct (physical) and indirect (functional) associations [28]. By importing CPA4 into the online tool STRING, protein–protein interaction (PPI) network information was compiled. Confidence scores > 0.4 were considered median significant.

2.6. Analysis of the Tumor Immune Estimation Resource (TIMER2)

The Tumor Immune Estimation Resource (TIMER2) is a comprehensive resource including 32 cancer types and incorporates 10,897 samples from the TCGA database for systematically analysis of immune infiltrates across diverse cancer types (http://cistrome.org/TIMER/, accessed on 7 September 2021) [29]. The TIMER2 database is used to evaluate the correlation of the expression of CPA4 in BLCA patients with the six types of infiltrating immune cells (B cells, dendritic cells, CD4 + T cells, CD8 + T cells, macrophages, and neutrophils) and displays the relationship between the expression of the CPA4 gene and the tumor purity.

2.7. Univariate and Multivariate Logistic Regression Analysis

Univariate Cox regression used to calculate the association between OS and patients' CPA4 expression in two cohorts aims at further researching the effect of CPA4 expression. A multivariate analysis was used to assess if CPA4 is an independent prognostic factor for

BLCA patient survival. CPA4 is statistically significant in the Cox regression analysis when the *p*-value is less than 0.05 [30].

2.8. Identification of CPA4 Coexpression Genes and Construction of a Prognostic Nomogram

cBiopor tal (https://www.cbioportal.org/, accessed on 7 Septtember 2021) (an online tool based on the TCGA database) was used to identify sets of coexpression genes. According to the *p*-value, we select the most relevant genes about CPA4. Then, the clinical factors (T, M, and N stages; radiation therapy; and primary therapy outcome) and the gene expression levels were used to construct a prognostic nomogram to evaluate the probability of 1-, 2-, and 3-year OS for BLCA patients via the R package (https://cran.r-project.org/web/packages/rms/, accessed on 7 September 2021) [31].

2.9. Immune Infiltration Analysis by ssGSEA

Single sample GSEA (ssGSEA) was performed to analyze the state of immune infiltration of BLCA from R package GSVA (version3.6) (http://www.bioconductor.org/packages/release/bioc/html/GSVA.Html, accessed on 8 September 2021), and we quantified the infiltration levels of 24 immune cell types from gene expression profiles in the literature [32]. In order to discover the correlation between CPA4 and the infiltration levels of 24 immune cells, adjusted *p*-values were established by the Spearman and Wilcoxon rank-sum tests.

2.10. Analysis of the Gene Expression Profiling Interactive Analysis 2

The Gene Expression Profiling Interactive Analysis2 (GEPIA2) (http://gepia.cancer-pku.cn/index.html, accessed on 7 September 2021) is an updated database used for analyzing the RNA sequencing expression data of 9736 tumors and 8587 normal samples from the TCGA and the GTEx projects, which include 60,498 genes and 198,619 isoforms [33]. GEPIA2 database investigated the expression level of CPA4 with various immune cells' markers. TIMER2 was used to identify the gene with a significant correlation with CPA4 expression in the GEPIA2 web.

2.11. Statistical Analysis

The expression of CPA4 for non-paired and paired samples was analyzed by the Wilcoxon rank-sum test and Wilcoxon signed-rank test, respectively. By using the pROC package, the ROC curve was generated to evaluate the CPA4 expression with diagnostic performance. The relations between the CPA expression and the clinical features were analyzed by the Kruskal–Wallis test, Chi-Squared test, and Wilcoxon signed rank test. The survival curves were generated via the long-rank test for the Kaplan–Meier analysis. $p < 0.05$ was considered statistically significant: * $p < 0.05$, ** $p < 0.01$, and *** $p < 0.001$; R software was used to process all kinds of statistical analyses (Version 4.0.2). In R, we use $p$adj = p.adjust (p, method = "BH", n = length(p)) to correct the *p*-value.

3. Results

3.1. Characteristics of BLCA Patients

In total, the information for 414 BLCA tumor tissues and 19 normal tissues were collected from the TCGA database including RNA-seq and relative clinical prognostic information in 414 patients. We grouped the BLCA patients into two sets: low ($n = 207$) and high expressions ($n = 207$) of CPA4. The clinical information of BLCA patients includes age, race, gender, pathologic stage, pathologic stage (T, N, or M), pathologic stage, primary therapy outcome, histologic grade, radiation therapy, subtype, smoking status, lymphovascular invasion, and OS event (Table 1).

Table 1. Clinical characteristics of two sets of patients with different expressions of CPA4 from the TCGA dataset.

Characteristic	Low Expression of CPA4	High Expression of CPA4	*p*
n	207	207	
Age, *n* (%)			0.921
≤70	116 (28%)	118 (28.5%)	
>70	91 (22%)	89 (21.5%)	
Race, *n* (%)			0.003
Asian	32 (8.1%)	12 (3%)	
Black or African American	8 (2%)	15 (3.8%)	
White	159 (40.1%)	171 (43.1%)	
Gender, *n* (%)			0.372
Female	50 (12.1%)	59 (14.3%)	
Male	157 (37.9%)	148 (35.7%)	
T stage, *n* (%)			0.004
T1	4 (1.1%)	1 (0.3%)	
T2	73 (19.2%)	46 (12.1%)	
T3	89 (23.4%)	107 (28.2%)	
T4	23 (6.1%)	37 (9.7%)	
N stage, *n* (%)			0.494
N0	120 (32.4%)	119 (32.2%)	
N1	18 (4.9%)	28 (7.6%)	
N2	39 (10.5%)	38 (10.3%)	
N3	3 (0.8%)	5 (1.4%)	
M stage, *n* (%)			0.810
M0	109 (51.2%)	93 (43.7%)	
M1	5 (2.3%)	6 (2.8%)	
Pathologic stage, *n* (%)			0.014
Stage I	4 (1%)	0 (0%)	
Stage II	76 (18.4%)	54 (13.1%)	
Stage III	63 (15.3%)	79 (19.2%)	
Stage IV	63 (15.3%)	73 (17.7%)	
Radiation therapy, *n* (%)			0.369
No	181 (46.6%)	186 (47.9%)	
Yes	13 (3.4%)	8 (2.1%)	
Primary therapy outcome, *n* (%)			<0.001
PD	18 (5%)	52 (14.6%)	
SD	14 (3.9%)	17 (4.8%)	
PR	12 (3.4%)	10 (2.8%)	
CR	136 (38.1%)	98 (27.5%)	
Histologic grade, *n* (%)			<0.001
High Grade	186 (45.3%)	204 (49.6%)	
Low Grade	19 (4.6%)	2 (0.5%)	
Lymphovascular invasion, *n* (%)			0.666
No	62 (21.9%)	68 (24%)	
Yes	78 (27.6%)	75 (26.5%)	
Subtype, *n* (%)			0.003
Non-Papillary	124 (30.3%)	151 (36.9%)	
Papillary	82 (20%)	52 (12.7%)	
OS event, *n* (%)			<0.001
Alive	139 (33.6%)	92 (22.2%)	
Dead	68 (16.4%)	115 (27.8%)	
Age, meidan (IQR)	69 (60, 76)	68 (61, 76)	0.990

3.2. Tumor Tissues Express Higher CPA4 Than Normal Tissue

The expression of CPA4 in pan-cancer was analyzed between tumor and normal tissues. From the TCGA + GETx database, the expression level of CPA4 in non-matched patients ($p = 1.6 \times 10^{-5}$) was significantly higher than that in normal people (Figure 1). The analysis of the correlation between CPA4 expression in BLCA patients and relative clinical information shows that a higher DLEU1 expression level is correlated with OS events and the subtype papillary. No statistically significant differences were found between the expression levels of CPA4 in BLCA and age; gender; pathological T, N, or M stages; and pathologic stage.

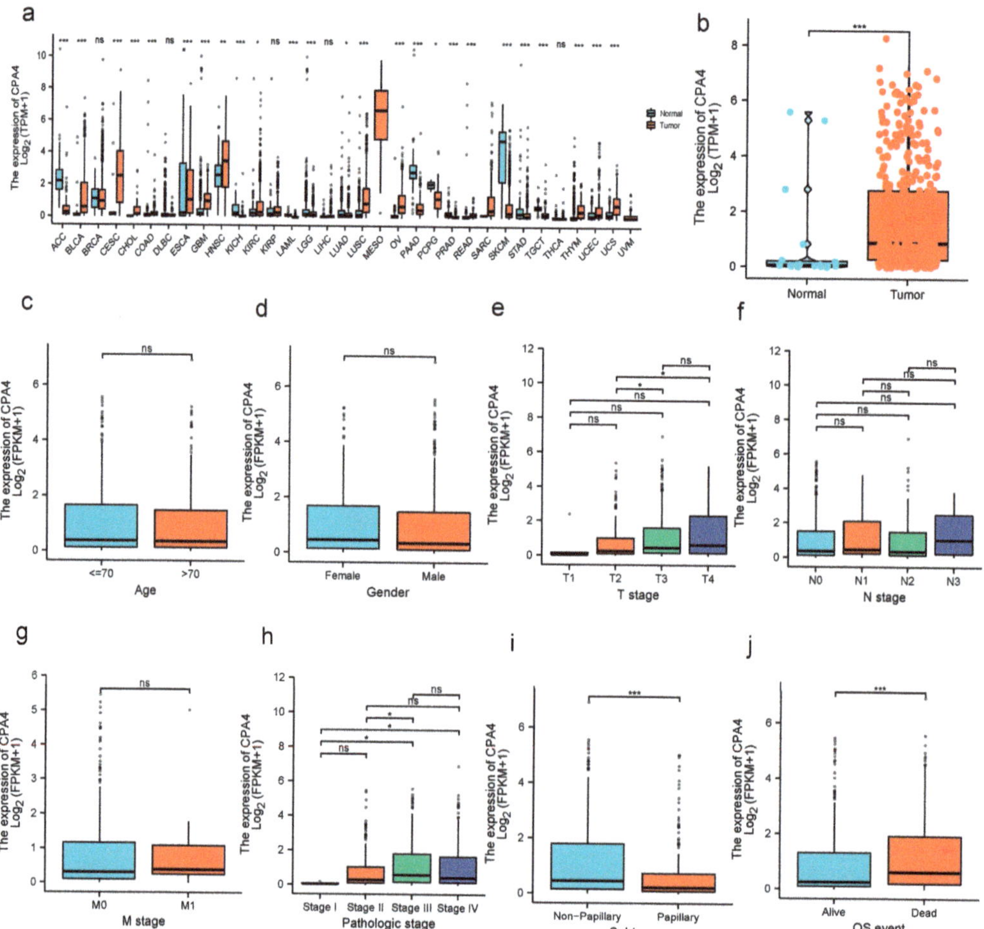

Figure 1. CPA4 expression and clinicopathological features in BLCA. (**a**) human CPA4 expression levels in different cancer tissues and corresponding normal tissues. (**b**) The expression level of CPA4 in BLCA tissue was significantly higher compared with the normal tissues from the TCGA + GTEx database. (**c–g**) No statistically significant differences were found between the expression levels of CPA4 in BLCA and age; gender; and pathological T, N, or M stage. (**h–j**) High pathologic stage, higher dead event, and nonpapillary were associated with higher expressions of CPA4 in BLCA. * $p < 0.05$; ** $p < 0.01$; *** $p < 0.001$; ns: no significance.

3.3. Impact of High CPA4 Expression on the Detection and Prognosis of BLCA Patients

The expression of CPA4 indicated a significant discriminative power in identifying tumors from normal cells with an AUC value of 0.798 (Figure 2d). The Kaplan–Meier survival analysis showed that BLCA patients with higher CPA4 expressions have shorter overall survival, disease-specific survival, and progress-free intervals (Figure 2a–c). The KM plots show that a higher expression of CPA4 had a worse prognosis than a lower expression. Promoter methylation of CPA4 in the TCGA-BLCA data was significantly lower than that of normal tissues adjacent to cancer in the UALCAN webpage ($p < 0.001$; Figure 2e).

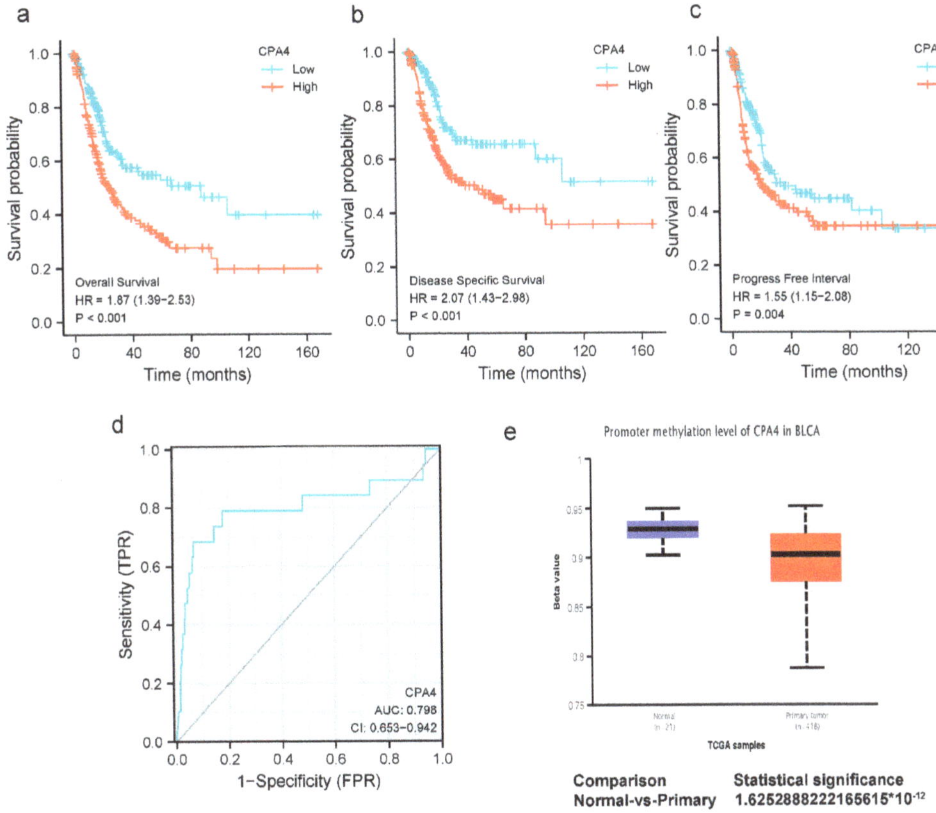

Figure 2. (**a–c**) Kaplan–Meier survival curves comparing high and low expressions of CPA4 in BLCA patients. (**a**) overall survival; (**b**) disease-specific survival; (**c**) progression-free interval; (**d**) ROC analysis of CPA4 indicates promising discrimination power between tumor and normal tissues; (**e**) the promoter methylation of CPA4 in tumor tissues ($n = 418$) and normal tissues ($n = 21$) from TCGA-BLCA data.

3.4. Differentially Expressed Genes and GO Enrichment Analysis in High- and Low CPA4 Expression Samples

We analyzed the DEGs in altered expressions of CPA4 including in low and high samples to explore the potential mechanisms of CPA4 that promote tumor progression. There were 529 DEGs identified, of which 349 genes were upregulated and 180 were downregulated ($|\log2(FC)| > 2$ and $p.adj < 0.05$). The DEGs's expression is shown in a heat map and volcano plot (Figure 3) using GO enrichment analysis to predict the co-expression functions in patients with BLCA. The top GO enrichment items in the biological process (BP), molecular function (MF), and cellular component (CC) groups were epidermal cell differentiation, keratinocyte differentiation, keratinization, intermediate filament cytoskeleton, intermediate filament, cornified envelope, endopeptidase inhibitor activity, peptidase inhibitor activity, peptidase inhibitor activity, serine-type endopeptidase inhibitor activity, metabolism of xenobiotics by cytochrome P450, drug metabolism-cytochrome P450, and retinol metabolism (Figure 4a).

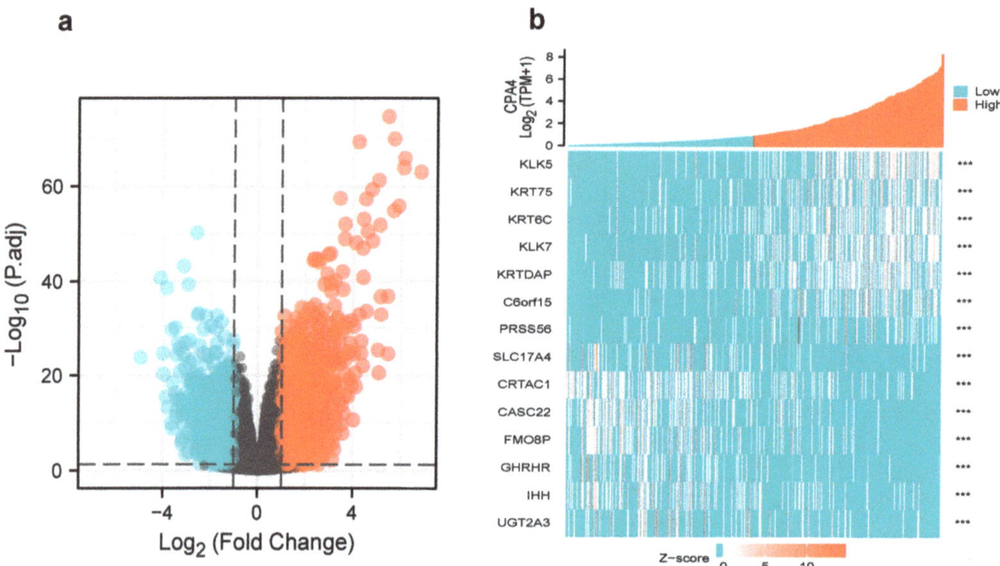

Figure 3. (**a**) Volcano plot of differentially expressed genes (DEGs) connected with the expression of CPA4; (**b**) heatmap of differentially expressed genes (DEGs) connected with the expression of CPA4. *** $p < 0.001$.

3.5. Gene Set Enrichment Analysis for CPA4-Related Signaling Pathways

By the enrichment of MSigDB Collection (c2.all.v7.0.symbols.gmt (curated)), we used the GSEA to identify signaling pathways associated with CPA4 between the different expression levels of CPA4 with significant differences (adjusted *p*-value < 0.05 and FDR < 0.25). The eight pathways included the formation of the cornified envelope, keratinization, immunoregulatory interactions between a lymphoid and a non-lymphoid cell, wp hair follicle development cytodifferentiation part 3 of 3, antigen processing and presentation, assembly of collagen fibrils and other multimeric structures, graft versus host disease, and cytokine–cytokine receptor interaction (Figure 4).

3.6. CPA4 Expression Predicts Poor Prognosis in Different Cancer Stages

Univariate cox proportional-hazards model analysis showed that high CPA4 expression, high pathologic grade and stage (T, N, and M), and subtype papillary were negative predictors for OS in BLCA patients. Meanwhile, in the multivariate regression analysis, CPA4 expression was an independent factor correlated with OS both in the low-expression set and high-expression set ($p = 0.003$) (Figure 5).

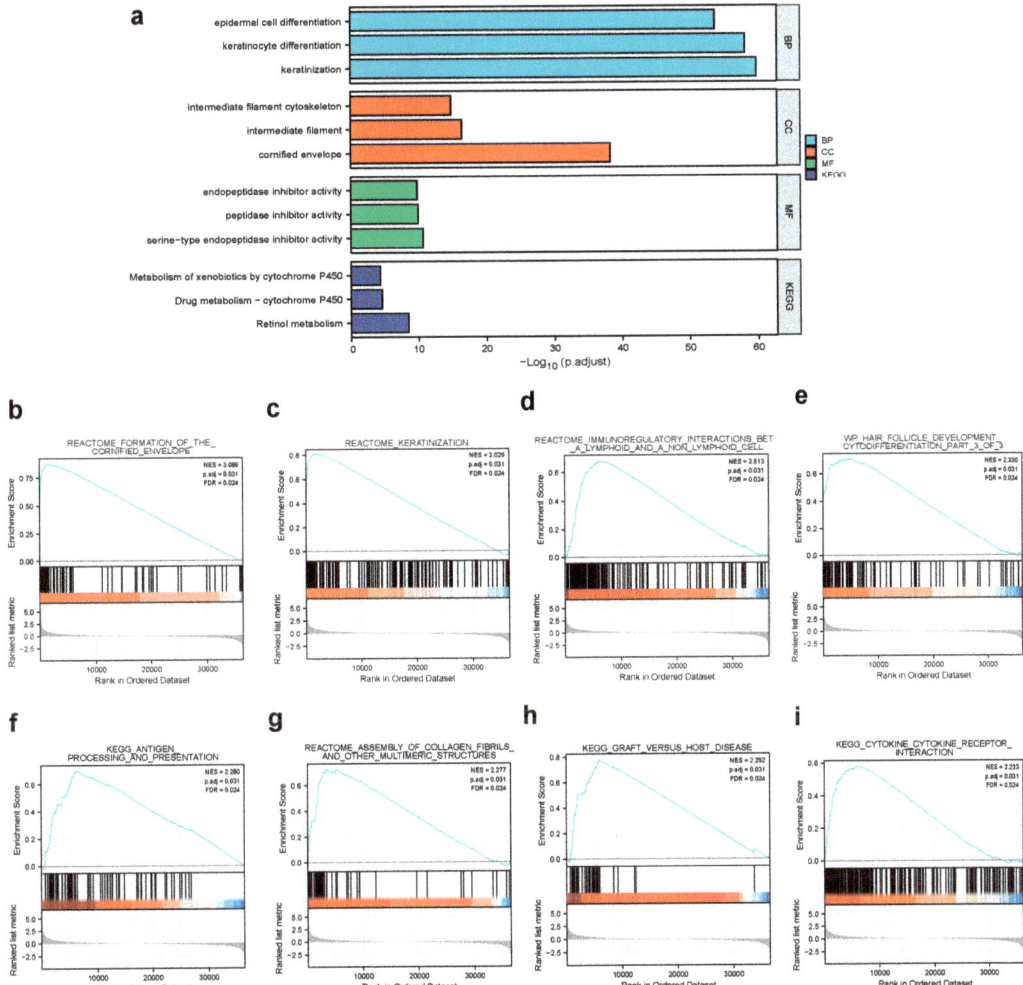

Figure 4. (**a**) GO enrichment analysis of differentially expressed genes (DEGs) in high- and low-CPA4 expression samples; (**b**,**c**) enrichment plots from GSEA. Several pathways were differentially enriched in BLCA patients according to different CPA4 expressions; (**b**) formation of the cornified envelope; (**c**) keratinization; (**d**) immunoregulatory interactions between a lymphoid and a non-lymphoid cell; (**e**) WP hair follicle development cytodifferentiation part 3 of 3; (**f**) antigen processing and presentation; (**g**) assembly of collagen fibrils and other multimeric structures; (**h**) graft versus host disease; (**i**) cytokine–cytokine receptor interaction. ES, enrichment score; NES, normalized enrichment score; ADJ *p*-Val, adjusted *p*-value; FDR, false discovery rate.

a

Characteristics	Total(N)	HR(95% CI) Univariate analysis	P value Univariate analysis
Age	413		
<=70	234	Reference	
>70	180	1.421 (1.063–1.901)	0.018
T stage	379		
T1&T2	124	Reference	
T3&T4	256	2.199 (1.515–3.193)	<0.001
N stage	369		
N0	239	Reference	
N1&N2&N3	131	2.289 (1.678–3.122)	<0.001
M stage	213		
M0	202	Reference	
M1	11	3.136 (1.503–6.544)	0.002
Pathologic stage	411		
Stage I&Stage II	134	Reference	
Stage III&Stage IV	278	2.310 (1.596–3.342)	<0.001
Subtype	408		
Non-Papillary	275	Reference	
Papillary	134	0.690 (0.488–0.976)	0.036
CPA4	413		
Low	207	Reference	
High	207	1.749 (1.296–2.359)	<0.001

b

Characteristics	Total(N)	HR(95% CI) Multivariate analysis	P value Multivariate analysis
Age	413		
<=70	234		
>70	180	1.311 (0.807–2.131)	0.275
T stage	379		
T1&T2	124		
T3&T4	256	3.346 (0.918–12.200)	0.067
N stage	369		
N0	239		
N1&N2&N3	131	2.094 (1.198–3.659)	0.009
M stage	213		
M0	202		
M1	11	1.518 (0.564–4.090)	0.409
Pathologic stage	411		
Stage I&Stage II	134		
Stage III&Stage IV	278	0.451 (0.110–1.843)	0.267
Subtype	408		
Non-Papillary	275		
Papillary	134	0.899 (0.505–1.600)	0.717
CPA4	413		
Low	207		
High	207	2.202 (1.309–3.702)	0.003

Figure 5. Univariate (**a**) and multivariate (**b**) regression analyses of CPA4 and other clinicopathologic parameters with OS in BLCA patients.

3.7. Construction of Nomogram for Predicting OS and Validation by Calibration

We constructed a nomogram for predicting the prognosis of BLCA with relative clinical situation, which integrates the clinical characteristics associated with the survival of BLCA. Based on the multivariate Cox analysis, a nomogram was assigned to the clinical characteristics of a point and the sum of points awarded to each characteristic is a point from 0 to 100. All of the points are accumulated and recorded as the total points. Using the absolute point axis down to the outcome axis, the probability of BLCA survival at 1, 3 and 5 years can be determined (Figure 6a). From the nomogram, the expression of CPA4 contributes many points compared with other relative clinical situations including the T, N, and M stages; radiation therapy; and primary therapy outcome. Meanwhile, the calibration plot indicates great agreement between the predicted and observed values, which are close to the 45-degree line, which is the ideal curve (Figure 6b).

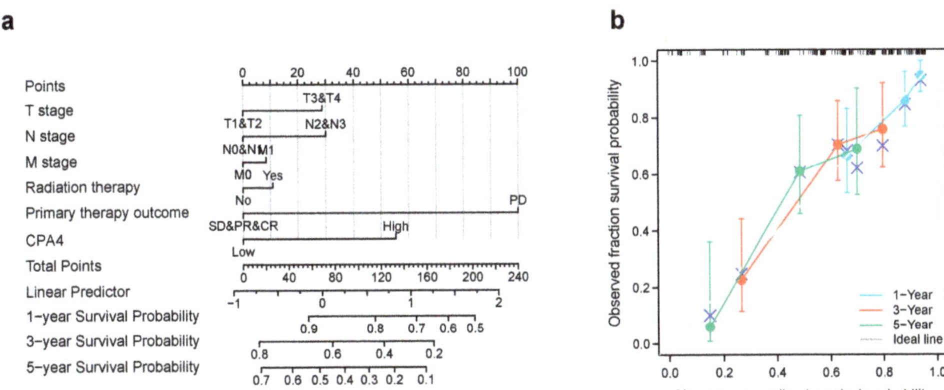

Figure 6. The relationship of CPA4 expression with other clinical factors and overall survival (OS). (**a**) Nomogram for predicting the probability of 1-, 3-, and 5-year OS for BLCA patients; (**b**) calibration plot of the nomogram for predicting the OS likelihood.

3.8. CPA4-Interaction Protein Networks in BLCA Tissue

CPA4-interaction protein networks were constructed to further explore the necessary proteins for metabolism and the molecular mechanism used by STRING. The PPI network of the CPA4 protein showed the relationship of the CPA4 protein in the progression of BLCA. Ten proteins and corresponding gene names were listed with their annotation scores (Figure 7). The top 10 genes included LXN, CMA1, SGCE, TPSAB1, AGBL2, TPSB2, PEG10, GRB10, TSGA13, and MEST, and LXN had the highest score.

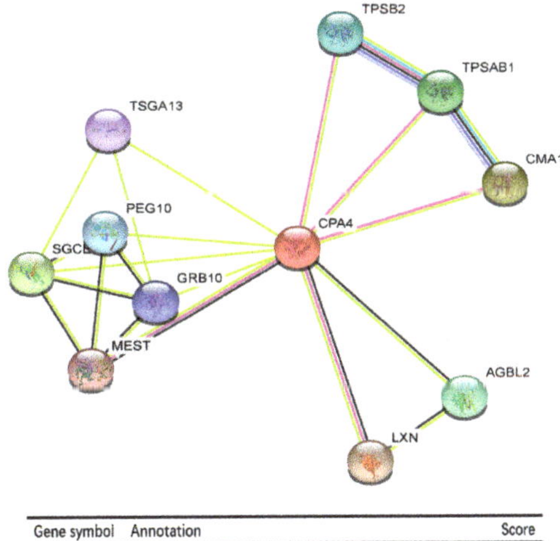

Gene symbol	Annotation	Score
LXN	Latexin	0.993
CMA1	Chymase	0.777
SGCE	Epsilon-sarcoglycan	0.72
TPSAB1	Tryptase alpha/beta-1	0.715
AGBL2	Cytosolic carboxypeptidase 2	0.67
TPSB2	Tryptase beta 2	0.652
PEG10	Retrotransposon-derived protein PEG10	0.646
GRB10	Growth factor receptor-bound protein 10	0.64
TSGA13	Testis-specific gene 13 protein	0.609
MEST	Mesoderm-specific transcript homolog protein	0.594

Figure 7. CPA4-interaction proteins in BLCA tissue; annotation of CPA4-interacting proteins and their co-expression scores.

3.9. Correlation Analysis between CPA4 Expression and Infiltrating Immune Cells

The survival of patients with different cancers including BLCA is associated with the tumor-infiltrating immune cells. From the result, the expression level of CPA4 had significant correlations with CD8+ T cells (r = 0.287, $p = 2.29 \times 10^{-8}$), B cells (r = 0.218, $p = 8.65 \times 10^{-10}$), neutrophils (r = 0.196, $p = 1.76 \times 10^{-4}$), and dendritic cells (r = 0.356, $p = 2.5 \times 10^{-12}$). $p < 0.05$ was considered significant (Figure 8a). Furthermore, we analyzed 24 immune cells including pDC, NK CD56bright cells, DC, cytotoxic cells, TFH, B cells, CD8 T cells, Th17 cells, Treg, T cells, mast cells, iDC, NK cells, Tem, aDC, neutrophils, Th1 cells, NK CD56dim cells, macrophages, eosinophils, Tgd T helper cells, Th2 cells, and Tcm. We analyzed the correlation between the expression of CPA4 and immune infiltration by ssGSEA using Spearman's R. From the result, the expression level of CPA4 was negatively correlated with the infiltration levels of NK CD56bright cells ($p < 0.001$) and positively correlated with cytotoxic cells, T cells, NK cells, idc, Tem, Treg, aDC, Neutrophils, NK CD56dim cells, macrophages, Th2 cells, and Th1 cells (Figure 8).

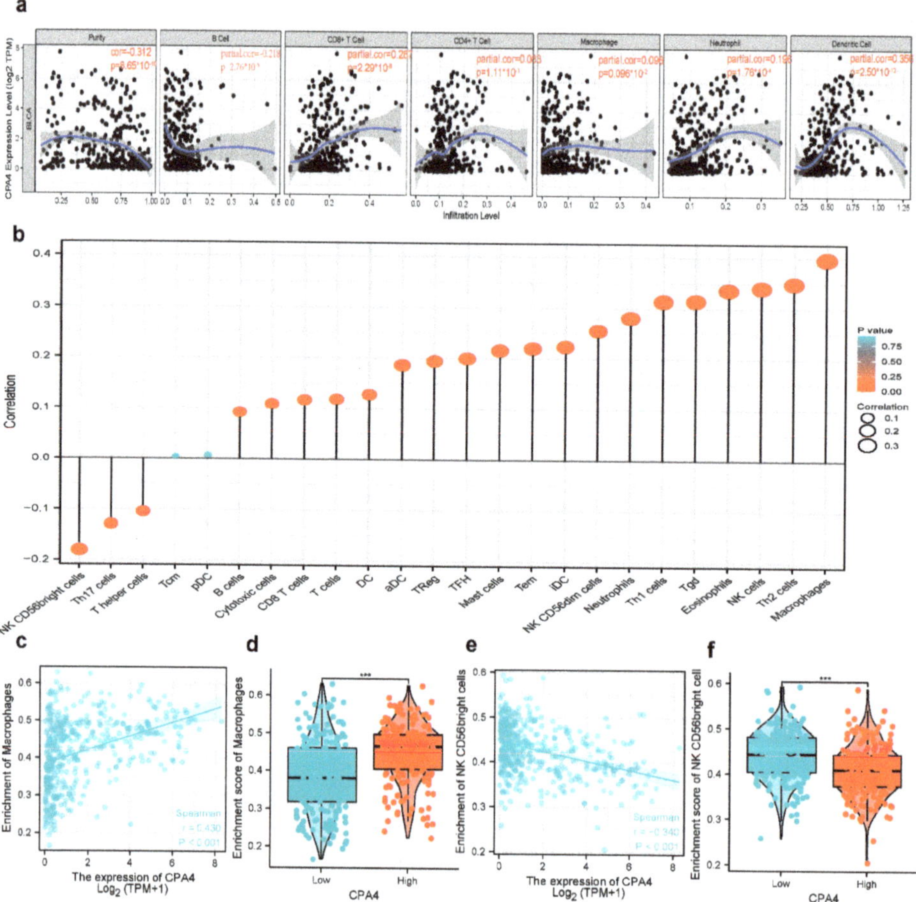

Figure 8. The expression level of CPA4 was related to immune infiltration in the tumor microenvironment. (**a**) Correlation of CPA4 expression with infiltrating immune infiltration in BLCA (**b**) The forest plot shows the correlation between CPA4 expression level and 24 immune cells. The size of the dots indicates the absolute value of Spearman's R. (**c**,**d**) The Wilcoxon rank sum test was used to analyze the difference in the macrophage cell infiltration levels between the CPA4 high- and low-expression groups; (**e**,**f**) the correlation between CPA4 expression and NK CD56 bright cell infiltration levels. *** $p < 0.001$.

3.10. Possible Role of the Expression of CPA4 in Various Infiltrating Immune Cells

We used the TIMER2 and GEPIA2 databases to further identify the possible role of the expression of CPA4 in various infiltrating immune cells including T cells (general), M1/M2 macrophages, tumor-associated macrophages, B cells, neutrophils, monocytes, NK, CD8+ T cells, and functional DCs as well as T cells such as Th1, Th2, Th9, Th17, Th22, Tfh, exhausted T cells, and Treg. From the results, Th1, T cell exhaustion, and TAM sets marking were greatly connected with the expression of CPA4 in BLCA (Table 2).

Table 2. Correlation analysis between CPA4 and markers of immune cells in BLCA patients found in the TIMER2 and GEPIA2.

Cell Type	Gene Marker	None Cor	*p*	Purity Cor	*p*	Tumor R	*p*	Normal R	*p*
B cell	CD19	−0.042	0.397	−0.138	**	−0.032	0.52	−0.033	0.89
	CD20(KRT20)	−0.314	***	−0.226	***	−0.13	*	−0.18	0.47
	CD38	0.301	***	0.148	**	0.1	*	−0.032	0.9
CD8+ T cell	CD8A	0.267	***	0.12	*	−0.032	0.9	−0.074	0.76
	CD8B	0.15	**	0.018	0.727	0.0031	0.95	−0.096	0.69
Tfh	BCL6	−0.247	***	−0.214	***	−0.14	**	−0.28	0.25
	ICOS	−0.307	***	0.154	**	0.13	**	−0.095	0.7
	CXCR5	0.109	*	−0.095	0.0677	0.075	0.13	0.039	0.87
Th1	T-bet(TBX21)	0.227	***	0.046	0.375	0.2	***	0.041	0.87
	STAT4	0.37	***	0.223	***	0.16	**	0.0068	0.98
	IL12RB2	0.403	***	0.327	***	0.24	***	−0.22	0.37
	WSX1(IL27RA)	0.39	***	0.291	***	0.18	***	0.027	0.91
	STAT1	0.386	***	0.282	***	0.24	***	−0.14	0.56
	IFN-γ(IFNG)	0.278	***	0.161	**	0.13	**	−0.085	0.73
	TNF-α(TNF)	0.287	***	0.194	***	0.098	*	0.27	0.26
Th2	GATA3	−0.484	***	−0.402	***	−0.26	***	−0.26	0.28
	CCR3	0.188	***	0.131	*	0.67	*	−0.14	0.58
	STAT6	−0.228	***	−0.209	***	−0.13	**	−0.32	0.18
	STAT5A	0.004	0.936	−0.158	**	−0.015	0.76	−0.53	*
Th9	TGFBR2	0.087	0.079	−0.014	0.792	0.038	0.45	−0.45	0.056
	IRF4	0.188	***	−0.03	0.571	0.043	0.39	−0.12	0.63
	PU.1(SPI1)	0.356	***	0.181	***	0.15	**	−0.17	0.49
Th17	STAT3	0.325	***	0.232	***	0.15	**	−0.11	0.64
	IL-21R	0.318	***	0.132	*	0.073	0.14	−0.1	0.68
	IL-23R	−0.003	0.945	−0.076	0.143	−0.0048	0.92	−0.019	0.94
	IL-17A	−0.019	0.705	−0.057	0.274	−0.051	0.31	−0.18	0.47
Th22	CCR10	−0.025	0.626	−0.068	0.195	0.029	0.57	−0.34	0.16
	AHR	−0.271	***	−0.195	***	−0.11	*	−0.29	0.23
Treg	FOXP3	0.287	***	0.15	**	0.16	**	0.037	0.88
	CD25(IL2RA)	0.369	***	0.22	***	0.037	0.88	−0.066	0.79
	CCR8	0.218	***	0.083	0.113	0.083	0.094	−0.0059	0.98
T cell exhaustion	PD-1(PDCD1)	0.255	***	0.089	*	0.089	0.073	−0.099	0.69
	CTLA4	0.311	***	0.16	**	0.23	***	−0.11	0.64
	LAG3	0.362	***	0.227	***	0.22	***	−0.19	0.45
	TIM-3(HAVCR2)	0.375	***	0.218	***	0.21	***	−0.097	0.69
Macrophage	CD68	0.316	***	0.193	***	0.14	**	0.49	*
	CD11b(ITGAM)	0.303	***	0.119	*	0.72	*	−0.29	0.23
M1	INOS(NOS2)	−0.033	0.511	−0.092	0.0774	−0.0068	0.89	−0.14	0.57
	IRF5	−0.123	*	−0.116	*	−0.063	0.2	−0.026	0.92
	COX2(PTGS2)	0.209	***	0.164	**	0.057	0.25	−0.24	0.32
M2	CD16	0.408	***	0.273	***	0.17	***	−0.2	0.42
	ARG1	−0.049	0.322	−0.007	0.894	0.076	0.13	0.68	**
	MRC1	0.334	***	0.164	**	0.042	0.4	−0.23	0.34
	MS4A4A	0.353	***	0.199	***	0.12	**	−0.23	0.34
TAM	CCL2	0.26	***	0.113	*	0.022	0.66	−0.12	0.62
	CD80	0.413	***	0.285	***	0.17	***	−0.18	0.46
	CD86	0.396	***	0.244	***	0.17	***	−0.074	0.76
	CCR5	0.29	***	0.101	0.0522	0.13	*	−0.08	0.75
Monocyte	CD14	0.406	***	0.253	***	0.11	*	−0.21	0.38
	CD16(FCGR3B)	0.316	***	0.22	***	0.15	**	−0.073	0.77
	CD115(CSF1R)	0.353	***	0.178	***	0.14	**	−0.28	0.24

Table 2. *Cont.*

Cell Type	Gene Marker	None Cor	p	Purity Cor	p	Tumor R	p	Normal R	p
Neutrophil	CD66b(CEACAM8)	0.089	0.0721	0.098	0.0609	−0.031	0.53	−0.084	0.73
	CD15(FUT4)	0.141	**	0.047	0.369	0.0041	0.93	−0.33	0.17
	CD11b(ITGAM)	0.303	***	0.119	*	0.018	0.72	−0.29	0.23
Natural killer cell	XCL1	−0.01	0.844	−0.005	0.93	−0.06	0.23	0.13	0.59
	CD7	0.304	***	0.131	*	0.15	**	−0.029	0.91
	KIR3DL1	0.136	**	0.049	0.346	0.075	0.13	0.19	0.44
Dendritic cell	CD1C(BDCA-1)	0.086	0.0823	−0.054	0.305	−0.023	0.65	−0.02	0.93
	CD141(THBD)	0.356	***	0.322	***	0.055	0.27	0.37	0.12
	CD11c(ITGAX)	0.35	***	0.181	***	0.099	*	−0.2	0.41

BLCA, Bladder Urothelial Carcinoma; Tfh, Follicular helper T cell; Th, T helper cell; Treg, Regulatory T cell; TAM, tumor-associated macrophage;.None, correlation without adjustment correlation; Purity, correlation adjusted by purity; Tumor, correlation analysis in the tumor tissue of TCGA; Normal, correlation analysis in normal tissue of TCGA; Cor, R value of Spearman's correlation * $p < 0.05$; ** $p < 0.01$; *** $p < 0.001$.

4. Discussion

CPA4 (carboxypeptidase A4) is a member of the metallocarboxypeptidase family and is a zinc-containing exopeptidase that catalyzes the release of carboxy-terminal amino acids [34]. In recent years, CPA4 has shown the potential to be a biomarker in the early diagnosis with clinical benefit for certain cancers. Some studies revealed that CPA4 is connected with various cancer cells in its differentiation and growth, including non-small-cell lung cancer and gastric cancer [35,36]. Furthermore, it is reported that CPA4 is located on chromosome 7q32 in a region linked to prostate cancer aggressiveness [11], and Sun suggested that CPA4 is closely associated with colorectal cancer liver metastasis [37]. Although CPA4 expression has been confirmed to have potential significance in multiple types of cancer, no studies have shown the expression level and clinical significance of CPA4 in BLCA. In this study, based on a pan-cancer analysis, we demonstrated that human CPA4 expression levels were highly expressed in 11 types of cancer with corresponding normal tissues (Figure 1), which are consistent with the findings in the previous study reported by Sun and Handa et al. [17,35,38]. We also confirmed that CPA4 is significantly upregulated in BLCA (Figure 1b). Moreover, a previous study has shown that CPA4 expression was detected specifically in the cytoplasm of cancer tissue cells, and in the CPA4-suppressed triple-negative breast cancer (TNBC), viability, and migration were decreased [38]. It can act as a potential biomarker of poor prognosis in TNBC. It is reported that CPA4 might be used as an independent poor prognostic factor in esophageal squamous cell carcinoma [39]. In our study, the results in BLCA are consistent. However, one trial showed that CPA4 is a protective factor in muscle-invasive bladder cancer, contrary to the role of CPA4 in most cancers [40]. A potential reason for the difference is due to updates in the TCGA database and different objects. We studied BLCA, while that study investigated muscle-invasive bladder cancer. We compared the different expression levels of BLCA with age; gender; T, N, and M stage; pathologic stage; subtype; and OS. Surprisingly, we found that higher dead events, higher pathologic stages, and the subtype non-papillary were associated with higher expressions of CPA4 in BLCA, with statistical differences (Figure 2). These findings suggest that CPA4 may be a potential biomarker of poor prognosis in identifying BLCA with poor clinical outcome.

Currently, the function of CPA4 in tumors had not been fully reported. Previous trials suggested that the inhibition of CPA4 could reduce the number of breast cancer cells with stemness properties and may be a potential target for TNBC therapy [41]. The CircCPA4 sponge let-7 regulates the expression of CPA4 and glioma progression [42]. All of these results suggest that CPA4 could be regarded as an emerging target or promising biomarker for cancer therapy. Since the mRNA expression of CPA4 in BLCA was significantly higher than that in normal bladder tissue, we speculated that CPA4 can be regarded as a biomarker to detect BLCA from normal controls. To verify the clinical value of CPA4, an ROC curve analysis was performed to verify the clinical value of CPA4 in the diagnosis of BLCA; our results showed that CPA4 may be a potential diagnostic biomarker between bladder cancer and normal tissues, with an AUC of 0.798 (Figure 2d).

Many studies have shown that CPA4 is a significant biomarker of poor prognosis in lots of cancers and is associated with the upregulation of CPA4 with poor overall survival. In hepatocellular carcinoma, AC10364 inhibited cell proliferation and viability through the abnormal expression of genes including CPA4 associated with tumorigenesis or growth [43]. A paper from Yan suggested that the inhibition of CPA4 might be of great significance for improving early stage non-small cell lung cancer survival after ablation [43]. However, the prognostic value in BLCA of CPA4 has not been investigated. With the increased level of CPA4 related to a higher number of dead events and higher pathologic stages, we speculated that CPA4 is involved in the development of BLCA. In light of the Kaplan–Meier curves, we confirmed that the overexpression of CPA4 is associated with shorter overall survival (OS), disease-specific survival (DSF), and progress-free intervals (PFIs) (Figure 2). Moreover, by univariate and multivariate regression analysis, we found that high CPA4 expression; high pathologic stage; T, N, and M stage; and the subtype papillary were negative predictors for OS in BLCA patients and that CPA4 can be an independent factor correlated with OS (Figure 5). The nomogram more accurately predicted 1-, 3-, and 5-year OS in BLCA patients and could help to screen and determine those high-risk patients (Figure 6).

Through GSEA, CPA4 was found to be involved in epidermal cell differentiation, keratinocyte differentiation, keratinization, etc., indicating CPA4 potentially playing a role in cell metabolism and protein synthesis (Figure 4).

The PPI network of CPA4 protein, which were constructed by STRING, showed the relationship of CPA4 in the progression of BLCA such as LXN, CMA1, SGCE, TPSAB1, etc. (Figure 7). It has been reported that latexin (LXN) can inhibit human CPA4, in which the expression is induced in prostate cancer cells after treatment with histone deacetylase inhibitors [44]. The level of CMA1, a key gene, is significantly correlated with gastric cancer prognosis and infiltration level [45]. SGCE promotes breast cancer stem cell self-renewal, chemoresistance, and metastasis both in vitro and in vivo by stabilizing EGFR [45]. Thus, it is speculated that a high expression of CPA4 may increase the degree of malignancy of tumors through CPA4 interacting proteins, leading to the deterioration of patients' conditions.

Moreover, CPA4 plays a specific role in immune infiltration in bladder cancer. Compellingly, we unraveled that several infiltrating immune cells (Th1cell, Th2 cell, T cell exhaustion, and TAM) were correlated with the expression of CPA4 in bladder cancer using TIMER2 and GEPIA2. Type 1 T helper (Th1) cells produce interferon-gamma [46] (Figure 8, Table 2). The dual inhibition of STAT1 and STAT3 activation downregulates the expression of PD-L1 in cancer cells [47]. T-cell exhaustion is a state of T-cell dysfunction that occurs in many chronic infections and cancers [48]. Scholars have observed that CTLA4 was identified as a crucial negative regulator of the immune system, which transmits an inhibitory signal [49].

There are some limitations in our study. First, basic experiments are needed to verify the results, which were conducted with online public databases. Second, in vivo/vitro experiments are needed to further investigate the potential mechanism of the effect of CPA4 on immune invasion in BLCA.

5. Conclusions

In conclusion, our study first demonstrated that CPA4 expression increased in BLCA, and univariate and multivariate regression analyses and a nomogram were used to prove that increased CPA4 is correlated with shorter overall survival, which means high risk factors in BLCA patients. Moreover, we illustrated that a high level of CPA4 was positively related to a high pathologic grade; high T, N, and M stages; and the subtype papillary. The immune infiltration in the tumor microenvironment has also been shown to be associated with CPA4. Collectively, this study partially unveiled that CPA4 in BLCA could be regarded as a potential biomarker for diagnosis and prognosis and may play a special role in immune infiltration.

Author Contributions: C.W. designed the original idea, organized the model of TCGA data with relative pictures, and wrote part of the article. Y.Z. was responded for the modification of manuscript and wrote part of the manuscript. Q.X. was responded for the statistics problems and grammar check. M.X. was responsible for the language, bioinformation problems and correcting them. Z.C. and X.Y. was a supporter of this manuscript. Y.H. edited the whole manuscript. All of the authors of this manuscript are responsible for the accuracy, appropriation, and completing any part of this research and agree to open this manuscript. All authors have read and agreed to the published version of the manuscript.

Funding: This research received no external funding.

Institutional Review Board Statement: Not applicable.

Informed Consent Statement: Not applicable.

Data Availability Statement: All of the data in this manuscript are available and approved by the institution.

Acknowledgments: The authors would like to thank all who supported this manuscript as well as provided any useful suggestions and help.

Conflicts of Interest: All of the authors declare no conflict of interest. The funders had no role in the design of the study; in the collection, analyses, or interpretation of data; in the writing of the manuscript; or in the decision to publish the results.

References

1. Bray, F.; Ferlay, J.; Soerjomataram, I.; Siegel, R.L.; Torre, L.A.; Jemal, A. Global cancer statistics 2018: GLOBOCAN estimates of incidence and mortality worldwide for 36 cancers in 185 countries. *CA Cancer J. Clin.* **2018**, *68*, 394–424. [CrossRef] [PubMed]
2. Lenis, A.T.; Lec, P.M.; Chamie, K.; Mshs, M.D. Bladder Cancer: A Review. *JAMA* **2020**, *324*, 1980–1991. [CrossRef] [PubMed]
3. McConkey, D.J.; Choi, W. Molecular Subtypes of Bladder Cancer. *Curr. Oncol. Rep.* **2018**, *20*, 77. [CrossRef]
4. Bhanvadia, S.K. Bladder Cancer Survivorship. *Curr. Urol. Rep.* **2018**, *19*, 111. [CrossRef] [PubMed]
5. Babjuk, M.; Burger, M.; Compérat, E.M.; Gontero, P.; Mostafid, A.H.; Palou, J.; van Rhijn, B.W.G.; Rouprêt, M.; Shariat, S.F.; Sylvester, R.; et al. European Association of Urology Guidelines on Non-muscle-invasive Bladder Cancer (TaT1 and Carcinoma In Situ)—2019 Update. *Eur. Urol.* **2019**, *76*, 639–657. [CrossRef] [PubMed]
6. Berdik, C. Unlocking bladder cancer. *Nature* **2017**, *551*, S34–S35. [CrossRef]
7. James, A.C.; Gore, J.L. The costs of non-muscle invasive bladder cancer. *Urol. Clin. N. Am.* **2013**, *40*, 261–269. [CrossRef]
8. Abdollah, F.; Gandaglia, G.; Thuret, R.; Schmitges, J.; Tian, Z.; Jeldres, C.; Passoni, N.M.; Briganti, A.; Shariat, S.F.; Perrotte, P.; et al. Incidence, survival and mortality rates of stage-specific bladder cancer in United States: A trend analysis. *Cancer Epidemiol.* **2013**, *37*, 219–225. [CrossRef]
9. Grayson, M. Bladder cancer. *Nature* **2017**, *551*, S33. [CrossRef]
10. Tanco, S.; Zhang, X.; Morano, C.; Avilés, F.X.; Lorenzo, J.; Fricker, L.D. Characterization of the substrate specificity of human carboxypeptidase A4 and implications for a role in extracellular peptide processing. *J. Biol. Chem.* **2010**, *285*, 18385–18396. [CrossRef]
11. Ross, P.L.; Cheng, I.; Liu, X.; Cicek, M.S.; Carroll, P.R.; Casey, G.; Witte, J.S. Carboxypeptidase 4 gene variants and early-onset intermediate-to-high risk prostate cancer. *BMC Cancer* **2009**, *9*, 69. [CrossRef]
12. Kayashima, T.; Yamasaki, K.; Yamada, T.; Sakai, H.; Miwa, N.; Ohta, T.; Yoshiura, K.; Matsumoto, N.; Nakane, Y.; Kanetake, H.; et al. The novel imprinted carboxypeptidase A4 gene (CPA4) in the 7q32 imprinting domain. *Hum. Genet.* **2003**, *112*, 220–226. [CrossRef] [PubMed]
13. Sun, L.; Burnett, J.; Guo, C.; Xie, Y.; Pan, J.; Yang, Z.; Ran, Y.; Sun, D. CPA4 is a promising diagnostic serum biomarker for pancreatic cancer. *Am. J. Cancer Res.* **2016**, *6*, 91–96. [PubMed]
14. Hong, W.; Xue, M.; Jiang, J.; Zhang, Y.; Gao, X. Circular RNA circ-CPA4/ let-7 miRNA/PD-L1 axis regulates cell growth, stemness, drug resistance and immune evasion in non-small cell lung cancer (NSCLC). *J. Exp. Clin. Cancer Res. CR* **2020**, *39*, 149. [CrossRef] [PubMed]
15. Shao, Q.; Zhang, Z.; Cao, R.; Zang, H.; Pei, W.; Sun, T. CPA4 Promotes EMT in Pancreatic Cancer via Stimulating PI3K-AKT-mTOR Signaling. *OncoTargets Ther.* **2020**, *13*, 8567–8580. [CrossRef] [PubMed]
16. Huang, H.; Reed, C.P.; Zhang, J.S.; Shridhar, V.; Wang, L.; Smith, D.I. Carboxypeptidase A3 (CPA3): A novel gene highly induced by histone deacetylase inhibitors during differentiation of prostate epithelial cancer cells. *Cancer Res.* **1999**, *59*, 2981–2988.
17. Sun, L.; Guo, C.; Burnett, J.; Pan, J.; Yang, Z.; Ran, Y.; Sun, D. Association between expression of Carboxypeptidase 4 and stem cell markers and their clinical significance in liver cancer development. *J. Cancer* **2017**, *8*, 111–116. [CrossRef]

18. Fu, Y.; Su, L.; Cai, M.; Yao, B.; Xiao, S.; He, Q.; Xu, L.; Yang, L.; Zhao, C.; Wan, T.; et al. Downregulation of CPA4 inhibits non small-cell lung cancer growth by suppressing the AKT/c-MYC pathway. *Mol. Carcinog.* **2019**, *58*, 2026–2039. [CrossRef]

19. Tomczak, K.; Czerwińska, P.; Wiznerowicz, M. The Cancer Genome Atlas (TCGA): An immeasurable source of knowledge. *Contemp. Oncol. (Pozn. Pol.)* **2015**, *19*, A68–A77. [CrossRef]

20. Blum, A.; Wang, P.; Zenklusen, J.C. SnapShot: TCGA-Analyzed Tumors. *Cell* **2018**, *173*, 530. [CrossRef] [PubMed]

21. Chandrashekar, D.S.; Bashel, B.; Balasubramanya, S.A.H.; Creighton, C.J.; Ponce-Rodriguez, I.; Chakravarthi, B.; Varambally, S. UALCAN: A Portal for Facilitating Tumor Subgroup Gene Expression and Survival Analyses. *Neoplasia (N. Y.)* **2017**, *19*, 649–658. [CrossRef]

22. He, J.; Chen, D.L.; Samocha-Bonet, D.; Gillinder, K.R.; Barclay, J.L.; Magor, G.W.; Perkins, A.C.; Greenfield, J.R.; Yang, G.; Whitehead, J.P. Fibroblast growth factor-1 (FGF-1) promotes adipogenesis by downregulation of carboxypeptidase A4 (CPA4)—A negative regulator of adipogenesis implicated in the modulation of local and systemic insulin sensitivity. *Growth Factors (Chur Switz.)* **2016**, *34*, 210–216. [CrossRef] [PubMed]

23. Zhou, Y.; Zhou, B.; Pache, L.; Chang, M.; Khodabakhshi, A.H.; Tanaseichuk, O.; Benner, C.; Chanda, S.K. Metascape provides a biologist-oriented resource for the analysis of systems-level datasets. *Nat. Commun.* **2019**, *10*, 1523. [CrossRef] [PubMed]

24. Love, M.I.; Huber, W.; Anders, S. Moderated estimation of fold change and dispersion for RNA-seq data with DESeq2. *Genome Biol.* **2014**, *15*, 550. [CrossRef] [PubMed]

25. Subramanian, A.; Tamayo, P.; Mootha, V.K.; Mukherjee, S.; Ebert, B.L.; Gillette, M.A.; Paulovich, A.; Pomeroy, S.L.; Golub, T.R.; Lander, E.S.; et al. Gene set enrichment analysis: A knowledge-based approach for interpreting genome-wide expression profiles. *Proc. Natl. Acad. Sci. USA* **2005**, *102*, 15545–15550. [CrossRef]

26. Yu, G.; Wang, L.G.; Han, Y.; He, Q.Y. clusterProfiler: An R package for comparing biological themes among gene clusters. *Omics J. Integr. Biol.* **2012**, *16*, 284–287. [CrossRef]

27. Dinu, I.; Potter, J.D.; Mueller, T.; Liu, Q.; Adewale, A.J.; Jhangri, G.S.; Einecke, G.; Famulski, K.S.; Halloran, P.; Yasui, Y. Improving gene set analysis of microarray data by SAM-GS. *BMC Bioinform.* **2007**, *8*, 242. [CrossRef]

28. Szklarczyk, D.; Gable, A.L.; Lyon, D.; Junge, A.; Wyder, S.; Huerta-Cepas, J.; Simonovic, M.; Doncheva, N.T.; Morris, J.H.; Bork, P.; et al. STRING v11: Protein-protein association networks with increased coverage, supporting functional discovery in genome-wide experimental datasets. *Nucleic Acids Res.* **2019**, *47*, D607–D613. [CrossRef] [PubMed]

29. Li, T.; Fu, J.; Zeng, Z.; Cohen, D.; Li, J.; Chen, Q.; Li, B.; Liu, X.S. TIMER2.0 for analysis of tumor-infiltrating immune cells. *Nucleic Acids Res.* **2020**, *48*, W509–W514. [CrossRef]

30. Liu, J.; Lichtenberg, T.; Hoadley, K.A.; Poisson, L.M.; Lazar, A.J.; Cherniack, A.D.; Kovatich, A.J.; Benz, C.C.; Levine, D.A.; Lee, A.V.; et al. An Integrated TCGA Pan-Cancer Clinical Data Resource to Drive High-Quality Survival Outcome Analytics. *Cell* **2018**, *173*, 400–416. [CrossRef]

31. Shen, C.; Liu, J.; Wang, J.; Zhong, X.; Dong, D.; Yang, X.; Wang, Y. Development and validation of a prognostic immune-associated gene signature in clear cell renal cell carcinoma. *Int. Immunopharmacol.* **2020**, *81*, 106274. [CrossRef]

32. Hänzelmann, S.; Castelo, R.; Guinney, J. GSVA: Gene set variation analysis for microarray and RNA-seq data. *BMC Bioinform.* **2013**, *14*, 7. [CrossRef]

33. Li, C.; Tang, Z.; Zhang, W.; Ye, Z.; Liu, F. GEPIA2021: Integrating multiple deconvolution-based analysis into GEPIA. *Nucleic Acids Res.* **2021**, *49*, W242–W246. [CrossRef]

34. Wei, S.; Segura, S.; Vendrell, J.; Aviles, F.X.; Lanoue, E.; Day, R.; Feng, Y.; Fricker, L.D. Identification and characterization of three members of the human metallocarboxypeptidase gene family. *J. Biol. Chem.* **2002**, *277*, 14954–14964. [CrossRef]

35. Sun, L.; Wang, Y.; Yuan, H.; Burnett, J.; Pan, J.; Yang, Z.; Ran, Y.; Myers, I.; Sun, D. CPA4 is a Novel Diagnostic and Prognostic Marker for Human Non-Small-Cell Lung Cancer. *J. Cancer* **2016**, *7*, 1197–1204. [CrossRef]

36. Sun, L.; Guo, C.; Yuan, H.; Burnett, J.; Pan, J.; Yang, Z.; Ran, Y.; Myers, I.; Sun, D. Overexpression of carboxypeptidase A4 (CPA4) is associated with poor prognosis in patients with gastric cancer. *Am. J. Transl. Res.* **2016**, *8*, 5071–5075.

37. Sun, L.; Guo, C.; Burnett, J.; Yang, Z.; Ran, Y.; Sun, D. Serum carboxypeptidaseA4 levels predict liver metastasis in colorectal carcinoma. *Oncotarget* **2016**, *7*, 78688–78697. [CrossRef] [PubMed]

38. Handa, T.; Katayama, A.; Yokobori, T.; Yamane, A.; Fujii, T.; Obayashi, S.; Kurozumi, S.; Kawabata-Iwakawa, R.; Gombodorj, N.; Nishiyama, M.; et al. Carboxypeptidase A4 accumulation is associated with an aggressive phenotype and poor prognosis in triple-negative breast cancer. *Int. J. Oncol.* **2019**, *54*, 833–844. [CrossRef] [PubMed]

39. Sun, L.; Cao, J.; Guo, C.; Burnett, J.; Yang, Z.; Ran, Y.; Sun, D. Associations of carboxypeptidase 4 with ALDH1A1 expression and their prognostic value in esophageal squamous cell carcinoma. *Dis. Esophagus Off. J. Int. Soc. Dis. Esophagus* **2017**, *30*, 1–5. [CrossRef] [PubMed]

40. Abudurexiti, M.; Xie, H.; Jia, Z.; Zhu, Y.; Zhu, Y.; Shi, G.; Zhang, H.; Dai, B.; Wan, F.; Shen, Y.; et al. Development and External Validation of a Novel 12-Gene Signature for Prediction of Overall Survival in Muscle-Invasive Bladder Cancer. *Front. Oncol.* **2019**, *9*, 856. [CrossRef]

41. Wang, Y.; Xie, Y.; Niu, Y.; Song, P.; Liu, Y.; Burnett, J.; Yang, Z.; Sun, D.; Ran, Y.; Li, Y.; et al. Carboxypeptidase A4 negatively correlates with p53 expression and regulates the stemness of breast cancer cells. *Int. J. Med Sci.* **2021**, *18*, 1753–1759. [CrossRef] [PubMed]

42. Peng, H.; Qin, C.; Zhang, C.; Su, J.; Xiao, Q.; Xiao, Y.; Xiao, K.; Liu, Q. circCPA4 acts as a prognostic factor and regulates the proliferation and metastasis of glioma. *J. Cell. Mol. Med.* **2019**, *23*, 6658–6665. [CrossRef]

43. Wu, J.; Qu, J.; Cao, H.; Jing, C.; Wang, Z.; Xu, H.; Ma, R. Monoclonal antibody AC10364 inhibits cell proliferation in 5-fluorouracil resistant hepatocellular carcinoma via apoptotic pathways. *OncoTargets Ther.* **2019**, *12*, 5053–5067. [CrossRef]
44. Pallarès, I.; Bonet, R.; García-Castellanos, R.; Ventura, S.; Avilés, F.X.; Vendrell, J.; Gomis-Rüth, F.X. Structure of human carboxypeptidase A4 with its endogenous protein inhibitor, latexin. *Proc. Natl. Acad. Sci. USA* **2005**, *102*, 3978–3983. [CrossRef]
45. Shi, S.; Ye, S.; Mao, J.; Ru, Y.; Lu, Y.; Wu, X.; Xu, M.; Zhu, T.; Wang, Y.; Chen, Y.; et al. CMA1 is potent prognostic marker and associates with immune infiltration in gastric cancer. *Autoimmunity* **2020**, *53*, 210–217. [CrossRef] [PubMed]
46. Romagnani, S. Th1/Th2 cells. *Inflamm. Bowel Dis.* **1999**, *5*, 285–294. [CrossRef]
47. Sasidharan Nair, V.; Toor, S.M.; Ali, B.R.; Elkord, E. Dual inhibition of STAT1 and STAT3 activation downregulates expression of PD-L1 in human breast cancer cells. *Expert Opin. Ther. Targets* **2018**, *22*, 547–557. [CrossRef]
48. Wherry, E.J. T cell exhaustion. *Nat. Immunol.* **2011**, *12*, 492–499. [CrossRef]
49. Walker, L.S.; Sansom, D.M. The emerging role of CTLA4 as a cell-extrinsic regulator of T cell responses. *Nat. Rev. Immunol.* **2011**, *11*, 852–863. [CrossRef] [PubMed]

biology

MDPI

Article

The Identification of RNA Modification Gene *PUS7* as a Potential Biomarker of Ovarian Cancer

Huimin Li [1], Lin Chen [1], Yunsong Han [1], Fangfang Zhang [1], Yanyan Wang [1], Yali Han [1], Yange Wang [1], Qiang Wang [2,*] and Xiangqian Guo [1,*]

[1] Cell Signal Transduction Laboratory, Bioinformatics Center, Henan Provincial Engineering Center for Tumor Molecular Medicine, Institute of Biomedical Informatics, School of Basic Medical Sciences, Henan University, Kaifeng 475001, China; lihuimin0202@163.com (H.L.); cyy050067@163.com (L.C.); 18298352872@163.com (Y.H.); zff11211281010@163.com (F.Z.); w18237833015@163.com (Y.W.); hangreen5508@163.com (Y.H.); wangyange3@henu.edu.cn (Y.W.)
[2] School of Software, Henan University, Kaifeng 475001, China
* Correspondence: qiangwang@henu.edu.cn (Q.W.); xqguo@henu.edu.cn (X.G.)

Simple Summary: RNA modifications are involved in a variety of diseases, including cancers. Given the lack of efficient and reliable biomarkers for early diagnosis of ovarian cancer (OV), this study was designed to explore the role of RNA modification genes (RMGs) in the diagnosis of OV. The study first selected PUS7 (Pseudouridine Synthase 7) as a diagnostic biomarker candidate through the analysis of differentially expressed genes using TCGA and GEO data. Then, we evaluated its specificity and sensitivity using Receiver Operating Characteristic (ROC) analysis in TCGA and GEO data. The protein expression, mutation, protein interaction networks, correlated genes, related pathways, biological processes, cell components, and molecular functions were analyzed for PUS7 as well. The upregulation of PUS7 protein in OV was confirmed by the staining images in HPA and tissue arrays. In conclusion, the findings of the present study point towards the potential of PUS7 as the diagnostic marker and therapeutic target for ovarian cancer.

Citation: Li, H.; Chen, L.; Han, Y.; Zhang, F.; Wang, Y.; Han, Y.; Wang, Y.; Wang, Q.; Guo, X. The Identification of RNA Modification Gene *PUS7* as a Potential Biomarker of Ovarian Cancer. *Biology* **2021**, *10*, 1130. https://doi.org/10.3390/biology 10111130

Academic Editors: Shibiao Wan, Yiping Fan, Chunjie Jiang and Shengli Li

Received: 29 September 2021
Accepted: 30 October 2021
Published: 3 November 2021

Publisher's Note: MDPI stays neutral with regard to jurisdictional claims in published maps and institutional affiliations.

Abstract: RNA modifications are reversible, dynamically regulated, and involved in a variety of diseases such as cancers. Given the lack of efficient and reliable biomarkers for early diagnosis of ovarian cancer (OV), this study was designed to explore the role of RNA modification genes (RMGs) in the diagnosis of OV. Herein, 132 RMGs were retrieved in PubMed, 638 OV and 18 normal ovary samples were retrieved in The Cancer Genome Atlas (TCGA), and GSE18520 cohorts were collected for differential analysis. Finally, *PUS7* (Pseudouridine Synthase 7) as differentially expressed RMGs (DEGs-RMGs) was identified as a diagnostic biomarker candidate and evaluated for its specificity and sensitivity using Receiver Operating Characteristic (ROC) analysis in TCGA and GEO data. The protein expression, mutation, protein interaction networks, correlated genes, related pathways, biological processes, cell components, and molecular functions of *PUS7* were analyzed as well. The upregulation of PUS7 protein in OV was confirmed by the staining images in HPA and tissue arrays. Collectively, the findings of the present study point towards the potential of PUS7 as a diagnostic marker and therapeutic target for ovarian cancer.

Keywords: DEGs; diagnosis; ovarian cancer; PUS7; RMGs

1. Introduction

Ovarian cancer (OV) is the leading cause of death among gynecologic malignancies in most developed countries [1,2]. It accounts for an estimated 239,000 new cases and 152,000 deaths worldwide annually [3]. The risk of having ovarian cancer during the lifetime of a woman is approximately 1 in 78, and the lifetime chance of dying of ovarian cancer is approximately 1 in 108 [4]. Four out of five OV patients are diagnosed with advanced stage [5], and out of these, only 30% of patients survive more than 5 years [4].

The lack of a practical screening strategy and the asymptomatic characteristic of OV contribute to the late presentation of the disease. Hence, the efficient and early detection of OV is pivotal to improving the survival of ovarian cancer patients.

Post-transcriptional modifications affect RNA stability, localization, structure, splicing, or function [6]. Different RNAs have been detected to contain numerous types of modifications [7,8]. For example, mRNA modifications include N6-methyladenosine (m6A), inosine (I), 5-methylcytosine (m5C), and 5-hydroxymethylcytosine (hm5C). Deregulated RNA modifications are reported to be associated with several pathological processes such as tumorigenesis, cardiovascular diseases, and neurological disorders [9]. RNA modification enzymes have been generally considered important decorations for RNAs [10], and dysregulation and mutation in RNA modification genes are involved in the development of numerous cancers including lung cancer, bladder cancer, leukemia, prostate cancer, breast cancer, etc. [11]. For example, Alpha-Ketoglutarate Dependent Dioxygenase (*FTO*) was deciphered as a prognosticator for lung squamous cell carcinoma and promoted cell proliferation and invasion [12]. Methyltransferase Like 3 (*METTL3*), acting as an oncogene in lung cancer, upregulated *EGFR* and *TAZ* expression and promoted growth, survival, and invasion of human lung cancer cells [13]. NOP2/Sun RNA Methyltransferase 2 (*NSUN2*) was reported to be overexpressed in breast cancer and to be associated with cancer progression [14]. Elongator Acetyltransferase Complex Subunit 3 (*ELP3*), responsible for mcm5s2 modification, has been found to be upregulated in breast cancer and to facilitate cancer cell metastasis [15]. tRNA methyltransferase 9B (*TRM9L/TRMT9B*) has been shown to be downregulated in breast cancer [16]. Similarly, in renal cell carcinomas, G3BP Stress Granule Assembly Factor 1 (*G3BP1*) has been shown to promote tumor progression and metastasis [17]. Taken together, RNA modification genes play pivotal roles in human cancers.

Pseudouridine synthases (PUS) are divided into six families (TruA, TruB, TruD, RsuA, RluA, and Pus10) [18]. PUS7 is the only member of the TruD family that is involved in the modification of tRNAs, at position Tyr35 in pre-tRNA, at position 13 in cytoplasmic tRNA, and at numerous nucleotides in mRNAs. PUS7 is the only pseudouridine synthase to possess a consensus sequence (UGUAR) for substrate recognition [19]. *PUS7* was also reported to be associated with human myeloid malignancies in embryonic stem cells [20]. However, no reports have expounded the role of *PUS7* in OV, so far.

In this study, PUS7 was identified as a novel and potential biomarker for early diagnosis, using transcriptional profiles in the GEO and TCGA databases, ROC, HPA, and Oncomine analyses. Protein–protein interaction (PPI); GSEA pathway; and GO analyses, including the biological process (BP), cell component (CC), and molecular function (MF) terms, were also performed to provide in-depth insights into PUS7.

2. Materials and Methods

2.1. Data Collection

The RMGs were collected from PubMed according to the keywords "RNA modification". The transcriptome profiles, including datasets GSE18520 and TCGA, were obtained from GEO (https://www.ncbi.nlm.nih.gov/gds, accessed on 15 October 2019) [21] and UCSC Xena (https://xena.ucsc.edu/, accessed on 16 October 2019) [22], respectively. A total of 53 OV and 10 normal cases were enrolled in GSE18520 (platform: GPL570), and 585 OV and 8 normal cases in TCGA (Affymetrix Human Genome U133 Plus 2.0 Array) were adopted to carry out the following analyses.

2.2. Differential Expression Analysis

The GEO2R, an interactive web tool that facilitates users to compare the gene expression between different groups of samples in a GEO dataset, was used to identify the differentially expressed genes (DEGs). The SangerBox was adopted to analyze the TCGA expression profile of ovarian cancer. A p value < 0.05 and $|\log_2 FC| > 1$ were used as the cut-off criteria to screen out DEGs. The DEGs of the two datasets were listed in Supplementary Table S1. Subsequently, the RMGs and DEGs that overlapped between GSE18520 and TCGA were selected using Venny 2.1 and were used for further analysis. The analysis of the volcano plot of DEGs in GSE18520 and TCGA, and the heatmaps of DEGs-RMGs in GSE18520 and TCGA were obtained through the SangerBox web tool.

2.3. PUS7 Protein Level Analysis of OV Tissues in HPA and Tissue Array

The protein expression of PUS7 was analyzed using HPA data [23]. A tissue chip (HOvaC070PT01) was purchased from SHANGHAI OUTDO BIOTECH CO., LTD. A total of 12 OV samples and 2 healthy ovary samples, and 65 OV samples and 5 healthy ovary samples were retrieved from HPA and tissue array, respectively. The one case with an equivocal staining result was excluded, and the baseline characteristics of the remaining 64 cases of OV tissues in tissue array are described in Supplementary Table S2. The immunohistochemistry (IHC) staining intensity was graded from 0 to 3 (0, negative; 1, weak; 2, moderate; and 3, strong). The staining quantity was graded from 0 to 3 (0, none; 1, <25%; 2, 25–75%; and 3, >75%) according to the percentage of positive cells in the HPA database. The staining quantity was graded from 0 to 4 (0, none; 1, <25%; 2, 25–50%; 3, 50–75%; and 4, >75%) in the tissue assay. The staining scores were calculated by multiplying the staining intensity with the staining quantity.

2.4. PUS7 Gene Expression Analysis Using TCGA and GEO Datasets

The *PUS7* expression analysis was carried out using TCGA and GSE119056 expression profiling data. An ROC analysis (the method frequently used for binary assessment) was subsequently performed to evaluate the effectiveness of the expression level of any gene of interest in discriminating between OV and healthy samples. The area under the curve (AUC) value ranged from 0.5 to 1.0, which indicates 50 to 100% discrimination ability.

2.5. PUS7 Gene Expression Analysis Using Oncomine Database

The gene expression of *PUS7* was explored using the Oncomine database (https://www.oncomine.org/resource/main.html, accessed on 25 October 2019) [24]. The Oncomine database applies a combination of threshold values (p-value) and fold change (FC, tumors vs. controls) with $p \leq 0.05$ and fold change >1.

2.6. Protein–Protein Interaction (PPI) Network Analyses

STRING (https://stringdb.org/, accessed on 22 October 2019) [25] is a database used to predict and analyze functional interactions between proteins and was used to identify the functional protein–protein interactions (PPIs) of *PUS7*. GeneMANIA (http://genemania.org/, accessed on 24 October 2019) [26] was used to identify gene networks embracing *PUS7*.

2.7. The Mutation and Correlation Analyses of PUS7

The *PUS7* mutation was performed through cBioPortal (https://www.cbioportal.org/, 27 October 2019) [27]. The Gene Expression Profiling Interactive Analysis (GEPIA) database (http://gepia.cancer-pku.cn/, accessed on 27 October 2020) [28] was employed to analyze the PUS7 correlated genes based on TCGA data.

2.8. Pathways and BP, CC, and MF Analyses

Gene set enrichment analysis (GSEA) was carried out to identify potential cellular pathways involved with PUS7. The TCGA-OV dataset was divided into a high (25%) and a low group (75%) based on the PUS7 mRNA expression. Nominal p-value < 0.01 and false discovery rate (FDR) q-value < 0.05 were considered significant for enriched gene set analysis. Using 312 genes positively correlated (R > 0.3, p < 0.05) with *PUS7* derived from the cBioPortal analysis, the BP, CC, and MF analyses were carried out through the Database for Annotation, Visualization, and Integrated Discovery (DAVID, https:

//david.ncifcrf.gov/, 19 November 2020) [29] and visualized with bubble diagrams based on *p* values < 0.05.

2.9. Statistical Analysis

The statistical analyses were performed using SPSS ver. 26.0. The Student's *t*-test and the rank-sum test were used to evaluate the difference in PUS7 expression between the OV and normal samples. The ROC curve was constructed using *PUS7* expression profiles in the OV and normal samples by GraphPad Prism 8.0. A *p* value at < 0.05 was taken as a measure of statistically significant difference.

3. Results

3.1. The Identification of DEGs-RMGs of OV Data in TCGA and GEO

A total of 132 RMGs (Supplementary Table S3) were retrieved from PubMed. TCGA AffyU133a expression profiles and GSE18520 cohorts of ovarian cancer were downloaded from UCSC Xena and the GEO databases, respectively. A total of 1142 and 5215 DEGs (Supplementary Table S1) were obtained in the TCGA dataset and GSE18520 dataset between the OV and normal samples through DEO2R and SangerBox-limma analysis, respectively, and the volcano plots of DEGs are presented in Figure 1A,B. The RMGs and DEGs from the two cohorts were intersected to screen out the overlapping RMGs and DEGs for diagnostic biomarker analysis. As a result, two genes named *WDR77* and *PUS7* were identified as differentially expressed RMGs (Figure 1B). *WDR77* was excluded since it exhibited a contrary expression trend between OV and normal in TCGA and GSE18520 (Figure 2A,B). However, *PUS7* showed a consistent high expression in OV rather than normal cases; thus, *PUS7* could be a potential diagnostic biomarker and is subject to further analyses.

3.2. Expression Validation and Mutation Analysis for PUS7 in Ovarian Cancer

To validate the overexpression of *PUS7* in OV rather than normal samples, an Oncomine analysis was performed on ovarian cancer with different pathological types, and found that the *PUS7* expression is highly elevated in OV samples with fold change >1 and *p* < 0.05 (as presented in Figure 3A,B and Table 1). Moreover, Figure 3C,D displays the corresponding ROC curve of *PUS7* in the TCGA and GSE18520 datasets, indicating the remarkable potential of *PUS7* to discriminate OVs from normal tissue. The IHC analytic results showed the overexpression of PUS7 at the protein level (Figure 4A,B). To further explore the overexpression of PUS7 at the protein level in OV samples, a tissue array was performed. Typical staining images in the tissue array are exhibited in Figure 4C, confirming the protein upregulation of PUS7 in OV tissues (Figure 4D). Since mutations in RNA modification genes have been reported to be associated with several types of human cancers, the mutation analysis of PUS7 was performed in cBioPortal, demonstrating the fusion of PUS7 with SRSF Protein Kinase 2 (SRPK2) in serous ovarian cancer (Table 2).

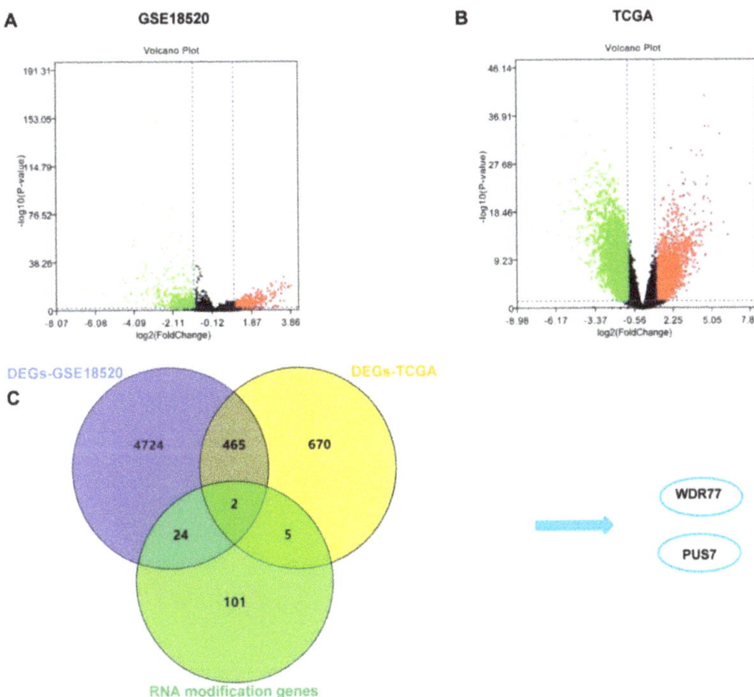

Figure 1. The identification of DEGs-RMGs using OV data in TCGA and GEO. (**A,B**) The volcano plot of DEGs between OV and normal samples in GSE18520 and TCGA data. (**C**) *WDR77* and *PUS7* were identified as the overlapping genes of DEGs in both datasets.

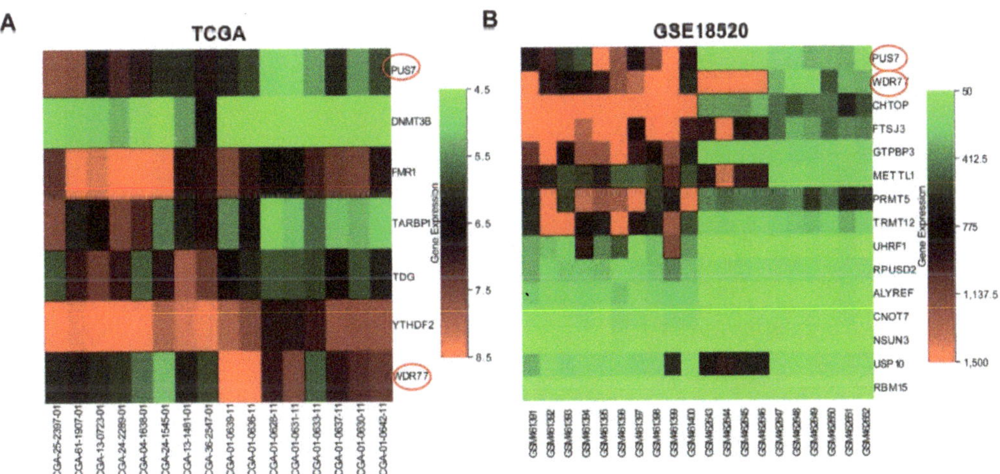

Figure 2. The heatmaps of differentially expressed RMGs. (**A**) The heatmap of the expression profile of overlapping genes of RMGs and DEGs in normal tissues and OV tissues in the TCGA dataset. (**B**) The heatmaps of the expression profile for overlapping genes of RMGs and DEGs in normal tissues and OV tissues in the GSE18520 dataset.

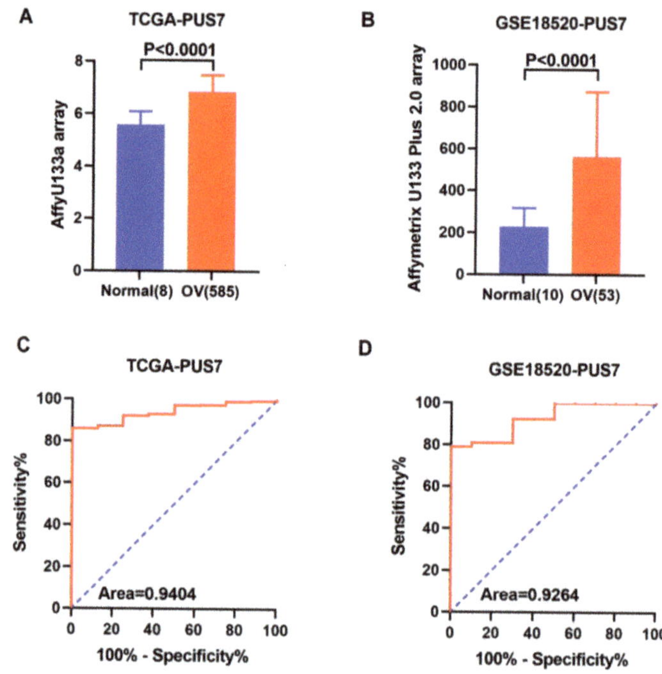

Figure 3. The differential expression analysis and ROC analysis of *PUS7* in OV and normal tissues. (**A**,**B**) The expression analysis of *PUS7* in TCGA and GSE18520 cohorts, respectively. (**C**,**D**) The ROC analysis of PUS7 between OV and normal samples in TCGA and GSE18520 cohorts. AUC is plotted as sensitivity% vs. 100-specificity%. A $p < 0.05$ was considered a significant difference.

Table 1. The comparison analysis of *PUS7* in ovarian cancer and normal tissue in different cohorts (Oncomine).

Dataset	Tumor (Cases)	Normal (Cases)	Fold Change	*t*-Test	*p*-Value
Lu Ovarian	Ovarian Serous Adenocarcinoma (20)	Ovarian Surface Epithelium (5)	1.913	9.134	2.28×10^{-9}
Lu Ovarian	Ovarian Endometrioid Adenocarcinoma (9)	Ovarian Surface Epithelium (5)	1.808	5.846	0.0000904
Lu Ovarian	Ovarian Mucinous Adenocarcinoma (9)	Ovarian Surface Epithelium (5)	1.405	4.275	0.000692
Lu Ovarian	Ovarian Clear Cell Adenocarcinoma (7)	Ovarian Surface Epithelium (5)	1.457	2.64	0.017
Hendrix Ovarian	Ovarian Mucinous Adenocarcinoma (13)	Ovary (4)	1.216	4.26	0.003
Hendrix Ovarian	Ovarian Clear Cell Adenocarcinoma (8)	Ovary (4)	1.275	4.44	0.000934
Hendrix Ovarian	Ovarian Endometrioid Adenocarcinoma (37)	Ovary (4)	1.299	6.012	0.00098
Hendrix Ovarian	Ovarian Serous Adenocarcinoma (37)	Ovary (4)	1.301	6.304	0.001
Yoshihara Ovarian	Ovarian Serous Adenocarcinoma (43)	Peritoneum (10)	1.537	3.171	0.003

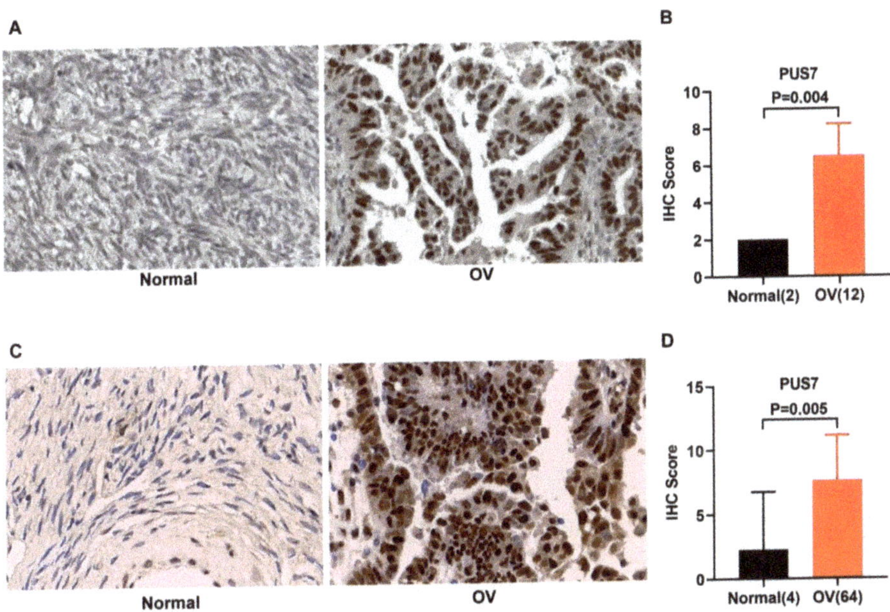

Figure 4. PUS7 protein expression was significantly higher in OV tissues than normal tissues. (**A**) Representative IHC images of PUS7 in normal (**left**) and OV (**right**) tissues in HPA. (**B**) Statistical analysis of the protein expression of PUS7 according to the staining scores of OV and normal tissues. (**C**) Representative IHC images of PUS7 in normal (**left**) and OV (**right**) tissues according to tissue microarray. (**D**) Statistical analysis of the protein expression of PUS7 according to the staining scores of OV and normal tissues. $p < 0.05$ was considered significant.

Table 2. The mutation distribution of PUS7 in ovarian cancer according to cBioPortal.

Cancer Type	Sample ID	Fusion Partner	Copy	Mutation in Sample
Serious Ovarian Cancer	TCGA-24-1469-01	Fusion, SRPK2-PUS7	ShallowDel	223
Serious Ovarian Cancer	TCGA-31-1953-01	Fusion, SRPK2-PUS7	Gain	56
Serious Ovarian Cancer	TCGA-61-1740-01	Fusion, SRPK2-PUS7	Gain	183

SRPK 2. SRSF Protein Kinase 2; PUS7: Pseudouridine Synthase 7.

3.3. The Interaction Network of PUS7

To explore the PPI and gene networks of PUS7 and its partner, an analysis using the String and GeneMANIA tools was performed. The PPI analysis results showed that a total of 10 proteins including NSUN2, NOP2, NOC3L, RBM28, BRIX1, TRUB1, WDR12, PUS1, DKC1, and NMD3 have interactions with PUS7 (Figure 5A). In the GeneMANIA analysis, a total of 20 genes named *ETFDH, WDR74, THUMPD1, NOC3L, CXXC4, DPYSL2, HSPA4L, RAD21, STAT3, MRPS2, HMBS, IDE, UBC, HSPH1, HDDC2, CMTR2, ATP6V0A1,* and *DRG1* were demonstrated to have physical interactions or genetic interactions or to share protein domains with *PUS7* or were co-expressed or co-located with *PUS7* (Figure 5B). The shared genes of the above two analyses are *PUS1* and *NOC3L* (Figure 5C), where *PUS1* was co-expressed with *PUS7* [30,31] and *NOC3L* physically interacted with *PUS7* [31–33] in GeneMANIA, both of which were known to interact with PUS7, according to the String results. In addition, GEPIA analysis showed that the expression of *PUS7* is significantly correlated with *PUS1* (R = 0.57, *p*-value = 0) and *NOC3L* (R = 0.61, *p*-value = 0) (Figure 5D).

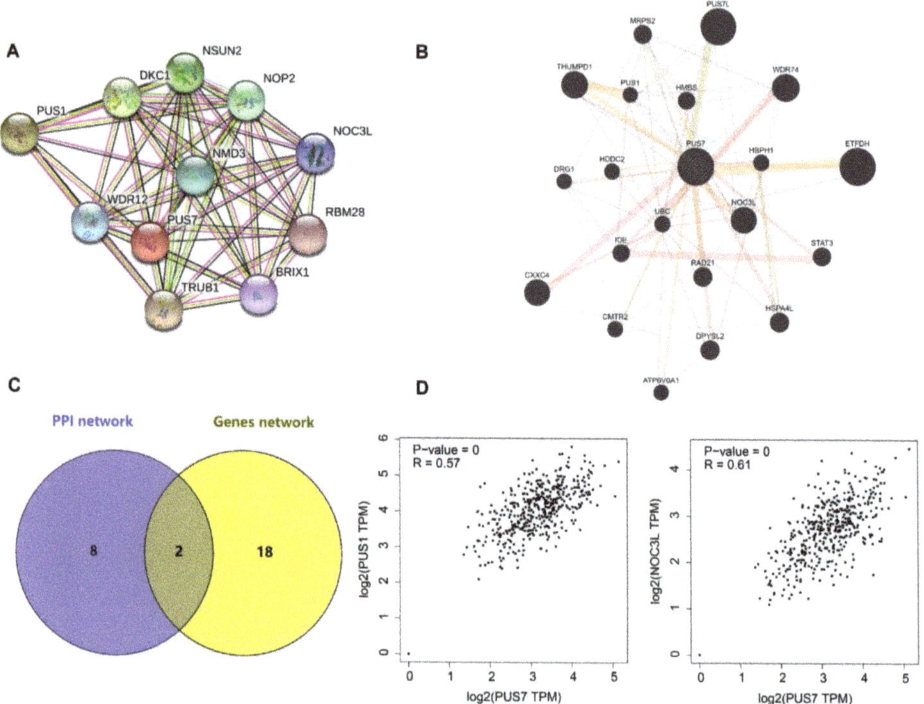

Figure 5. The interaction network and correlation of *PUS7*. (**A**) The protein interaction network of PUS7. Ten proteins including NSUN2, NOP2, NOC3L, RBM28, BRIX1, TRUB1, WDR12, PUS1, DKC1, and NMD3 physically/functionally interact with PUS7. (**B**) Twenty genes named *ETFDH, WDR74, THUMPD1, NOC3L, CXXC4, DPYSL2, HSPA4L, RAD21, STAT3, MRPS2, HMBS, IDE, UBC, HSPH1, HDDC2, CMTR2, ATP6V0A1,* and *DRG1* have physical interactions or genetic interactions, share protein domains with *PUS7*, or co-express or co-localize with *PUS7*. (**C**) Two genes were shared by the two networks. (**D**) The correlation analysis of *PUS7* with *PUS1* and *NOC3L*. R > 0.5 plus *p* < 0.05 was regarded as a significant correlation.

3.4. The Pathway Enrichment Analysis of PUS7 in Ovarian Cancer

To investigate the pathways that *PUS7* may be involved in or may regulate in ovarian cancer, a GSEA pathway analysis was performed using TCGA data, which was separated into a high (top 25%) *PUS7* group and a low (down 75%) *PUS7* group. The top eight pathways in which PUS7 participates are DNA replication, the cell cycle, mismatch repair, spliceosomes, homologous recombination, RNA polymerase, aminoacyl tRNA biosynthesis, and one carbon pool by folate in ovarian cancer (Figure 6). Among the eight pathways, the top two pathways are DNA replication and the cell cycle, both of which are linked to ovarian cancer cell proliferation. These results may imply that the overexpression of *PUS7* in ovarian cancer might promote ovarian cancer proliferation via regulation of DNA replication and the cell cycle.

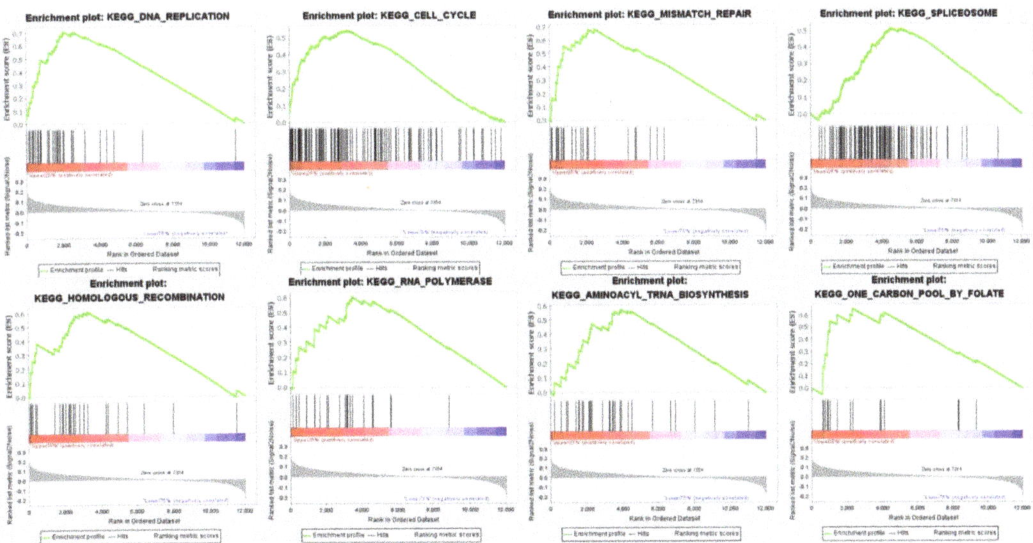

Figure 6. The pathway enrichment analysis of PUS7 in ovarian cancer. GSEA pathway analysis using TCGA ovarian cancer data, which was separated to a high (top25%) *PUS7* group and a low (down75%) *PUS7* group. Eight top pathways in which *PUS7* participates were DNA replication, the cell cycle, mismatch repair, spliceosomes, homologous recombination, RNA polymerase, aminoacyl, tRNA biosynthesis, and one carbon pool by folate in ovarian cancer.

3.5. Gene Ontology (GO) Analyses of PUS7 in Ovarian Cancer

To further clarify the GO terms of BP (biological processes), CC (cellular components) and MF (molecular functions) of *PUS7*, a total of 312 genes (Supplementary Table S4) positively related to *PUS7* (R > 0.3, p < 0.0001) according to the TCGA ovarian cancer data through the cBioPortal database were subjected to DAVID analysis. The results showed that biological processes in which *PUS7* mainly participates include the regulation of DNA templates and transcription, rRNA processing, tRNA export from nuclei, the regulation of glucose transport, the intracellular transport of viruses, mitotic nuclear envelope disassembly, viral processes, RNA processing, the regulation of cellular response to heat, gene silencing by RNA, and the positive regulation of gene expression (Figure 7A). The cellular components affected by *PUS7* include the nucleoplasm, nucleolus, nucleus, small subunit processsomes, nuclear envelope, and nuclear membrane (Figure 7B). The molecular functions of *PUS7* include poly(A) RNA binding, nucleic acid binding, helicase activity, ATP binding, ATP-dependent RNA helicase activity, structural constituents of a nuclear pore, DNA binding, RNA binding, protein binding, single-stranded DNA binding, nucleocytoplasmic transporter activity, DNA replication origin binding, ATP-dependent helicase activity, and nucleotide binding (Figure 7C).

5. Lindsey, A.T.; Britton, T.; Carol, E.D.; Kimberly, D.M.; Goli, S.; Carolyn, D.R.; Mia, M.G.; Ahmedin, J.; Rebecca, L.S. Ovarian Cancer Statistics, 2018. *CA Cancer J. Clin.* **2018**, *68*, 284–296.
6. Roundtree, I.A.; Evans, M.E.; Pan, T.; He, C. Dynamic RNA Modifications in Gene Expression Regulation. *Cell* **2017**, *169*, 1187–1200. [CrossRef]
7. Machnicka, M.A.; Olchowik, A.; Grosjean, H.; Bujnicki, J.M. Distribution and frequencies of post-transcriptional modifications in tRNAs. *RNA Biol.* **2014**, *11*, 1619–1629. [CrossRef]
8. Cantara, W.A.; Crain, P.F.; Rozenski, J.; McCloskey, J.A.; Harris, K.A.; Zhang, X.; Vendeix, F.A.P.; Fabris, D.; Agris, P.F. The RNA Modification Database, RNAMDB: 2011 update. *Nucleic Acids Res.* **2011**, *39*, D195–D201. [CrossRef]
9. Carell, T.; Brandmayr, C.; Hienzsch, A.; Muller, M.; Pearson, D.; Reiter, V.; Thoma, I.; Thumbs, P.; Wagner, M. Structure and function of noncanonical nucleobases. *Angew. Chem. Int. Ed. Engl.* **2012**, *51*, 7110–7131. [CrossRef]
10. Karijolich, J.; Yu, Y.T. Spliceosomal snRNA modifications and their function. *RNA Biol.* **2010**, *7*, 192–204. [CrossRef]
11. Nicky, J.; Julia, T.; Martin, A.S.; Nicole, S.; John, S.M.; Eva, M.N. The RNA modification landscape in human disease. *RNA* **2017**, *23*, 1754–1769.
12. Liu, J.; Ren, D.; Du, Z.; Wang, H.; Zhang, H.; Jin, Y. m6A demethylase FTO facilitates tumor progression in lung squamous cell carcinoma by regulating MZF1 expression. *Biochem. Biophys. Res. Commun.* **2018**, *502*, 456–464. [CrossRef]
13. Lin, S.; Junho, C.; Du, P.; Robinson, T.; Richard, I.G. The m6A methyltransferase METTL3 promotes translation in human Cancer cells. *Mol. Cell* **2016**, *62*, 335–345. [CrossRef]
14. Yi, J.; Gao, R.; Chen, Y.; Yang, Z.; Han, P.; Zhang, H.; Dou, Y.; Liu, W.; Wang, W.; Du, G.; et al. Overexpression of NSUN2 by DNA hypomethylation is associated with metastatic progression in human breast cancer. *Oncotarget* **2016**, *8*, 20751–20765. [CrossRef]
15. Sylvain, D.; Francesca, R.; Lars, T.; Zhou, Z.; Lukas, H.; Martin, T.; Kateryna, S.; Iva, K.; Alexandra, F.; Hadrien, D.; et al. Elp3 links tRNA modification to IRES-dependent translation of LEF1 to sustain metastasis in breast cancer. *J. Exp. Med.* **2016**, *213*, 2503–2523.
16. Begley, U.; Sosa, M.S.; Avivar-Valderas, A.; Patil, A.; Endres, L.; Estrada, Y.; Chan, C.T.; Su, D.; Dedon, P.C.; Aguirre-Ghiso, J.A.; et al. A human tRNA methyltransferase 9-like protein prevents tumour growth by regulating LIN9 and HIF1-α. *EMBO Mol. Med.* **2013**, *5*, 366–383. [CrossRef]
17. Wang, Y.; Fu, D.; Chen, Y.; Su, J.; Wang, Y.; Li, X.; Zhai, W.; Niu, Y.; Yue, D.; Geng, H. G3BP1 promotes tumor progression and metastasis through IL-6/G3BP1/STAT3 signaling axis in renal cell carcinomas. *Cell Death Dis.* **2018**, *9*, 501. [CrossRef]
18. Tomoko, H.; Adrian, R.F. Pseudouridine synthases. *Chem. Biol.* **2006**, *13*, 1125–1135.
19. Rintala-Dempsey, A.C.; Kothe, U. Eukaryotic stand-alone pseudouridine synthases-RNA modifying enzymes and emerging regulators of gene expression? *RNA Biol.* **2017**, *14*, 1185–1196. [CrossRef]
20. Guzzi, N.; Cieśla, M.; Ngoc, P.C.T.; Lang, S.; Arora, S.; Dimitriou, M.; Pimková, K.; Sommarin, M.N.E.; Munita, R.; Lubas, M.; et al. Pseudouridylation of tRNA-Derived Fragments Steers Translational Control in Stem Cells. *Cell* **2018**, *73*, 1204–1216.e25. [CrossRef]
21. Ron, E.; Michael, D.; Alex, E.L. Gene Expression Omnibus: NCBI gene expression and hybridization array data repository. *Nucleic Acids Res.* **2002**, *30*, 207–210.
22. Goldman, M.; Craft, B.; Swatloski, T.; Cline, M.; Morozova, O.; Diekhans, M.; David Haussler, D.; Zhu, J. The UCSC cancer genomics browser: Update 2015. *Nucleic Acids Res.* **2015**, *43*, D812–D817. [CrossRef]
23. Pontén, F.; Jirström, K.; Uhlen, M. The Human Protein Atlas-a tool for pathology. *J. Pathol.* **2008**, *216*, 387–393. [CrossRef]
24. Rhodes, D.R.; Yu, J.; Shanker, K.; Deshpande, N.; Varambally, R.; Ghosh, D.; Barrette, T.; Pandey, A.; Chinnaiyan, A.M. ONCOMINE: A cancer microarray database and integrated data-mining platform. *Neoplasia* **2004**, *6*, 1–6. [CrossRef]
25. Snel, B.; Lehmann, G.; Bork, P.; Huynen, M.A. STRING: A web-server to retrieve and display the repeatedly occurring neighbour-hood of a gene. *Nucleic Acids Res.* **2000**, *28*, 3442–3444. [CrossRef]
26. Sara, M.; Debajyoti, R.; David, W.-F.; Chris, G.; Quaid, M. GeneMANIA: A real-time multiple association network integration algorithm for predicting gene function. *Genome Biol.* **2008**, *9*, S4.
27. Cerami, E.; Gao, J.; Dogrusoz, U.; Gross, B.E.; Sumer, S.O.; Aksoy, B.A.; Jacobsen, A.; Byrne, C.J.; Heuer, M.L.; Larsson, E.; et al. The cBio cancer genomics portal: An open platform for exploring multidimensional cancer genomics data. *Cancer Discov.* **2012**, *2*, 401–404. [CrossRef]
28. Tang, Z.; Li, C.; Kang, B.; Gao, G.; Li, C.; Zhang, Z. GEPIA: A web server for cancer and normal gene expression profiling and interactive analyses. *Nucleic Acids Res.* **2017**, *45*, W98–W102. [CrossRef]
29. Huang, D.W.; Sherman, B.T.; Lempicki, R.A. Systematic and integrative analysis of large gene lists using DAVID bioinformatics resources. *Nat. Protoc.* **2009**, *4*, 44–57. [CrossRef]
30. Innocenti, F.; Cooper, G.M.; Stanaway, I.B.; Gamazon, E.R.; Smith, J.D.; Mirkov, S.; Ramirez, J.; Liu, W.; Lin, Y.S.; Moloney, C.; et al. Identification, replication, and functional fine-mapping of expression quantitative trait loci in primary human liver tissue. *PLoS Genet.* **2011**, *7*, e1002078. [CrossRef]
31. Denis, A.S.; Michael, M.; Eunice, S.; Richard, S.S.; Vivian, G.C. Genetic analysis of radiation-induced changes in human gene expression. *Nature* **2009**, *459*, 587–591.
32. Wang, I.X.; Ramrattan, G.; Cheung, V.G. Genetic variation in insulin-induced kinase signaling. *Mol. Syst. Biol.* **2015**, *11*, 820. [CrossRef] [PubMed]

33. Wan, C.; Borgeson, B.; Phanse, S.; Tu, F.; Drew, K.; Clark, G.; Xiong, X.; Kagan, O.; Kwan, J.; Bezginov, A.; et al. Panorama of ancient metazoan macromolecular complexes. *Nature* **2015**, *525*, 339–344. [CrossRef] [PubMed]
34. Siegel, R.L.; Miller, K.D.; Jemal, A. Cancer statistics. *CA Cancer J. Clin.* **2019**, *69*, 7–34. [CrossRef]
35. Chen, X.; Zhang, J.; Zhu, J. The role of m6A RNA methylation in human cancer. *Mol. Cancer* **2019**, *18*, 103. [CrossRef]
36. Liu, T.; Wei, Q.; Jin, J.; Luo, Q.; Liu, Y.; Yang, Y.; Cheng, C.; Li, L.; Pi, J.; Si, Y.; et al. The m6A reader YTHDF1 promotes ovarian cancer progression via augmenting EIF3C translation. *Nucleic Acids Res.* **2020**, *48*, 3816–3831. [CrossRef]
37. Shaheen, R.; Tasak, M.; Maddirevula, S.; Abdel-Salam, G.; Sayed, I.; Alazami, A.; Al-Sheddi, T.; Eman Alobeid, E.; Phizicky, E.; Alkuraya, F. PUS7 Mutations Impair Pseudouridylation in Humans and Cause Intellectual Disability and Microcephaly. *Hum. Genet.* **2019**, *138*, 231–239. [CrossRef]
38. Hu, G.; Wang, R.; Wei, B.; Wang, L.; Yang, Q.; Kong, D.; Du, C. Prognostic Markers Identification in Glioma by Gene Expression Profile Analysis. *J. Comput. Biol.* **2020**, *27*, 81–90. [CrossRef]
39. Obuchowski, N.A.; Bullen, J.A. Receiver operating characteristic (ROC) curves: Review of methods with applications in diagnostic medicine. *Phys. Med. Biol.* **2018**, *63*, 07tr01. [CrossRef]
40. Zhe, H.; Jane, C.; Dawn, T. What is an ROC curve? *Emerg. Med. J.* **2017**, *34*, 357–359.
41. Yan, C.; Zhu, M.; Ding, Y.; Yang, M.; Wang, M.; Li, G.; Ren, C.; Huang, T.; Yang, W.; He, B.; et al. Meta-analysis of genome-wide association studies and functional assays decipher susceptibility genes for gastric cancer in Chinese populations. *Gut* **2020**, *69*, 641–651. [CrossRef]
42. Yuan, L.; Jin, T.; Yin, J.; Du, X.; Wang, Q.; Dong, R.; Wang, S.; Cui, Y.; Chen, C.; Lu, G. Polymorphisms of tumor-related genes IL-10, PSCA, MTRR and NOC3L are associated with the risk of gastric cancer in the Chinese Han population. *Cancer Epidemiol.* **2012**, *36*, e366–e372. [CrossRef]
43. Tesarova, M.; Vondrackova, A.; Stufkova, H.; Veprekova, L.; Stranecky, V.; Berankova, K.; Hansikova, H.; Magner, M.; Galoova, N.; Honzik, T.; et al. Sideroblastic anemia associated with multisystem mitochondrial disorders. *Pediatr. Blood Cancer* **2019**, *66*, e27591. [CrossRef]

Article

R-Score: A New Parameter to Assess the Quality of Variants' Calls Assessed by NGS Using Liquid Biopsies

Roberto Serna-Blasco [1], Estela Sánchez-Herrero [1,2], María Berrocal Renedo [1], Silvia Calabuig-Fariñas [3,4,5], Miguel Ángel Molina-Vila [6], Mariano Provencio [7] and Atocha Romero [1,7,*]

[1] Liquid Biopsy Laboratory, University Hospital Puerta de Hierro, 28222 Madrid, Spain; rserna@idiphim.org (R.S.-B.); esanchez@idiphim.org (E.S.-H.); mariaberrocal18@gmail.com (M.B.R.)

[2] Atrys Health, I+D Department, 08025 Barcelona, Spain

[3] CIBERONC, Liquid Biopsy WM, 28029 Madrid, Spain; calabuix_sil@gva.es

[4] Mixed Unit TRIAL, Príncipe Felipe Research Center & General University Hospital of Valencia Research Foundation, 46012 Valencia, Spain

[5] Department of Pathology, Universitat de València, 46010 Valencia, Spain

[6] Laboratory of Oncology/Pangaea Oncology, Quirón-Dexeus University Hospital, 08028 Barcelona, Spain; mamolina@panoncology.com

[7] Medical Oncology, University Hospital Puerta de Hierro, 28222 Madrid, Spain; mariano.provencio@salud.madrid.org

* Correspondence: aromero@idiphim.org; Tel.: +34-1917769

Citation: Serna-Blasco, R.; Sánchez-Herrero, E.; Berrocal Renedo, M.; Calabuig-Fariñas, S.; Molina-Vila, M.Á.; Provencio, M.; Romero, A. R-Score: A New Parameter to Assess the Quality of Variants' Calls Assessed by NGS Using Liquid Biopsies. *Biology* **2021**, *10*, 954. https://doi.org/10.3390/biology10100954

Academic Editors: Shibiao Wan, Yiping Fan, Chunjie Jiang, Shengli Li and Dirk Rades

Received: 3 August 2021
Accepted: 4 September 2021
Published: 24 September 2021

Publisher's Note: MDPI stays neutral with regard to jurisdictional claims in published maps and institutional affiliations.

Simple Summary: Circulating tumor DNA profiling by next-generation sequencing (NGS) is becoming essential for guiding targeted therapies. However, it remains challenging. Here, we show that variant allele fraction and the median of absolute values of all pairwise differences impact the agreement between digital PCR and NGS calls. Therefore, we propose a new parameter, named R-score, which integrates both variables, and we evaluate its usefulness for optimizing NGS variant calling.

Abstract: Next-generation sequencing (NGS) has enabled a deeper knowledge of the molecular landscape in non-small cell lung cancer (NSCLC), identifying a growing number of targetable molecular alterations in key genes. However, NGS profiling of liquid biopsies risk for false positive and false negative calls and parameters assessing the quality of NGS calls remains lacking. In this study, we have evaluated the positive percent agreement (PPA) between NGS and digital PCR calls when assessing *EGFR* mutation status using 85 plasma samples from 82 *EGFR*-positive NSCLC patients. According to our data, variant allele fraction (VAF) was significantly lower in discordant calls and the median of the absolute values of all pairwise differences (MAPD) was significantly higher in discordant calls ($p < 0.001$ in both cases). Based on these results, we propose a new parameter that integrates both variables, named R-score. Next, we sought to evaluate the PPA for *EGFR* mutation calls between two independent NGS platforms using a subset of 40 samples from the same cohort. Remarkably, there was a significant linear correlation between the PPA and the R-score ($r = 0.97$; $p < 0.001$). Specifically, the PPA of samples with an R-score ≤ -1.25 was 95.83%, whereas PPA falls to 81.63% in samples with R-score ≤ 0.25. In conclusion, R-score significantly correlates with PPA and can assist laboratory medicine specialists and data scientists to select reliable variants detected by NGS.

Keywords: NGS; ctDNA; VAF; liquid biopsy; filtering; variant calling

1. Introduction

The analysis of circulating tumor DNA (ctDNA) has become an attractive approach for non-invasive biomarker testing as well as for real-time monitoring of cancer patients; its usefulness is especially remarkable in lung cancer patients [1–4]. These tumors are mostly diagnosed at advanced stages, in elderly patients with a median age at diagnosis of approximately 65 years [5], and they are difficult to access owing to their anatomical location, which makes it sometimes difficult to obtain sufficient material for molecular analysis [6]. Moreover, in the last decades, there has been a major paradigm shift in the management of metastatic non-small cell lung cancer (NSCLC), with the advent of targeted therapies for patients harbouring druggable alterations such as *EGFR* or *BRAF* mutations, as well as *ALK*, *ROS*, and *RET* rearrangements, and so on [7]. Furthermore, novel *KRAS* inhibitors constitute a promising therapeutic approach for advanced NSCLC patients [8,9]. Specifically, 30% of NSCLC tumours harbour activating mutations in the *EGFR* gene, which identify patient candidates to receive tyrosine kinase inhibitors (TKIs) [10]. For this subset of patients, liquid biopsy has been proven to be extremely useful, saving time in the process of diagnosis. In this way, guidelines recommend testing for the T790M *EGFR* mutation in the blood after progression to an *EGFR* TKI as a first choice, and re-biopsies are suggested in the case of a negative result in order to identify patients that can benefit from osimertinib (a third-generation TKI) [11]. Moreover, ctDNA plasma levels have been shown to be of prognostic significance for these patients, and monitoring *EGFR* mutation levels in the plasmas has been proven useful for response to treatment monitoring [5,12,13].

Next-generation sequencing (NGS) enables simultaneous detection of multiple alterations in a single test. Incorporation of unique molecular identifiers (UMIs), random nucleotide sequences assigned to each DNA fragment prior to PCR amplification during library preparation, enables the detection, quantification, and sequencing of unique DNA fragments with high-resolution, allowing the identification and removal of amplification artifacts arising from library preparation and the reduction of the limit of detection (LOD) [14,15]. Nonetheless, ctDNA is present at very low levels in the plasma and its profiling is still challenging with working conditions sometimes close to the edge of this technology. Therefore, there is a need to develop new parameters assessing the quality of the reads in order to avoid false positive and false negative calls.

Here, we assess the impact of two parameters, namely, variant allele fraction (VAF) and median of the absolute values of all pairwise differences (MAPD), separately and together on variant calls when using the Oncomine Pan-Cancer Cell-Free Assay™ (ThermoFisher Scientific®, Palo Alto, CA, USA) by evaluating the agreement between digital PCR (dPCR) and NGS for the assessment of *EGFR* mutation status. Based on our data, we propose a new parameter named R-score and, finally, we evaluate the agreement in NGS calls between two independent NGS methods according to R-score.

2. Materials and Methods

2.1. Patients and Samples

A total of 85 samples from advanced *EGFR*-positive NSCLC patients were recruited upon disease progression to a first-line with a TKI, between February 2016 and March 2019. The study was approved by the Hospital Puerta de Hierro Ethics Committee. All patients provided the appropriate written informed consent to participate in the study prior to enrolment. Briefly, eligible patients were both male and female, age >18 years, with a pathologically confirmed diagnosis of stage IV NSCLC harbouring an *EGFR* mutation. Blood samples were collected in 8.5 mL PPT™ tubes (Becton Dickinson, Franklin Lakes, NJ, USA).

2.2. Laboratory Procedures

Two independent laboratories were involved in this study: laboratory 1 (L1) and laboratory 2 (L2). Samples for which we did not have available at least 8 mL of plasma (N = 45) were processed only by L1 exclusively, and they were used to test the agreement between dPCR and NGS exclusively. For 40 plasma samples, we had available at least 8 mL of plasma, and samples were divided into two aliquots, which were then distributed to L1 and L2. L1 carried out dPCR assays and NGS analysis using the Oncomine Pan-Cancer Cell-Free Assay and an Ion S5 sequencer (ThermoFisher Scientific®, Palo Alto, CA, USA), whereas L2 carried out NGS with QIAact Lung DNA UMI Panel using the GeneRead Platform (QIAgen, Valencia, CA, USA).

Isolation of plasma was achieved by two consecutive centrifugations at room temperature, the first one at $1500 \times g$ for 10 min and the second at $5000 \times g$ for 20 min. cfDNA was extracted with the QIAamp Circulating Nucleic Acid Kit (QIAgen, Valencia, CA, USA) according to the manufacturer's protocol (QIAamp Circulating Nucelic Acid Handbook 10/2013). DNA concentration was measured by Qubit 2.0 Fluorometer with Qubit 1X dsDNA HS Assay Kit (ThermoFisher Scientific®, Palo Alto, CA, USA) and fragment length and sample quality were evaluated using the Agilent High Sensitivity DNA Kit using Agilent 2100 Bioanalyzer (Agilent Technologies, Santa Clara, CA, USA). Supplementary Figure S1 shows the observed size of the cfDNA fragments, which was approximately 180 bp. cfDNA was stored at -80 °C until further analysis.

In order to detect somatic mutation in the *EGFR* gene, dPCRs were performed using predesigned TaqMan® dPCR assays in a QuantStudio® 3D Digital PCR (Applied Biosystems®, South San Francisco, CA, USA). dPCR reaction was carried out in a final volume of 18 µL; this reaction included 8.55 µL of template cfDNA, 9 µL of 20X QuantStudio® Master Mix, and 0.45 µL 40X TaqMan assay. Subsequently, 14.5 µL of final reaction volume was loaded to QuantStudio® 3D digital PCR 20K chip. The thermal cycler conditions were as follows: initial denaturalization at 96 °C for 10 min, 40 cycles at 56 °C for 2 min, 98 °C for 30 s, 72 °C for 10 min, and finally samples were maintained at 22 °C for at least 30 min. Chips were read using QuantStudio® 3D Digital PCR instrument. The results were analysed with QuantStudio® 3D AnalysisSuite™ Cloud. Default call assignments for each data cluster were manually adjusted when needed. A positive and a negative control were included in every run. The LOD and limit of quantitation of the dPCR TaqMan® assays were estimated based on the standard deviation of the response and the slope according to the recommendations of The International Council for Harmonisation of Technical Requirements for Pharmaceuticals for Human Use; ICH Q2 (R1) guidelines (validation of analytical procedures: text and methodology), and they have been published elsewhere [13]. The sensitivity and specificity of the assays, considering tissue genotyping to be the gold standard, have also been reported [16].

The presence of *EGFR* mutations was evaluated in parallel by two independent NGS platforms, Ion S5™ XL and GeneReader™, and using two different gene panels, Oncomine™ Pan-Cancer Cell-Free Assay (ThermoFisher Scientific®, Palo Alto, CA, USA) and the QIAact Lung DNA UMI Panel (QIAgen, Valencia, CA, USA), respectively. The comparison was performed using 40 samples.

For NGS analysis using the Oncomine Pan-Cancer Cell-Free Assay (NGS-Oncomine), library preparation was performed with a minimum input of 10 ng of cfDNA according to manufacturer's instructions. The final pool was loaded in an Ion 550™ Chip using Ion Chef™ Instrument (ThermoFisher Scientific®, Palo Alto, CA, USA). Finally, loaded chips were sequenced on an Ion GeneStudio™ S5 Sequencer (ThermoFisher Scientific®, Palo Alto, CA, USA). Torrent Suite Software (v5.12) was used to perform raw sequencing data analysis. The CoverageAnalysis (v. 5.12.0.0) plugin was used for sequencing coverage analysis (ThermoFisher Scientific®, Palo Alto, CA, USA). As recommended by the manufacturer, a median read coverage >25,000 and median molecular coverage >2500 were required to detect a variant with a VAF of 0.1%. Raw reads were aligned to the human reference genome hg19. Variant calling, annotation, and filtering were performed on the Ion Reporter (v5.10) platform using the OncomineTagSeq Pan-Cancer Liquid Biopsy workflow (v2.1). Briefly, sequencing reads were mapped to defined target regions (Oncomine Pan-Cancer DNA Regions v1.0 (5.10)) and subjected to variant calling using Oncomine Pan-Cancer Annotations v1r.0.

For NGS analysis using the QIAact Lung DNA UMI Panel (NGS-GeneReader), libraries were performed with an input of 16.75 µL and ~10–70 ng of purified cfDNA, according to manufacturer's instructions. Then, libraries were quantified using a QIAxcel Advanced System and Qubit dsDNA HS Assay kit in order to pool in batches of six samples.

GeneRead Clonal Amp Q Kit was used to clonal amplification of pooled libraries. After bead enrichment, pooled libraries were sequenced using the GeneRead UMI Advanced Sequencing Q kit in a GeneReader instrument. Finally, FASTQ files alignment was performed using hg19 as reference genome, and variant calling and report generation of sequencing results were performed by QIAGEN Clinical Insight Analyze software.

2.3. Parameters

VAF was defined as the number of mutant molecules at a specific nucleotide location relative to the sum of total DNA molecules (mutant + wild type). VAF was provided for each detected mutation after dPCR and NGS analysis. In dPCR analysis, VAF was calculated, following the next equation, by QuantStudio® 3D AnalysisSuite™ Cloud:

$$VAF = (FAMcopies/\mu L)/(FAMcopies/\mu L + VICcopies/\mu L) \times 100 \tag{1}$$

where FAM copies = number of reads of mutated sequences and VIC copies = number of reads of wild-type sequences.

In the case of NGS-Oncomine, VAF was calculated, using the CoverageAnalysis (v. 5.12.0.0) plugin. Likewise, using NGS-GeneReader, VAF was calculated with QIAGEN Clinical Insight Analyze software in the same way as NGS-Oncomine.

NGS-Oncomine platform also provides a quality sequencing parameter, MAPD, as a pair is defined as adjacent amplicons in terms of genomic distances. Assuming that adjacent amplicons in the genome most likely have the same underlying copy number in a sample, the difference between the log2c(read count ratio) values against the reference baseline for all adjacent amplicons contains information for the noise level of the data. The MAPD is an estimation of coverage variability between adjacent amplicons. The default threshold is 0.5 [17]. As a result, sample results with an MAPD above this value should be reviewed with caution

$$MAPD = median(\,|\,x_{i+1} - x_i\,|\,) \tag{2}$$

where x_i = log2 ratio for marker i.

2.4. Statistical Analysis

The primary objective was to evaluate the impact of VAF and MAPD parameters, separately and together, firstly on the positive percent agreement (PPA) between dPCR and NGS (NGS-Oncomine) and secondly on the PPA between two independent NGS platforms (NGS-Oncomine and NGS-GeneReader).

Each mutation was treated as a separate measurement for statistical analysis; therefore, 137 measurements were used in this study.

The correlation between VAFs measured by dPCR and NGS was assessed with simple linear regression analysis, using the concordance correlation coefficient (p) and Spearman's coefficient (r). For comparisons between numerical variables, Mann–Whitney U test was used. Comparisons between categorical variables were made using Fisher's exact test or chi-squared test, whichever was most appropriate.

To describe how often NGS and dPCR methods agreed on *EGFR* calls, as well as concordance between the two different NGS platforms, we calculated the PPA.

The threshold of $p < 0.05$ was considered as statistically significant. Statistical software used was Stata v16.0 (StataCorp 2019. Stata Statistical Software Release 16. College station, TX: StataCorp LLC) and R version 3.6.3. (R core team 2020. The R Foundation for Statistical Computing Platform, Vienna, Austria) URL https://www.R-project.org/ (last accessed on 26 July 2021).

3. Results

3.1. EGFR Mutation Detection by dPCR and NGS-Oncomine

EGFR mutation status was evaluated in 85 plasma samples from 82 *EGFR*-positive NSCLC patients in parallel by dPCR and NGS-Oncomine. All samples used in this study had detectable *EGFR* driver mutations by dPCR. The mutation detected by dPCR was always concordant with the *EGFR* mutation detected on the pre-treatment tissue sample as reported by pathologists. Among the total number of detected *EGFR* mutations (N = 137), 62% were activating mutations, among which the most common mutations were exon 19 deletions (55.3%) or L858R (36.5%). The rest of the *EGFR* driver mutations were L861Q (3.5%), G719A (2.3%), S768I (1.2%), and exon 20 insertions (1.2%). Regarding T790M resistance mutation, 61.2% of samples were identified as T790M positive by dPCR. Data for T790M status in tissue samples were not available. Of note, 42 (30.6%) mutations detected by dPCR were not found using NGS-Oncomine. When analysing mutations separately, a lower PPA was measured in L858R mutation (67.74%; 95%CI 50.31–85.17) compared with exon 19 deletion (76.60%; 95%CI 64.03–89.16). Less common *EGFR* mutations such as L861Q, S768I, and G719A were detected by both methods. It should be noted that the exon 20 insertion (c.2310_2311insGGT; p.D770_N771insG) was not found by NGS-Oncomine.

Finally, regarding T790M resistance mutation, 52 (61.2%) samples were identified as T790M positive by dPCR, whereas only 28 (33%) samples were T790M positive using NGS-Oncomine (53.85% of agreement; 95% CI 39.83–67.86).

3.2. VAF and MAPD Involvement in the Agreement between dPCR and NGS-Oncomine Calls

Overall, there were 91 concordant calls by both technologies and 46 discordant calls with a PPA of 66.42% (95% CI 58.42–74.43).

First, we evaluated the overall correlation between VAF values assessed by dPCR and NGS-Oncomine when the mutation was detected by both methods. According to our data, VAFs measured by NGS-Oncomine were significantly correlated to VAFs assessed by dPCR ($r = 0.89$; $p < 0.001$) (Figure S2). Next, VAFs values and MAPD scores were compared between discordant and concordant calls. Overall, dPCR VAFs values were significantly lower in discordant calls compared with concordant calls ($p < 0.001$) (Figure 1A). Specifically, 1.1% and 10.9% of concordant calls have VAF $\leq 0.1\%$ and $\leq 0.5\%$, respectively, compared with 8.9% and 46.7% in discordant calls. Likewise, MAPD score was significantly higher in discordant samples compared with concordant samples ($p < 0.001$) (Figure 1B).

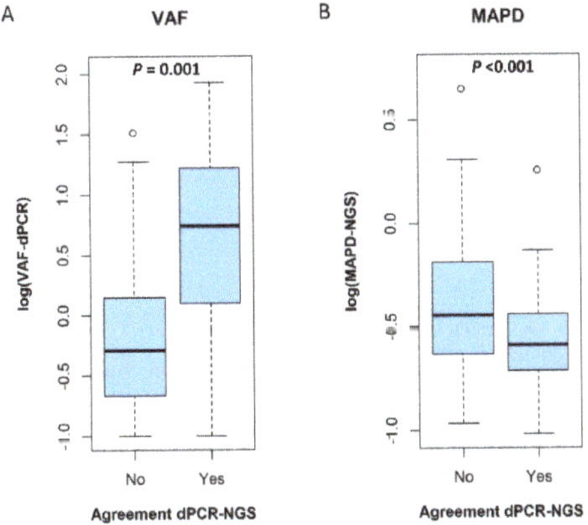

Figure 1. Boxplot. (**A**) VAF values in concordant and discordant calls (dPCR-NGS-Oncomine) in logarithmic scale. (**B**) MAPD values in concordant and discordant calls (dPCR-NGS-Oncomine) in logarithmic scale.

Next, we sought to evaluate the combined effect of VAF and MAPD parameters. Dot plots in Figure 2 show the concordance between dPCR and NGS-Oncomine on variant calls according to VAF and MAPD parameters. Discordant calls are coloured in red and concordant calls are coloured in blue. Figure 2A is divided into four quadrants using as cut-offs the logarithmic median values of VAF and MAPD according to our data set.

As shown, the highest PPA (96.9%; 95%CI: 83.8–99.9%) was observed in the lower-right quadrant. Conversely, the PPA descended as much as 27.6% (95%CI: 12.7–47.2%) for calls clustered in the upper-left quadrant, meaning that, the higher the VAF and the lower the MAPD, the higher the PPA. Similar results were obtained when quadrants were divided using thresholds according to technical specifications for each parameter (Figure 2B). As illustrated, PPA between NGS and dPCR calls was 0% (95% CI: 0–60.2%) when using a cut off of ≤-0.301 for VAF and >-0.301 for MAPD, whereas in the opposite conditions, the PPA increased to 84.9% (95% CI: 74.5–90.9%).

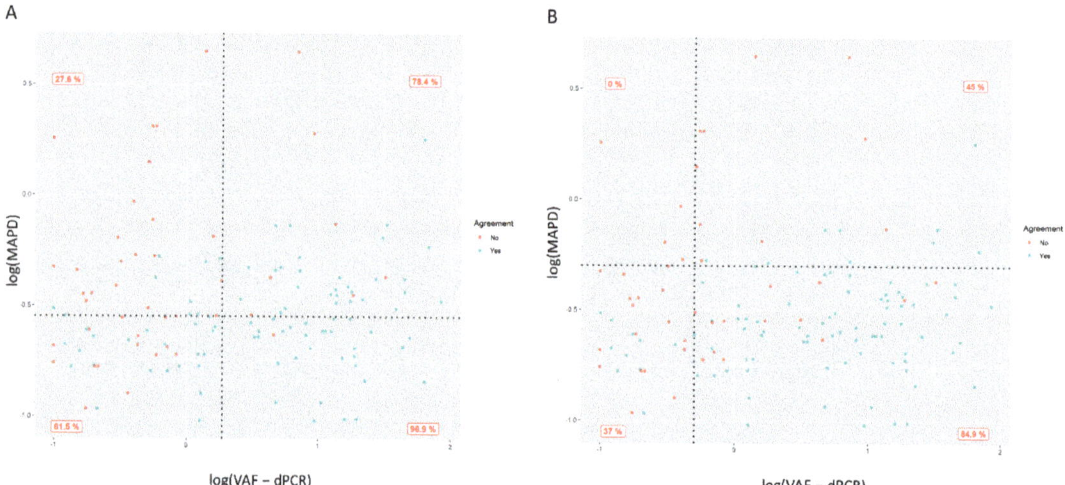

Figure 2. Dot plot showing the agreement in variant calls between dPCR—NGS-Oncomine, according to VAF and MAPD. VAF and MAPD values, both in logarithmic scale, are represented in the x and y-axis, respectively. Concordant calls are coloured in blue, while discordant calls are coloured in red. PPA for calls clustered in each quadrant is shown. (**A**) Dot plot divided into four quadrants using as cut-off the logarithmic median values of VAF and MAPD. In this way, the median VAF in our data set was 1.87, which is 0.272 on logarithmic scale, and the median MAPD was 0.28, which corresponds to −0.553 on logarithmic scale. (**B**) Dot plot divided into four quadrants according to technical specifications. MAPD threshold was selected following Ion Reporter recommendations [17]. According to the manufacturer, a value of MAPD above 0.5 is considered too high. Samples with high MAPD values have low coverage uniformity, which can result in missed or erroneous variant calls. The VAF threshold was chosen based on results from previous studies [18]. Therefore, both axes were divided using −0.301 value for VAF and MAPD (log(0.5)).

3.3. R-Score Is a Useful Parameter to Select Reliable Variant Calls

Based on previous observations, we proposed a new parameter, named R-score, which is defined as follows:

$$R\text{-score} = \log(MAPD/VAF) \tag{3}$$

In order to evaluate the utility of R-score for assessing the quality of an *EGFR* variant call, we evaluated the PPA between NGS-Oncomine and dPCR and NGS-Oncomine and NGS-GeneReader according to R-score.

First, we assessed the correlation between VAF values from NGS-Oncomine and NGS-GeneReader when the mutation was detected by both methods. According to our data, VAFs values from NGS-Oncomine significantly correlated with VAFs from NGS-GeneReader ($r = 0.80$; $p < 0.001$).

R-score was then calculated for each variant detected by NGS-Oncomine using the VAF and MAPD provided by the corresponding analysis software. MAPD and R-score values were significantly higher in discordant calls between dPCR and NGS-Oncomine compared with concordant calls ($p < 0.001$) (Figure 3A and Table S1). Conversely, VAF values were significantly lower in discordant calls between dPCR and NGS-Oncomine (Table S1). Subsequently, the PPA for *EGFR* variant calling between both NGS platforms was evaluated using different arbitrary R-score cut-offs (-1.25, -1, -0.75, -0.5, -0.25, 0, and 0.25). As shown in Figure 3C, there was a clear linear correlation between the PPA and the R-score ($r = 0.97$; $p < 0.001$). Of note, the PPA of samples with an R-score ≤ -1.25 was 95.83%, whereas PPA falls to 81.63% in samples with an R-score ≤ 0.25 (Figure 3B). A complete list of all mutations detected according to the NGS platform is available in Table S2.

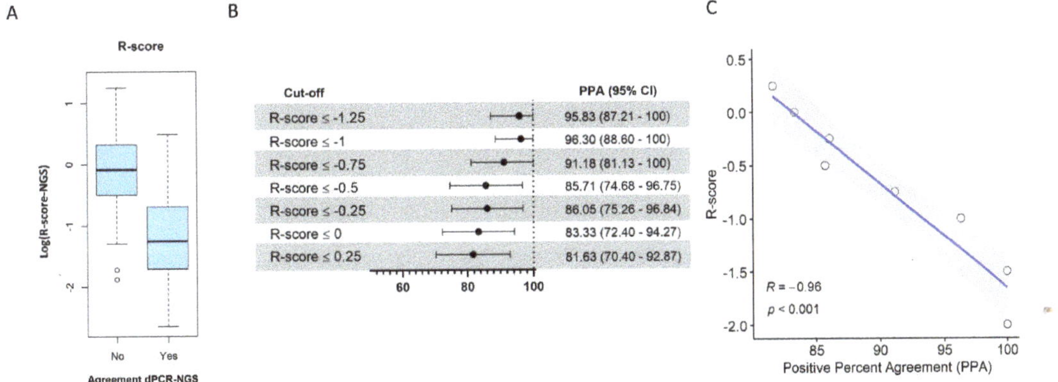

Figure 3. (**A**) R-score values in logarithmic scale in concordant and discordant samples (dPCR—NGS-Oncomine). R-score was calculated for each variant detected with Oncomine panel, using VAF and MAPD values from Ion Reporter (v5.10) analysis software. (**B**) Positive percent agreement (PPA) with corresponding 95% confidence interval (95%CI) between NGS-Oncomine and NGS-GeneReader according to R-score cut-off. The following arbitrary cut-offs for R-score were established: -1.25, -1, -0.75, -0.5, -0.25, 0, and 0.25, and PPA between both NGS platforms was estimated. (**C**) Correlation between PPA values and R-score cut-off values. As shown, there was a linear correlation between PPA and the R-score cut-off values; the lower the R-score, the greater the PPA. Abbreviations: R = Spearman correlation coefficient.

4. Discussion

Biomarker testing in NSCLC has been demonstrated to improve survival outcomes [19–21]. Of note, the number of biomarkers that need to be tested is constantly increasing in NSCLC as new targeted therapies are becoming available [7]. Unlike PCR-based platforms, which only allow a few mutations to be analyzed, NGS enables for interrogating multiple genomic alterations simultaneously in a single test. Indeed, National Comprehensive Cancer Network guidelines recommend that, when feasible, biomarker testing should be performed via a broad, panel-based approach by NGS [2]. However, NGS profiling of liquid biopsies, although feasible [22,23], remains challenging. On one hand, the sensitivity of the assays remains a major limitation [24], and approaches aimed to increase sensitivity might risk false positive calls. Moreover, it has been reported that tumor mutational burden (TMB) analysis, which has been proposed as a predictive biomarker for the identification of patients most likely to respond to immune checkpoint inhibitors, through liquid biopsies, is feasible [25]. TMB is optimally assessed by whole-exome sequencing (WES) [26], but targeted panels provide a time-effective and cost-effective alternative [27]. Nevertheless, TMB analysis requires sequencing over 0.5 Mb [28,29]. In this scenario, it is of particular interest to reduce as much as possible the risk of false-positive and false-negative calls. Thus, new parameters evaluating the quality of NGS calls are needed. A recent comprehensive study, in which several methodologies for the analysis of circulating tumor DNA were compared, revealed that the agreement between platforms significantly improved when discarding samples with VAF ≤ 0.5% [16]. Likewise, a study comparing BEAMing and droplet dPCR for ctDNA analysis using plasma samples from advanced breast cancer patients enrolled in the PALOMA-3 trial showed that discordant calls occurred at VAFs < 1% [30]. In the view of our findings, we hypothesized that the combination of VAF with MAPD could further improve the assessment of the reliability of a variant call. According to our data, MAPD was significantly higher in discordant samples compared with concordant calls ($p < 0.001$), while VAF values were significantly lower in discordant calls compared with concordant samples ($p < 0.001$). Remarkably, as shown in Figure 2, the highest PPA (96.9%; 95%CI: 83.8–99.9%) was observed in the lower-right quadrant. Conversely, the PPA descended as much as 27.6% (95%CI: 12.7–47.2%) for calls clustered in the upper-left quadrant.

Our results are limited to *EGFR* locus as the cohort included exclusively *EGFR*-positive NSCLC patients. Nonetheless, mutations in other key genes were found. Specifically, in our data set, there were two samples testing positive for *KRAS* mutations by both NGS platforms (data not shown). Larger cohorts assessing the utility R-score for assessing the reliability of variant calls in other loci different may be of particular interest.

Taken together, we propose the R-score defined as the log(MAPD/VAF). According to our results, *EGFR* variants with positive R-score are particularly sensitive to genotyping errors. As presented in Figure 3, a significant correlation was found between PPA and the R-score cut-off values, indicating that R-sore can be useful to discriminate between true and false calls in the *EGFR* locus.

5. Conclusions

VAF and MAPD have an impact on *EGFR* variant calling. Combining this information in a score (R-score) can further improve the assessment of the reliability of a variant call. Using a dataset of 85 *EGFR*-positive NSCLC patients, we find that *EGFR* variants with positive R-score are particularly sensitive to erroneous variant calls in the *EGFR* gene.

Supplementary Materials: The following are available online at https://www.mdpi.com/article/10.3390/biology10100954/s1, Figure S1: Two samples electropherogram showing a 180 bp peak corresponding to low fragment cfDNA and peaks corresponding to mono-, di-, and tri-nucleosomes. (B) Gel-like image of cfDNA samples an-alysed with Bioanalyzer 2100; Figure S2: Correlation between variant allele fraction assessed by dPCR and Oncomine-NGS. Linear regression line is shown in black and the 95% confidence interval is shaded in grey. Pearson's correlation coefficient and *p*-value are shown in the graph; Table S1: Effect size and *p*-values for VAF, MAPD, and R-score parameters when assessing significant differences between concordant and discordant calls; Table S2: Table with all *EGFR* mutations detected by dPCR, NGS-Oncomine, and NGS-GeneReader with VAF values for each assay and cfDNA concentration.

Author Contributions: All authors confirmed they have contributed to the intellectual content of this paper and have approved the submitted version and agree to be personally accountable for the author's own contributions and for ensuring that questions related to the accuracy or integrity of any part of the work, even ones in which the author was not personally involved, are appropriately investigated, resolved, and documented in the literature. All authors have read and agreed to the published version of the manuscript.

Funding: This paper is part of a project that has received funding from the European Union's Horizon 2020 research and innovation program under grant agreement No. 875160 and PI16/18018. Roberto Serna-Blasco is supported by Plan de Empleo Juvenil Comunidad de Madrid (PEJD-2018-PRE/BMD-8640). Estela Sánchez-Herrero is supported by the Consejería de Ciencia, Universidades e Innovación of the Comunidad de Madrid (Doctorados Industriales of the Comunidad de Madrid IND2019/BMD-17258).

Institutional Review Board Statement: The study was conducted according to the guidelines of the Declaration of Helsinki, and the study protocol was approved by the Hospital Puerta de Hierro Ethics Committee (internal code PI-154/17).

Informed Consent Statement: Informed consent was obtained from all subjects involved in the study.

Data Availability Statement: All data generated or analyzed during this study are included in this published article (and its additional files).

Acknowledgments: We wish to thank all the patients and staff who participated in the study.

Conflicts of Interest: Atocha Romero reports personal fees from Boehringer and Takeda during the conduct of the study. Mariano Provencio reports grants, personal fees, and non-financial support from BMS, ROCHE, and ASTRAZENECA, as well as personal fees from MSD and TAKEDA, outside the submitted work.

References

1. Provencio, M.; Torrente, M.; Calvo, V.; Gutiérrez, L.; Pérez-Callejo, D.; Pérez-Barrios, C.; Barquín, M.; Royuela, A.; Rodriguez-Alfonso, B.; Sotelo, M.; et al. Dynamic circulating tumor DNA quantificaton for the individualization of non-small-cell lung cancer patients treatment. *Oncotarget* **2017**, *8*, 60291–60298. [CrossRef]

2. Ettinger, D.S.; Wood, D.E.; Aisner, D.L.; Akerley, W.; Bauman, J.; Chirieac, L.R.; D'Amico, T.A.; Decamp, M.M.; Dilling, T.J.; Dobelbower, M.; et al. Non-small cell lung cancer, version 5.2017: Clinical practice guidelines in oncology. *JNCCN J. Natl. Compr. Cancer Netw.* **2017**, *15*, 504–535. [CrossRef]

3. Pérez-Callejo, D.; Romero, A.; Provencio, M.; Torrente, M. Liquid biopsy based biomarkers in non-small cell lung cancer for diagnosis and treatment monitoring. *Transl. Lung Cancer Res.* **2016**, *5*, 455–465. [CrossRef]

4. Romero, A.; Serna-Blasco, R.; Calvo, V.; Provencio, M. Use of Liquid Biopsy in the Care of Patients with Non-Small Cell Lung Cancer. *Curr. Treat. Options Oncol.* **2021**, *22*, 86. [CrossRef]

5. Provencio, M.; Serna-Blasco, R.; Franco, F.; Calvo, V.; Royuela, A.; Auglytė, M.; Sánchez-Hernández, A.; de Julián Campayo, M.; García-Girón, C.; Dómine, M.; et al. Analysis of circulating tumour DNA to identify patients with epidermal growth factor receptor–positive non-small cell lung cancer who might benefit from sequential tyrosine kinase inhibitor treatment. *Eur. J. Cancer* **2021**, *149*, 61–72. [CrossRef] [PubMed]

6. McLean, A.; Barnes, D.; Troy, L. Diagnosing Lung Cancer: The Complexities of Obtaining a Tissue Diagnosis in the Era of Minimally Invasive and Personalised Medicine. *J. Clin. Med.* **2018**, *7*, 163. [CrossRef]

7. Schrank, Z.; Chhabra, G.; Lin, L.; Iderzorig, T.; Osude, C.; Khan, N.; Kuckovic, A.; Singh, S.; Miller, R.J.; Puri, N. Current molecular-targeted therapies in NSCLC and their mechanism of resistance. *Cancers* **2018**, *10*, 224. [CrossRef] [PubMed]

8. Hallin, J.; Engstrom, L.D.; Hargi, L.; Calinisan, A.; Aranda, R.; Briere, D.M.; Sudhakar, N.; Bowcut, V.; Baer, B.R.; Ballard, J.A.; et al. The KRASG12C inhibitor MRTX849 provides insight toward therapeutic susceptibility of KRAS-mutant cancers in mouse models and patients. *Cancer Discov.* **2020**, *10*, 54–71. [CrossRef] [PubMed]

9. Hong, D.S.; Fakih, M.G.; Strickler, J.H.; Desai, J.; Durm, G.A.; Shapiro, G.I.; Falchook, G.S.; Price, T.J.; Sacher, A.; Denlinger, C.S.; et al. KRAS G12C Inhibition with Sotorasib in Advanced Solid Tumors. *N. Engl. J. Med.* **2020**, *383*, 1207–1217. [CrossRef] [PubMed]

10. Stewart, E.L.; Tan, S.Z.; Liu, G.; Tsao, M.S. Known and putative mechanisms of resistance to *EGFR* targeted therapies in NSCLC patients with *EGFR* mutations-a review. *Transl. Lung Cancer Res.* **2015**, *4*, 67–81. [PubMed]

11. Dal Maso, A.; Lorenzi, M.; Roca, E.; Pilotto, S.; Macerelli, M.; Polo, V.; Cecere, F.L.; Del Conte, A.; Nardo, G.; Buoro, V.; et al. Clinical Features and Progression Pattern of Acquired T790M-positive Compared With T790M-negative *EGFR* Mutant Non–small-cell Lung Cancer: Catching Tumor and Clinical Heterogeneity Over Time Through Liquid Biopsy. *Clin. Lung Cancer* **2020**, *21*, 1–14.e3. [CrossRef]

12. Romero, A.; Serna-Blasco, R.; Alfaro, C.; Sánchez-Herrero, E.; Barquín, M.; Turpin, M.C.; Chico, S.; Sanz-Moreno, S.; Rodrigez-Festa, A.; Laza-Briviesca, R.; et al. ctDNA analysis reveals different molecular patterns upon disease progression in patients treated with osimertinib. *Transl. Lung Cancer Res.* **2020**, *9*, 532–540. [CrossRef]

13. Provencio, M.; Torrente, M.; Calvo, V.; Pérez-Callejo, D.; Gutiérrez, L.; Franco, F.; Pérez-Barrios, C.; Barquín, M.; Royuela, A.; García-García, F.; et al. Prognostic value of quantitative ctDNA levels in non small cell lung cancer patients. *Oncotarget* **2018**, *9*, 488–494. [CrossRef] [PubMed]

14. Salk, J.J.; Schmitt, M.W.; Loeb, L.A. Enhancing the accuracy of next-generation sequencing for detecting rare and subclonal mutations. *Nat. Rev. Genet.* **2018**, *19*, 269–285. [CrossRef]

15. Kivioja, T.; Vähärautio, A.; Karlsson, K.; Bonke, M.; Enge, M.; Linnarsson, S.; Taipale, J. Counting absolute numbers of molecules using unique molecular identifiers. *Nat. Methods* **2012**, *9*, 72–74. [CrossRef] [PubMed]

16. Romero, A.; Jantus-Lewintre, E.; García-Peláez, B.; Royuela, A.; Insa, A.; Cruz, P.; Collazo, A.; Pérez Altozano, J.; Vidal, O.J.; Diz, P.; et al. Comprehensive cross-platform comparison of methods for non-invasive *EGFR* mutation testing: Results of the RING observational trial. *Mol. Oncol.* **2021**, *15*, 43–56. [CrossRef] [PubMed]

17. Ion Reporter™ Software 5.12 USER GUIDE. Available online: https://assets.thermofisher.com/TFS-Assets/LSG/manuals/MAN0018032_IonReporterSoftware_5_12_UG.pdf (accessed on 5 July 2021).

18. Deveson, I.W.; Gong, B.; Lai, K.; LoCoco, J.S.; Richmond, T.A.; Schageman, J.; Zhang, Z.; Novoradovskaya, N.; Willey, J.C.; Jones, W.; et al. Evaluating the analytical validity of circulating tumor DNA sequencing assays for precision oncology. *Nat. Biotechnol.* **2021**, 1–14. [CrossRef]

19. Le Tourneau, C.; Delord, J.P.; Gonçalves, A.; Gavoille, C.; Dubot, C.; Isambert, N.; Campone, M.; Trédan, O.; Massiani, M.A.; Mauborgne, C.; et al. Molecularly targeted therapy based on tumour molecular profiling versus conventional therapy for advanced cancer (SHIVA): A multicentre, open-label, proof-of-concept, randomised, controlled phase 2 trial. *Lancet Oncol.* **2015**, *16*, 1324–1334. [CrossRef]

20. Fisher, K.E.; Zhang, L.; Wang, J.; Smith, G.H.; Newman, S.; Schneider, T.M.; Pillai, R.N.; Kudchadkar, R.R.; Owonikoko, T.K.; Ramalingam, S.S.; et al. Clinical Validation and Implementation of a Targeted Next-Generation Sequencing Assay to Detect Somatic Variants in Non-Small Cell Lung, Melanoma, and Gastrointestinal Malignancies. *J. Mol. Diagn.* **2016**, *18*, 299–315. [CrossRef]

21. Cottrell, C.E.; Al-Kateb, H.; Bredemeyer, A.J.; Duncavage, E.J.; Spencer, D.H.; Abel, H.J.; Lockwood, C.M.; Hagemann, I.S.; O'Guin, S.M.; Burcea, L.C.; et al. Validation of a next-generation sequencing assay for clinical molecular oncology. *J. Mol. Diagn.* **2014**, *16*, 89–105. [CrossRef]

22. Provencio, M.; Pérez-Barrios, C.; Barquin, M.; Calvo, V.; Franco, F.; Sánchez, E.; Sánchez, R.; Marsden, D.; Cristóbal Sánchez, J.; Martin Acosta, P.; et al. Next-generation sequencing for tumor mutation quantification using liquid biopsies. *Clin. Chem. Lab. Med.* **2019**, *58*, 306–313. [CrossRef]

23. Sánchez-Herrero, E.; Serna-Blasco, R.; Ivanchuk, V.; García-Campelo, R.; Dómine Gómez, M.; Sánchez, J.M.; Massutí, B.; Reguart, N.; Camps, C.; Sanz-Moreno, S.; et al. NGS-based liquid biopsy profiling identifies mechanisms of resistance to ALK inhibitors: A step toward personalized NSCLC treatment. *Mol. Oncol.* **2021**, *15*, 2363–2376. [CrossRef]

24. Merker, J.D.; Oxnard, G.R.; Compton, C.; Diehn, M.; Hurley, P.; Lazar, A.J.; Lindeman, N.; Lockwood, C.M.; Rai, A.J.; Schilsky, R.L.; et al. Circulating tumor DNA analysis in patients with cancer: American society of clinical oncology and college of American pathologists joint review. *J. Clin. Oncol.* **2018**, *142*, 1242–1253.

25. Gandara, D.R.; Paul, S.M.; Kowanetz, M.; Schleifman, E.; Zou, W.; Li, Y.; Rittmeyer, A.; Fehrenbacher, L.; Otto, G.; Malboeuf, C.; et al. Blood-based tumor mutational burden as a predictor of clinical benefit in non-small-cell lung cancer patients treated with atezolizumab. *Nat. Med.* **2018**, *24*, 1441–1448. [CrossRef]

26. Rizvi, N.A.; Hellmann, M.D.; Snyder, A.; Kvistborg, P.; Makarov, V.; Havel, J.J.; Lee, W.; Yuan, J.; Wong, P.; Ho, T.S.; et al. Mutational landscape determines sensitivity to PD-1 blockade in non-small cell lung cancer. *Science* **2015**, *348*, 124–128. [CrossRef]

27. Kowanetz, M.; Zou, W.; Shames, D.S.; Cummings, C.; Rizvi, N.; Spira, A.I.; Frampton, G.M.; Leveque, V.; Flynn, S.; Mocci, S.; et al. Tumor mutation load assessed by FoundationOne (FM1) is associated with improved efficacy of atezolizumab (atezo) in patients with advanced NSCLC. *Ann. Oncol.* **2016**, *27*, vi23. [CrossRef]

28. Chalmers, Z.R.; Connelly, C.F.; Fabrizio, D.; Gay, L.; Ali, S.M.; Ennis, R.; Schrock, A.; Campbell, B.; Shlien, A.; Chmielecki, J.; et al. Analysis of 100,000 human cancer genomes reveals the landscape of tumor mutational burden. *Genome Med.* **2017**, *9*, 34. [CrossRef] [PubMed]

29. Merino, D.M.; McShane, L.M.; Fabrizio, D.; Funari, V.; Chen, S.J.; White, J.R.; Wenz, P.; Baden, J.; Barrett, J.C.; Chaudhary, R.; et al. Establishing guidelines to harmonize tumor mutational burden (TMB): In silico assessment of variation in TMB quantification across diagnostic platforms: Phase I of the Friends of Cancer Research TMB Harmonization Project. *J. Immunother. Cancer* **2020**, *8*, e000147. [CrossRef] [PubMed]

30. O'Leary, B.; Hrebien, S.; Beaney, M.; Fribbens, C.; Garcia-Murillas, I.; Jiang, J.; Li, Y.; Bartlett, C.H.; André, F.; Loibl, S.; et al. Comparison of beaming and droplet digital PCR for circulating tumor DNA analysis. *Clin. Chem.* **2019**, *65*, 1405–1413. [CrossRef]

Article

Role of Persistent Organic Pollutants in Breast Cancer Progression and Identification of Estrogen Receptor Alpha Inhibitors Using In-Silico Mining and Drug-Drug Interaction Network Approaches

Bibi Zainab [1], Zainab Ayaz [1], Umer Rashid [2], Dunia A. Al Farraj [3], Roua M. Alkufeidy [3], Fatmah S. AlQahtany [4], Reem M. Aljowaie [3] and Arshad Mehmood Abbasi [1,5,*]

[1] Department of Environmental Sciences, Abbottabad Campus, COMSATS University Islamabad, Abbottabad 22060, Pakistan; abbasizainab2@gmail.com (B.Z.); zainabayaz321@gmail.com (Z.A.)
[2] Department of Chemistry, Abbottabad Campus, COMSATS University Islamabad, Abbottabad 22060, Pakistan; umerrashid@cuiatd.edu.pk
[3] Department of Botany and Microbiology, College of Sciences, King Saud University, P.O. Box 22452, Riyadh 11495, Saudi Arabia; dfarraj@ksu.edu.sa (D.A.A.F.); ralqufaidi@ksu.edu.sa (R.M.A.); raljowaie@ksu.edu.sa (R.M.A.)
[4] Department of Pathology, College of Medicine, King Saud University, Medical City, Riyadh 11495, Saudi Arabia; fatma@ksu.edu.sa
[5] University of Gastronomic Sciences, 12042 Pollenzo, Italy
* Correspondence: arshad799@yahoo.com or amabbasi@cuiatd.edu.pk

Citation: Zainab, B.; Ayaz, Z.; Rashid, U.; Al Farraj, D.A.; Alkufeidy, R.M.; AlQahtany, F.S.; Aljowaie, R.M.; Abbasi, A.M. Role of Persistent Organic Pollutants in Breast Cancer Progression and Identification of Estrogen Receptor Alpha Inhibitors Using In-Silico Mining and Drug-Drug Interaction Network Approaches. *Biology* **2021**, *10*, 681. https://doi.org/10.3390/biology10070681

Academic Editors: Shibiao Wan, Yiping Fan, Chunjie Jiang, Shengli Li and Lucia Mangone

Received: 2 June 2021
Accepted: 8 July 2021
Published: 19 July 2021

Publisher's Note: MDPI stays neutral with regard to jurisdictional claims in published maps and institutional affiliations.

Simple Summary: The role of persistent organic pollutants (POPs) in breast cancer progression and their bioaccumulation in adipose tissue has been reported. We used a computational approach to study molecular interactions of POPs with breast cancer proteins and identified natural and synthetic compounds to inhibit these interactions. Moreover, for comparative analysis, standard drugs and screened compounds were also docked against estrogen receptor alpha (ERα) and identification of the finest inhibitor was performed using in-silico mining and drug-drug interaction (DDI) network approaches. Based on scoring values, short-chained chlorinated paraffins demonstrated strong interactions with ERα compared to organo-chlorines and PCBs. Synthetic and natural compounds demonstrating strong associations with the active site of the ERα protein could be potential candidates to treat breast cancer specifically caused by POPs and other organic toxins and can be used as an alternative to standard drugs.

Abstract: The strong association between POPs and breast cancer in humans has been suggested in various epidemiological studies. However, the interaction of POPs with the ERα protein of breast cancer, and identification of natural and synthetic compounds to inhibit this interaction, is mysterious yet. Consequently, the present study aimed to explore the interaction between POPs and ERα using the molecular operating environment (MOE) tool and to identify natural and synthetic compounds to inhibit this association through a cluster-based approach. To validate whether our approach could distinguish between active and inactive compounds, a virtual screen (VS) was performed using actives (627 compounds) as positive control and decoys (20,818 compounds) as a negative dataset obtained from DUD-E. Comparatively, short-chain chlorinated paraffins (SCCPs), hexabromocyclododecane (HBCD), and perfluorooctanesulfonyl fluoride (PFOSF) depicted strong interactions with the ERα protein based on the lowest-scoring values of −31.946, −18.916, −17.581 kcal/mol, respectively. Out of 7856 retrieved natural and synthetic compounds, sixty were selected on modularity bases and subsequently docked with ERα. Based on the lowest-scoring values, ZINC08441573, ZINC00664754, ZINC00702695, ZINC00627464, and ZINC08440501 (synthetic compounds), and capsaicin, flavopiridol tectorgenin, and ellagic acid (natural compounds) showed incredible interactions with the active sites of ERα, even more convening and resilient than standard breast cancer drugs Tamoxifen, Arimidex and Letrozole. Our findings confirm the role of POPs in breast cancer progression and suggest that natural and synthetic compounds with high binding affinity could be more efficient and appropriate candidates to treat breast cancer after validation through in vitro and in vivo studies.

Keywords: breast cancer; estrogen receptor alpha; persistent organic pollutants; drug-drug interaction networks; molecular docking

1. Introduction

POPs are the most common synthetic, lipophilic, toxic, bio-accumulative, and persistent pollutants in the environment. Most POPs are of anthropogenic origin, but some substances, i.e., dioxins and furans, are also produced naturally during volcanism. POPs are also used intentionally in pesticides and other industrial products and may be released accidentally as a by-product from industrial processes or fuel combustion, such as dioxins and furans [1]. POPs release in the environment through industrial and agricultural effluents, drainage systems, urban effluents and landfill leachate [2,3]. Contaminated soil, water, air, dust and processed goods like textiles and packaging materials contain considerable amounts of POPs. Importantly, at ambient temperatures, POPs have a tendency to enter the gas phase; as a result they may volatilize from soils, plants, and water bodies into the atmosphere. They preferentially partition to solids, particularly organic materials in aquatic systems and soils, avoiding the aqueous phase. Being hydrophobic in nature [4], rather than entering the aqueous milieu of cells, some major types of POPs, such as polychlorinated dibenzo-p-dioxins and furans (PCDD/PCFs), polychlorinated biphenyl (PCBs), organo-chlorinated pesticides (OCPs), perfluorooctane sulfonate (PFOS) and pentadecafluorooctanoic acid (PFOAs) are hydrophobic and accumulate in the fatty tissues of the living host. POPs may accumulate in food chains [5] and, from contaminated food such as fruits, vegetables, chicken, meat, milk and fish etc., may enter humans and other living organisms [6,7]. As a result, predatory species like humans often have the highest concentration of POPs, and their presence in humans, i.e., in adipose tissue and human milk, is associated with the up-regulation of hormone-dependent breast cancers [2].

The prevalence of breast cancer, one of the most common types of cancers, specifically in females, is increasing worldwide, which cannot be explained solely by the emergence of mammography screening [8]. In 2018, about two million cases of breast cancer were reported in women globally [9]. The survival rate was up to 26% in cases where distant metastases were present. About 25% of breast cancers have been reported in developed countries; furthermore, it is one of the leading causes of death in Western countries. Deregulation of estrogen balance is known to promote breast cancer, and in Asia, over 60% of breast cancer cases have been diagnosed as estrogen receptor alpha-positive (ERα) cancers [10]. The estrogen receptor 1 (ESR1) gene encodes the ERα protein, a ligand regulated transcription factor, which plays a central role in the proliferation of breast cancer. Production of testosterone enhances the synthesis of progesterone and estrogen receptors in breast glands. Particularly, ERα expressed in the mammary glands and uterus of women has binding ability with DNA and contributes significantly to apoptosis, homeostasis, metabolism, and in breast cancer. An estimated 60% pre- and 75% postmenopausal women are suffering from estrogen-dependent breast cancer [11]. Through disturbing the functioning of adipose tissue, POPs affect the production of estrogens by stimulating genotoxic enzymes and leading to cross-generational epigenetic modifications by modifying the epigenome [12]. Many in vitro studies have shown that certain POPs promote the development of estrogen-positive breast cancer cells by receptor (ER). Exposure to certain POPs, particularly in perinatal studies, can enhance the development of breast cancer and sensitivity to carcinogens and cancerous breast tumors in animal studies. Chemotherapy, hormone therapy, immunotherapy, radiotherapy, and surgery are among the common methods for breast cancer treatment [13], which eventually have multiple side effects. Therefore, it is necessary to find better natural and synthetic compounds for treatment.

In this context, extensive use of anticancer drugs and potential inhibitors with increasing resistance together with numerous side effects highlights an urgent need for novel cancer treatment methods. Therefore, VS methods including negative image-based screening, molecular docking and the pharmacophore hypothesis could be effective tools for identification and screening of the ligands against ER-α receptor. Recent studies have demonstrated that VS methods have the ability to provide structural insights into complex interactions for repositioning and remediation [14], specifically using natural and synthetic compounds [15]. At present, in-silico methods for drug designing, receptor mapping, molecular modeling, and homology modulation etc. are gaining tremendous popularity in drug development, molecular biology, nanotechnology and biochemistry domains. In addition, these methods are used to complement in vitro and in vivo toxicity assessments, particularly to reduce the need for animal monitoring, costs, and time [16]. Furthermore, in-silico cancer modelling opens up new avenues for research into oncogenesis in different biological dimensions and systems. These approaches can assist in expediting the development of diagnostic and therapeutic technologies for clinical care. With reliable digital representations of cancer, the consequences of therapeutic treatments at both the molecular and surgical scales may be anticipated in silico without exposing patients to danger. Previously, an in-silico drug discovery technique exposed that a potential ligand, 1,2,3,4,6-penta-O-galloyl-β-D-glucopyranose, which is a naturally occurring tannin, can inhibit the activity of Ror1 (protein) that contributes significantly to cancer growth and proliferation [17].

Many complementary resources, including microarray, protein-protein interaction, and protein complexes, are being used to discover enriched biological processes and pathways. One example of this is graph theory, which is being used to analyze the lung cancer protein-protein interaction network (PPIN), and to discover highly dense modules which are potential cancer-associated protein complexes [18]. Previously, flavonoids have been proven as potential anticancer agents by virtue of molecular binding to some key targets such as aromatase, fatty acid synthase, xanthine oxidase, cyclooxygenase, lipoxygenase, ornithine decarboxylase, protein tyrosine kinase, phosphoinositide 3-kinase, protein kinase C, topoisomerase II (ATP binding site), ATP binding cassette (ABC) transporter, and phospholipase A2 [19]. The present study was conducted with the aim of determining the molecular interactions between ERα (target) and POPs which were considered as key factors in breast cancer progression. Moreover, for comparative analysis, standard drugs and screened compounds were docked against ERα and the finest inhibitors (natural and synthetic compounds) were identified using in-silico mining and DDI network approaches.

2. Materials and Methods

2.1. Disease Selection

Breast cancer (BC) was the target disease because of its prevalence around the globe. Currently, more than two million cases of breast cancer have been diagnosed in women [9], while in Pakistan BC is diagnosed in over 90,000 women annually, out of which 40,000 will not survive [20].

2.2. Identification of the Mutated Gene

Gene identification was completely disease specific. The GeneCard (www.genecards. org/ (accessed on 20 November 2020)) was used along with a literature review to determine a list of mutated genes involved in breast cancer as reported earlier [21]. Based on GeneCard, the estrogen receptor gene (ERg) was identified as a mutated gene of breast cancer.

2.3. Selection and Preparation of Targeted Protein

The Protein Data Bank (PDB), a global database providing the 3-D structure of biological molecules like proteins, DNA, and RNA, was used to select and prepare the targeted protein following the method of Rose et al. [22]. Protein Bank RCSB (https://www.rcsb.org/ (accessed on 21 November 2020)). was used to get the 3-D structure of the ERα (ERα/pdb id: 5W9D) protein of breast cancer (Figure S1). The protein selection was entirely based on mutated genes and the MOE tool was used to prepare the protein file while removing water molecules and attached ligands while hydrogen atoms were added. Afterward, a discovery studio was used to visualize the protein structure.

2.4. Validation of Virtual Screening (VS) Protocol

To validate whether our approach can distinguish between active and inactive compounds, we have performed a virtual screen (VS) experiment using actives (627 ERα inhibitors, i.e. binders) as positive control and decoys (20,818 compounds, i.e. non-binders) as a negative dataset obtained from the database of Useful Decoys: Enhanced (DUD-E). All the dataset compounds were docked into the binding site of ERα (PDB ID: 5W9D). The ligand enzyme complexes with the lowest binding energy were analyzed by the MOE ligand interaction module. Finally, Discovery Studio Visualizer was used for the 3-D interaction plot.

2.5. Screening and Toxicity Detection of Pollutants

POPs, whose emissions and/or output can be eliminated, or at least reduced substantially, were screened from the list as demonstrated in the Stockholm Convention in 2001. The online server 'admetSAR' containing 27 predictive models [23] was used to check the toxicity of screened POPs, and all were lying under toxic classes I, II, and III.

2.6. Preparation of Ligand and Molecular Docking

PubChem offers free access to information and biological functions about chemical substances. The database contains chemical information from individual PubChem data providers and the integrated database contains a distinction between chemical structures and the database of substances [24]. PubChem (https://pubchem.ncbi.nlm.nih.gov/ (accessed on 12 June 2021)) was used to extract 3-D structures of screened ligands (POPs) while adopting the previously reported procedure [24]. Afterward, ligands were prepared using the molecular operating environment (MOE) tool.

The molecular docking (MD) technique for the identification and optimization of drug candidates was used to analyze and simulate molecular interactions between the ligand and targeted macromolecules as reported formerly [25]. The MOE tool was used to evaluate the mechanism of molecular interactions between ligands including POPs, approved drugs (positive control), progesterone and testosterone (negative control), and drug candidates (natural and synthetic compounds) with an ERα receptor protein, while Discovery Studio software (DS 4.1) was used to visualize the 3-D interactions following the method as reported earlier [26].

2.7. Collection and Mining of Natural and Synthetic Compounds

The ZINC database (zinc.docking.org/ (accessed on 12 July 2021)) was used for the collection of natural and synthetic compounds along with their chemical properties, including Zinc ID, molecular weight, hydrogen bond donors log p, polar dissociation, rotatable bonds, a-polar dissociation, and hydrogen bond acceptors. Lipinski's rule of five was applied to the collected drug dataset for mining the natural and synthetic compounds as described earlier [15].

2.8. Cluster Formation

Weka, a platform for clustering, association, pre-processing, regression, classification, and screening of data [27], was used for clustering of the drug dataset based on a k-means algorithm (k-mean) clustering system. According to the properties of drugs, this method tracks a modest and quick way to categorize a particular record "$x_1, x_2, x_3 \ldots x_n$" to numbers of k clusters (k $\leq$ n), where k represents clusters and the row is denoted by n.

2.9. Drug-Drug Interaction (DDI) Network

Gephi, a primary platform for data analysis and the fastest graphical visualization of large networks [28], was used to create DDI networks from k-means clustering to find a strong connection between drug networks within each cluster. DDI networks were generated based on statistical parameters such as modularity (Ml), path lengths (PL), average degree (AD), average weighted degree (AWD), degree distribution (DD), and graph density (GD). Each network has borders E, vertical V, average path length L and node D, as well as network density and modularity classes. Most strongly-associated natural and synthetic compounds were docked against the targeted ERα protein to identify scoring values/binding energies. Standard breast cancer drugs i.e., Tamoxifen, Arimidex, Letrozole were used as a control to compare the scoring value of screened drugs.

3. Results and Discussion

3.1. Validation of VS and Reliability of MD

To validate whether our approach can distinguish between active and inactive compounds, a virtual screen (VS) experiment was performed using actives (627 ERα inhibitors i.e. binders) as positive control and decoys (20,818 compounds i.e. non-binders) as negative dataset obtained from the database of Useful Decoys: Enhanced (DUD-E) [29]. All the dataset compounds were docked into the binding site of ERα (PDB ID: 5W9D). Computed binding energy values of the active compound dataset were in the range of 28.6573 to -355.9801 kcal/mol and chemical structures of the most active binders are given in Figure S2. The binding energy values of the decoy set were in the range of -1.0988 to -3.0371 kcal/mol. Therefore, these findings suggest that our VS protocol can distinguish between active and inactive compounds.

The reliability of docking accuracy was assessed in two steps. In the first step, redocking of the native ligand was performed (Table S1). In the second step, a cross-docking experiment was carried out (Table S2). Three-dimensional structures of five estrogen receptor-alpha (PDB accession codes = 1A52, 3ERT, 1GWQ, 1UOM and 5W9D) were retrieved from PDB. In self-docking experiments, all the native ligands were extracted from receptors and root means square deviation (RMSD) was calculated for each re-docked and experimental native ligand [30]. Docking was carried out using the Triangle matcher algorithm (placement stage) and scored by the London dG scoring function [31]. Subsequently, best-scored poses were submitted to a rigid receptor protocol (refinement stage). Throughout the validation of docking protocol, the best performance in terms of computed RMSD value, conformation, binding energy, position, and pose (orientation) was obtained with the Triangle matcher London dG scoring function. The final score was calculated with the ASE scoring function. The whole validation process is presented in Supporting Information (Tables S1 and S2).

3.2. Interactions between POPs and ERα

Carcinogenesis is not a simple process; it involves initiation, promotion, and progression [32] of malignancy. The ERα gene is more likely to be involved in cell proliferation and is considered the most popular target to treat breast cancer. As per previous reports, POPs may not directly cause cancer, but act as co-carcinogenic agents [33]. It has been reported that organo-chlorines such as dichlorodiphenyltrichloroethane (DDT), hexachlorocyclohexane (HCH), aldrin, dieldrin, and polychlorinated biphenyls (PCBs) have the potential to stimulate breast cancer cell proliferation through the estrogenic pathway [34]. The association between organochlorine and PCBs exposure and risk of breast cancer has been reported. In the present study, substantial data retrieved from the molecular operating environment (MOE) tool, including scoring values, root-mean-square distance, and (RMSD) values, are given in Table 1. As reported earlier [35], the more negative the free binding affinity/scoring values are, the better the bond stability it forms between the ligand and the receptor protein [35]. In this context, strong relations between POPs and ERα protein of breast cancer were assessed based on scoring values ranging from

−31.94 to −8.650 kcal/mol. Out of 27 POPs, short-chained chlorinated paraffins (SCCPs), hexabromocyclododecane (HBCD), and perfluorooctanesulfonyl fluoride (PFOSF) showed the strongest molecular interactions with the ERα protein based on their lowest-scoring values of −31.94, −18.91, −17.58 kcal/mol, respectively (Table 1). These findings revealed that SCCPs, HBCD, and PFOSF could be potentially involved in breast cancer prevailing compared to PCBs and organochlorine as reported in previous studies [34]. These POPs are more suspected to cause breast cancer and are widely used pesticides in developing countries of Asia due to their low cost and utility against various pests. The key non-occupational exposure routes of these high-potency contaminants include ingestion, both directly and by tainted food, and dermal interaction with the substance [36].

Table 1. Docking of POPs with ERα protein.

S. #	Chemical Name	Structures	B.E. (kcal/mol)	RMSD (Å)
1.	Short-chained chlorinated paraffin's		−31.95	1.975
2.	HBCD (Hexabromocyclododecane)		−18.92	0.831
3.	PFOSF (Perfluorooctanesulfonyl fluoride)		−17.58	1.920
4.	Dieldrin		−17.22	0.635
5.	DDT (dichloro-diphenyl-trichloroethane)		−17.15	1.123
6.	PFOS (perfluorooctanesulfonic acid)		−17.13	0.884
7.	Endrin		−17.01	0.979
8.	Aldrin		−16.19	1.312
9.	Hexa bromodiphenyl ethers		−15.72	1.099
10.	Hexabromobiphenyl		−15.67	1.541
11.	Penta-bromodiphenyl ethers		−15.39	1.261
12.	PCDDs (Polychlorinated dibenzodioxins)		−15.01	1.345
13.	Chlordane		−14.85	0.937
14.	Toxaphene		−14.55	0.818
15.	PCDFs (polychlorinated dibenzofurans)		−14.39	1.885
16.	Beta endosulfans		−14.26	1.791
17.	PCBs (Polychlorinated biphenyls)		−14.16	1.233
18.	α-endosulfans		−14.09	0.978

Table 1. *Cont.*

S. #	Chemical Name	Structures	B.E. (kcal/mol)	RMSD (Å)
19.	Heptachlor		−13.70	0.999
20.	Lindane		−12.91	0.631
21.	Chlorinated_naphthalenes		−12.58	0.903
22.	Mirex		−12.39	1.508
23.	Chlordecone		−11.87	1.207
24.	Pentachlorophenol		−9.566	1.215
25.	Hexachlorobenzene		−9.501	1.212
26.	Pentachlorobenzen		−9.500	1.596
27.	Hexachlorobutadiene		−8.650	1.090

B.E. Binding energy, RMSD. Root-mean-square distance, S. #. Serial number.

Above mentioned results revealed that based on scoring values, short-chain chlorinated paraffin (SCCPs) have the strongest interactions with the ERα protein (Figure 1A). SCCPs are commonly used in metalworking fluids, paints, sealants, adhesives, leather manufacturing chemicals, plastics, rubber, and as a plastic agent and flame retardant [37]. In addition, these pollutants have also been isolated from kidneys, adipose tissue, and breast milk of Inuit women [38]. As previously reported, SCCPs (C10–C13) belong to the third class of carcinogens and are highly toxic to aquatic organisms [39]. Moreover, previous reports have provided convincing evidence on the disruptive effect of SCCPs on thyroid hormones and glucocorticoids [40], as well as their role in the regulation of different signaling pathways and physiological mechanisms [41]. However, there are few studies on the health effects of SCCPs in humans, especially their contribution to endocrine disruption. In this context, the aforementioned results revealed the strong association of SCCPs with the ERα protein of breast cancer based on least binding energy or scoring value (−31.946 kcal/mol). This indicates that the strong binding capacity of SCCPs with ERα protein may be involved in the spread of breast cancer in women. Hexabromocyclododecane (HBCD) has also demonstrated strong interactions with the ERα protein (Figure 1B). Therefore, HBCD could also be one of the main contributors to breast cancer.

These POPs are extensively used in flame retardants, as a neurotoxin, and in xenobiotic chemicals. HBCD acts as a nuclear receptor agonist, is hepatotoxic, and is an endocrine disruptor that induces developmental neurotoxicity in animals [42]. HBCD concentration in human breast milk, blood serum, and the umbilical cord has already been investigated. The concentrations in human breast milk raise questions about possible lactation and prenatal uptake during important developmental stages of the fetus [43]. Likewise, several fluorinated compounds, specifically perfluorooctanesulfonyl fluoride (PFOSF), also showed strong interactions with the ERα protein, based on a low scoring value (−17.58 kcal/mol), which allows it to easily form a stable complex (Figure 1C). These compounds are used as impregnating and grading agents and corrosion inhibitors, like insecticides and flame retardants, in cosmetics, paper coatings, and surfactants. The possible association of PFOSF with hormone disorders, genotoxic potential, and tumor formation in rodents has been suggested previously [44]. Hormones can be indirectly imitated by endocrine disorders or hormone disorders. Strong associations of SCCPs, HBCD, and PFOSF with the ERα protein revealed their possible role in breast cancer; therefore the role of these POPs should be further investigated in detail.

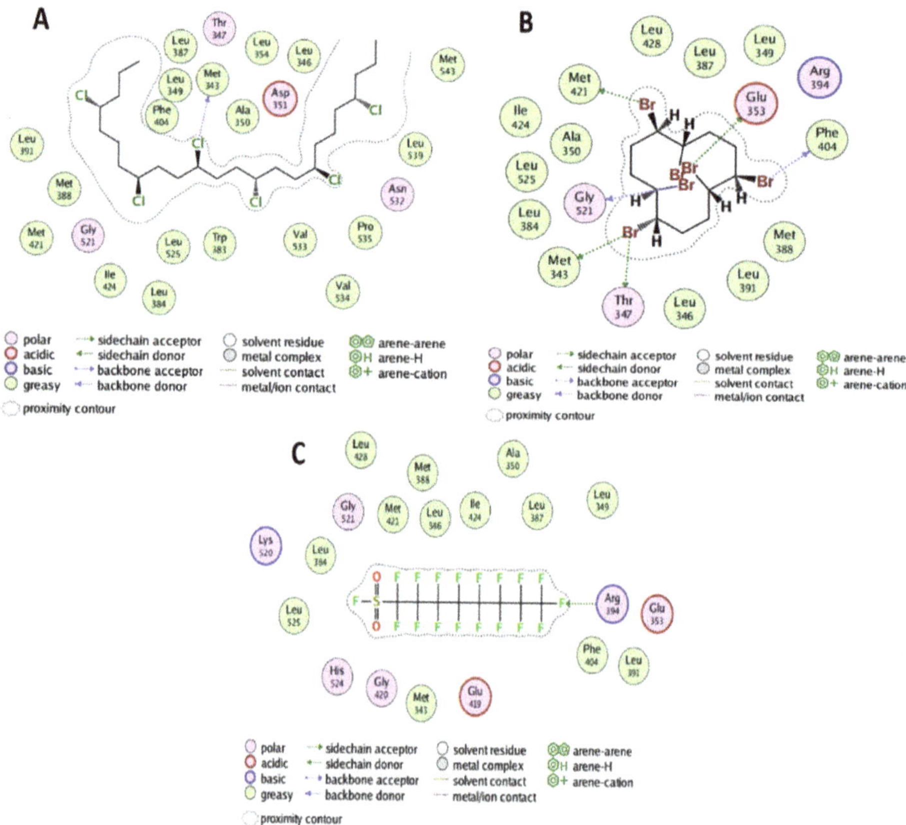

Figure 1. Two dimensional interactions of short chain chlorinated paraffins (**A**), Hexabromocyclododecane (**B**) and Perfluorooctanesulfonyl fluoride (**C**) with ERα protein of Breast Cancer.

3.3. Natural and Synthetic Compounds Collection, Mining, and Clustering

Approximately 7856 natural and synthetic compounds were collected with their structures and properties using the ZINC database (zinc.docking.org/ (accessed on 12 July 2020)). While applying the Lipinski rule of five on the drug dataset, 2390 compounds were chosen for further processing and the rest were discarded after mining. Afterward, 12 clusters were generated for natural and synthetic compounds based on the k-means clustering system using the Weka tool. The use of k-means clustering avoids the repetition of compounds, so natural and synthetic compounds exhibiting similar properties were placed in one group. Clusters possessing similar properties, including molecular weight, hydrogen bond donors log P, polar dissociation, rotatable bonds, a-polar dissociation, and hydrogen bond acceptors (HBA) are shown in a plot matrix with their attributes utilizing different color representations (Figure S3).

3.4. DDI Network

In total, 12 networks were generated using the k-means clustering algorithm and Gephi tool (Figure S4) based on statistical parameters as mentioned in Table 2. A strongly interacted network as illustrated in Figure 2 was generated from 457 compounds having higher modularity class in 12 networks based on the aforementioned parameters. Repulsion strength was set to 10,000, and Force Atlas and Fruchterman rein gold layouts were used to display and visualize the networks.

Table 2. Statistical parameters used to predict network interactions of various natural and synthetic compounds.

Networks	AD	AWD	ND	GD	Ml	APL	N	E
1.	2.306	2.860	1.000	0.005	0.518	1.000	507.0	1169
2.	2.120	2.519	1.000	0.008	0.524	1.000	266.0	564.0
3.	2.070	2.649	1.000	0.011	0.538	1.000	185.0	383.0
4.	1.531	2.266	1.000	0.024	0.613	1.000	64.00	98.00
5.	2.325	2.476	1.000	0.003	0.509	1.000	923.0	2146
6.	2.101	3.005	1.000	0.011	0.571	1.000	188.0	395.0
7.	2.275	3.028	1.000	0.005	0.525	1.000	469.0	1067
8.	1.629	2.056	1.000	0.001	0.552	1.000	107.0	181.0
9.	2.371	2.792	1.000	0.006	0.500	1.000	367.0	870.0
10.	2.155	2.662	1.000	0.004	0.501	1.000	541.0	1166
11.	1.518	1.789	1.000	0.007	0.528	1.000	218.0	331.0
12.	2.220	2.967	1.000	0.007	0.500	1.000	300.0	666.0
FSIN	2.325	2.476	1.000	0.003	0.503	1.000	923.0	2146

AD. Average degree, AWD. Average weighted degree, ND. Network diameter, GD. Graph density, Ml. modularity, APL. Average path length, N. nodes, E. edges, FSIN. Final strongly interacted network.

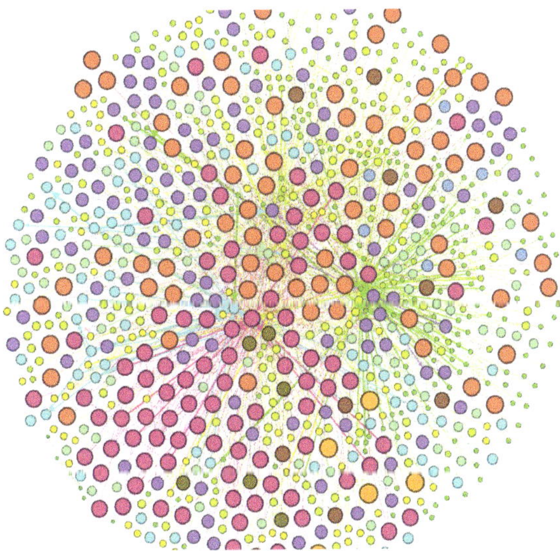

Figure 2. Final strongly drug-drug interaction network.

Nodes and edges in each network represent compounds and interactions between them, respectively. Darker color and larger size nodes show the strength of drug IDs/ZINC IDs. As shown in Table 2, network 1 comprises of 507.0 nodes and 1169 edges with modularity of 0.518; the second network consists of 266 nodes and 564 edges with 0.524 modularity. Network 3 showed 185.0 nodes and 383.0 edges with 0.538 modularity. Network 4 comprises 64.00 nodes and 98.00 edges with 0.613 modularity. Network 5 contains 923.0 nodes and 2146 edges with a modularity of 0.509. Network 6 consists of 188 nodes and 395.0 edges with a modularity of 0.571. Network 7 comprises 469.0 nodes and 1067 edges and possesses a modularity of 0.525. Network 8 consists of 107.0 nodes and 181.0 edges, and 0.001 graph density with a modularity of 0.552. Network 9 comprises 369.0 nodes and 870.0 edges with modularity of 0.500. Network 10 consists of 541.0 nodes and 1166 edges with a 0.501 modularity. Network 11 comprises 218.0 nodes and 331.0 edges and 0.528 modularity, while network 12 contains 300.0 nodes, 666.0 edges, and 0.500 modularity. Likewise, the finally generated strong network contains 923.0 nodes and 2166 edges with 0.503 modularity (Table 2).

3.5. Validation of Natural and Synthetic Compounds

ERα, the most common and effective target for breast cancer treatment, was docked against the screened natural and synthetic compounds [35]. For docking, the active site of the ERα protein was identified using standard drugs used to treat breast cancer i.e., Tamoxifen, Arimidex, and Letrozole as a positive control group, while progesterone and testosterone were utilized as a negative control group. As shown in Table 3, scoring values of Tamoxifen, Arimidex and Letrozole ranged from −31.26 to −20.97 kcal/mol. Tamoxifen showed pi-sulfur and pi-alkyl interactions; Arimidex also showed pi-sulfur and pi-alkyl interactions, while Letrozole showed conventional hydrogen bond and pi-sulfur interactions with the active site of ERα, as shown in Figure 3a–c.

Table 3. Docking results of positive control, negative control, synthetic and natural compounds with ERα protein.

S. #	Names	Structures	B.E. (kcal/mol)	RMSD (Å)
Positive control				
1	Tamoxifen		−31.26	1.006
2	Arimidex		−22.46	1.130
3	Letrozole		−20.97	1.385
Negative control				
1	Progestrone		−23.53	1.080
2	Testosterone		−23.95	0.859
Synthetic compounds				
1	5-(4-butoxyphenyl)-4-(2,3-dihydro-1,4-benzodioxin-6-ylcarbonyl)-1-(2-furylmethyl)-3-hydroxy-1,5-dihydro-2*H*-pyrrol-2-one [ZINC08441573]		−32.47	1.521
2	2-(4-*tert*-butylphenyl)-3*H*-quinazolin-4-one [ZINC00664754]		−31.38	1.575
3	(2*S*)-4-hydroxy-3-(5-methylfuran-2-carbonyl)-1-[[(2*R*)-oxolan-2-yl]methyl]-2-(3-phenoxyphenyl)-2*H*-pyrrol-5-one [ZINC00702695]		−30.35	1.820

Table 3. *Cont.*

S. #	Names	Structures	B.E. (kcal/mol)	RMSD (Å)
4	(1*R*,6*S*)-6-[[2-[4-(4-methylphenyl) piperazine-1-carbonyl]phenyl]carbamoyl]cyclohex-3-ene-1-carboxylic acid [ZINC00627464]		−30.31	1.650
5	(5*S*)-1-[3-(dimethylamino)propyl]-4-[hydroxy-(3-methyl-4-propan-2-yloxyphenyl)methylidene]-5-pyridin-4-ylpyrrolidine-2,3-dione [ZINC08440501]		−28.76	1.686
Natural compounds				
1	Capsaicin [ZINC01530575]		−24.90	0.905
2	Flavopiridol [ZINC21288966]		−24.76	1.875
3	Tectorigenin [ZINC00899915]		−21.71	0.994
4	Ellagic acid [ZINC03872446]		−16.36	0.996

B.E. Binding energy, RMSD. Root-mean-square distance, S. #. Serial number.

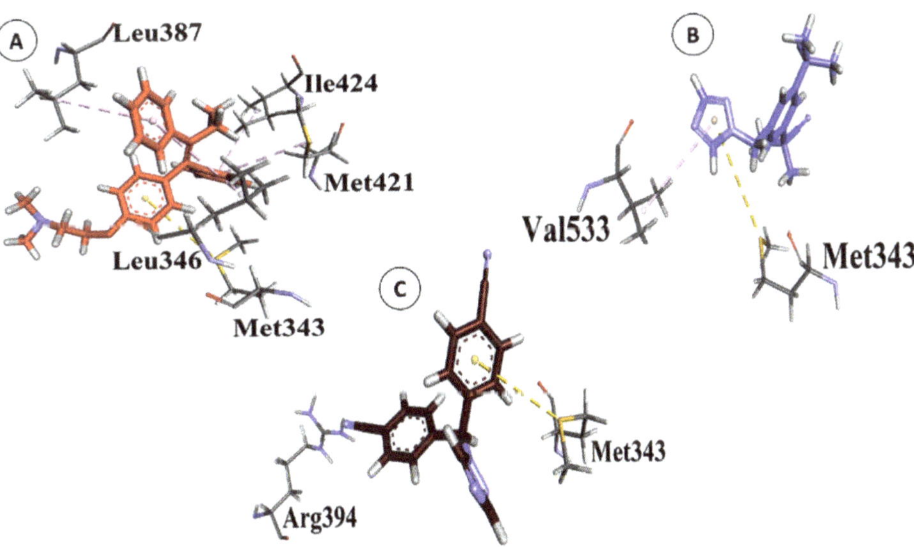

Figure 3. Three-dimensional interactions of Tamoxifen (**A**), Arimidex (**B**), Letrozole (**C**) with the binding site of ERα protein.

Progesterone and testosterone were used as a negative control group as their scoring values are −23.53 and −23.95 respectively, as shown in Table 3. Previous research reveals that adding an androgen to estrogen treatment lowers mammary epithelial proliferation and ER expression, implying that androgens may protect against breast cancer in the same way as progesterone does. In the past, androgens were used to treat breast cancer with moderate effectiveness [45]. There are mixed theories about these hormones, as some claim they down-regulate breast cancer progression, whereas some claims they up-regulate the process. However, according to our studies, both progesterone and testosterone have shown interactions with ER alpha protein. Furthermore, this study reveals that these hormones may up-regulate breast cancer development, as they interacted at the active site of ER-alpha with binding energies of −23.53 (progesterone) and −23.95 (testosterone). Up-regulation of breast cancer due to an imbalance of these hormones has already been reported. It has been reported that the effects of sex hormones on breast cancer development are clear from the advantages of hormone withdrawal treatment, with particular evidence of a relationship between completion of hormone withdrawal and clinical benefit [46].

The majority of clinical investigations that have utilized total testosterone as a measure of androgen exposure have found that greater total testosterone levels are related to an increased risk of breast cancer [47]. Progesterone metabolites may consequently be involved in the regulation of the generation of estradiol in the normal breast cell and so may be a multifaceted component in breast carcinogenesis [48]. Progesterone triggers normal human breast epithelial through paracrine mechanisms and is a risk factor for breast cancer because it promotes pre-neoplastic progression by stimulating cyclic proliferation of mammary stem cell pools or cell-initiating tumors in maturing breast epithelium. The development of cancer is therefore further promoted by progesterone signaling and a transition to autocrine proliferation regulation [49].

Natural and synthetic compounds possessing the highest binding affinity, RMSD less than 2 Å, and accurate binding sites were considered to form the most stable complexes. Data of 61 natural and synthetic compounds were collected from strong DDI networks and the top five synthetic (ZINC08441573, ZINC00664754, ZINC00702695, ZINC00627464, ZINC08440501) and four nature compounds (capsaicin, flavopiridol, tectorigenin, and ellagic acid) were docked with ERα (Table 3). Based on their scoring values and RMCD, both synthetic and natural compounds depicted strong binding capacity with the active site of the ERα protein.

Predicted scoring values of the top five synthetic compounds, ZINC08441573, ZINC00664754, ZINC00702695, ZINC00627464, and ZINC08440501, were −32.47, −31.38, −30.35, −30.31, and −29.350 kcal/mol, respectively. These findings were further confirmed by 3-D interactions with the active site of the ERα protein as shown in Figure 4a–e. The ZINC08441573 drug compound showed pi-pi, T shaped and pi-sigma interactions; ZINC00664754 exhibited conventional hydrogen bonding, pi-sigma, pi-sulphur and pi-alkyl bonding; ZINC00702695 had pi-sulfur, pi-lone pair and pi-alkyl associations; ZINC00627464 showed conventional hydrogen bonding, pi-pi T shaped, pi-lone pair and pi-alkyl associations; and ZINC08440501 had conventional hydrogen bond, Sulfur-X, pi-sulfur, and pi-alkyl interactions with the active sites of the ERα protein. Our findings suggest that, relatively, most of the synthetic compounds have scoring values even less than the standard breast cancer drugs, therefore showing strong interactions with the ERα protein. These compounds could be potential candidates to treat breast cancer. The inhibition potential of all synthetic compounds was greater than two standard breast cancer drugs i.e., Arimidex and letrozole. Two synthetic compounds including ZINC08441573 and ZINC00664754 exhibited a highly significant strong binding capacity (based on lowest scoring value) with the ERα protein. These two compounds possess the strong potential to inhibit interactions of all types of POPs (Table 1) with ERα protein and could be the best alternative to standard breast cancer drugs used currently.

Likewise, the binding capacity of the top four natural compounds from our dataset, namely capsaicin, flavopiridol, tectorigenin, and ellagic acid, with the ERα protein is demonstrated in Table 3. The scoring value of these compounds ranged from −25.30 to −16.36 kcal/mol, while their RMSD was between 0.905–1.875 Å. The association between these natural compounds and the ERα protein was further confirmed by 3-D networks as mentioned in Figure 5a–d. Capsaicin showed conventional hydrogen bonds and pi-lone pair interactions with ERα protein; flavopiridol had pi-sulfur, pi-sigma, and pi-alkylinteractions; tectorigenin exhibited conventional hydrogen bonding and pi-sulfur interactions; and ellagic acid showed conventional hydrogen bonding, pi-sulfur, and pi-alkyl interactions. Based on scoring values, it can be predicted that capsaicin and flavopiridol have a strong association with ERα protein, which is even stronger than standard drugs (Arimidex and letrozole). These natural compounds also have the potential to inhibit the binding of almost all POPs with ERα protein except short-chained chlorinated paraffins (Table 1). Based on scoring values and RMSD, the majority of the synthetic and natural compounds have the potential to inhibit various interactions between POPs and active sites of the ERα breast cancer protein. Such compounds could also be potential candidates for breast cancer drugs, and can also be useful as an alternative to standard drugs to treat breast cancer caused by POPs and other organic toxins.

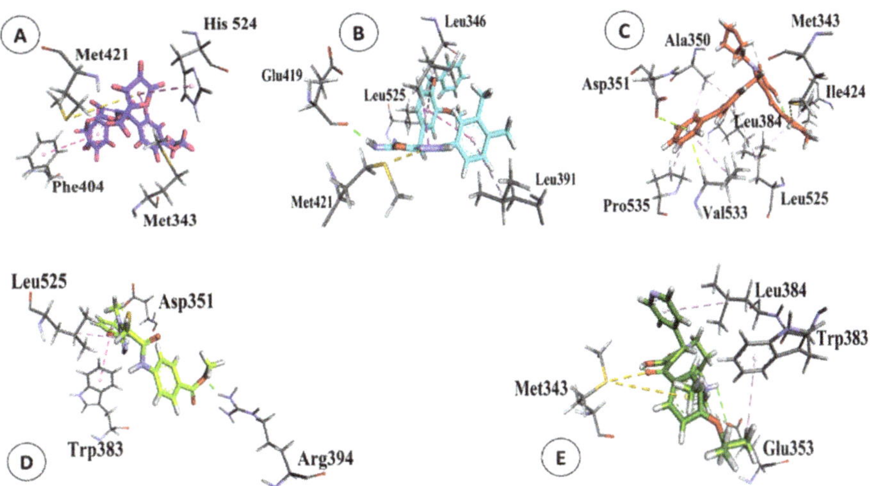

Figure 4. Three-dimensional interactions of synthetic compounds ZINC08441573 (**A**), ZINC00664754 (**B**), ZINC00702695 (**C**), ZINC00627464 (**D**), and ZINC08440501 (**E**) with the binding site of ERα protein.

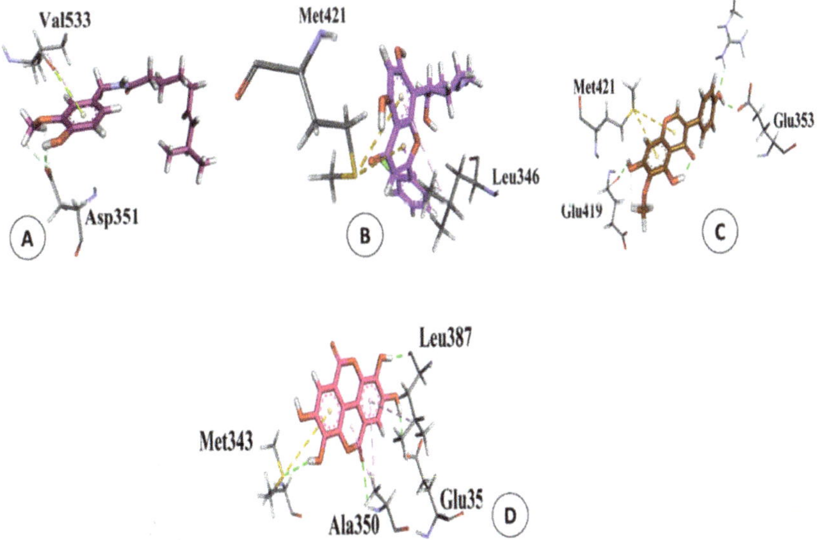

Figure 5. Three-dimensional interactions of natural compounds capsaicin (**A**), flavopiridol (**B**), tectorigenin (**C**), ellagic acid (**D**) with binding site of ERα protein.

4. Conclusions

The bioaccumulation of POPs in adipose tissue and their role in breast cancer development and/or progression was evaluated using the molecular interaction approach. Based on scoring values, short-chained chlorinated paraffins demonstrated strong interactions with the ERα breast cancer protein compared to organo-chlorines and PCBs. Both synthetic and natural compounds which demonstrated strong associations with the active site of the ERα protein could be potential candidates to treat breast cancer specifically caused by POPs and other organic toxins, and can be used as an alternative to standard drugs. Furthermore, our findings could be validated using in vitro and in vivo approaches.

Supplementary Materials: The following are available online at https://www.mdpi.com/article/10.3390/biology10070681/s1, Table S1: Results of re-docking of native inhibitors; Table S2: Cross-docking results for various PDB IDs from ERα; Figure S1: Structure of estrogen receptor alpha (ERα/pdb id = 5W9D) protein; Figure S2: Chemical structures and binding energies (B.E. in kcal/mol) of the active compounds obtained from CHEMBL; Figure S3. Plot matrix representation of drug compounds with their attributes; Figure S4. DDI networks generated using K means clustering algorithm and Gephi tool.

Author Contributions: Design, analysis, writing—B.Z. and Z.A.; conceptualization, supervision, resources, visualization, proof reading, validation—A.M.A. and U.R.; funding acquisition, visualization—D.A.A.F., R.M.A. (Roua M. Alkufeidy), F.S.A. and R.M.A. (Reem M. Aljowaie). All authors have read and agreed to the published version of the manuscript.

Funding: The authors extend their appreciation to the researchers supporting project number (RSP-2021/190) King Saud University, Riyadh, Saudi Arabia.

Institutional Review Board Statement: Not applicable.

Data Availability Statement: The data presented in this study are available within the article and supplementary materials.

Acknowledgments: The authors extend their appreciation to the researchers supporting project number (RSP-2021/190) King Saud University, Riyadh, Saudi Arabia. We are grateful to COMSATS University Islamabad, Abbottabad campus for lab facilitation.

Conflicts of Interest: All the authors declare no conflict of interest. The funders had no role in the design of the study; in the collection, analyses, or interpretation of data; in the writing of the manuscript, or in the decision to publish the results.

References

1. El-Shahawi, M.S.; Hamza, A.; Bashammakh, A.S.; Al-Saggaf, W.T. An overview on the accumulation, distribution, transformations, toxicity and analytical methods for the monitoring of persistent organic pollutants. *Talanta* **2010**, *80*, 1587–1597. [CrossRef]
2. Hinkebein, T. *Desalination: Limitations and Challenges*; National Academies Press (US): Washington, DC, USA, 2004; ISBN 0-309-53173-X.
3. Blanchard, M.; Teil, M.J.; Ollivon, D.; Legenti, L.; Chevreuil, M. Polycyclic aromatic hydrocarbons and polychlorobiphenyls in wastewaters and sewage sludges from the Paris area (France). *Environ. Res.* **2004**, *95*, 184–197. [CrossRef]
4. Bräuner, E.V.; Raaschou-Nielsen, O.; Gaudreau, E.; Leblanc, A.; Tjønneland, A.; Overvad, K.; Sørensen, M. Predictors of adipose tissue concentrations of organochlorine pesticides in a general Danish population. *J. Expo. Sci. Environ. Epidemiol.* **2012**, *22*, 52–59. [CrossRef]
5. Jones, K.C.; de Voogt, P. Persistent organic pollutants (POPs): State of the science. *Environ. Pollut.* **1999**, *100*, 209–221. [CrossRef]
6. Bonefeld-Jorgensen, E.C.; Long, M.; Bossi, R.; Ayotte, P.; Asmund, G.; Krüger, T.; Ghisari, M.; Mulvad, G.; Kern, P.; Nzulumiki, P.; et al. Perfluorinated compounds are related to breast cancer risk in greenlandic inuit: A case control study. *Environ. Health A Glob. Access Sci. Source* **2011**, *10*, 1–16. [CrossRef] [PubMed]
7. Ghosh, S.; Murinova, L.; Trnovec, T.; Christopher, A.L.; Washington, K.; Partha, S.M.; Sisir, K.D. Biomarkers linking PCB exposure and obesity. *Curr. Pharm. Biotechnol.* **2014**, *15*, 1058–1068. [CrossRef] [PubMed]
8. Soto, A.M.; Chung, K.L.; Sonnenschein, C. The pesticides endosulfan, toxaphene, and dieldrin have estrogenic effects on human estrogen-sensitive cells. *Environ. Health Perspect.* **1994**, *102*, 380–383. [CrossRef]
9. Bray, F.; Ferlay, J.; Soerjomataram, I.; Siegel, R.L.; Torre, L.A.; Jemal, A. Global cancer statistics 2018: GLOBOCAN estimates of incidence and mortality worldwide for 36 cancers in 185 countries. *CA. Cancer J. Clin.* **2018**, *68*, 394–424. [CrossRef] [PubMed]

10. Subramanian, A.; Salhab, M.; Mokbel, K. Oestrogen producing enzymes and mammary carcinogenesis: A review. *Breast Cancer Res. Treat.* **2008**, *111*, 191–202. [CrossRef]
11. Nagel, A.; Szade, J.; Iliszko, M.; Elzanowska, J.; Welnicka-Jaskiewicz, M.; Skokowski, J.; Stasilojc, G.; Bigda, J.; Sadej, R.; Zaczek, A.; et al. Clinical and biological significance of ESR1 gene alteration and estrogen receptors isoforms expression in breast cancer patients. *Int. J. Mol. Sci.* **2019**, *20*, 1881. [CrossRef]
12. Ennour-Idrissi, K.; Ayotte, P.; Diorio, C. Persistent organic pollutants and breast cancer: A systematic review and critical appraisal of the literature. *Cancers* **2019**, *11*, 1063. [CrossRef] [PubMed]
13. Tong, C.W.S.; Wu, M.; Cho, W.C.S.; To, K.K.W. Recent advances in the treatment of breast cancer. *Front. Oncol.* **2018**, *8*. [CrossRef]
14. Munir, A.; Elahi, S.; Masood, N. Clustering based drug-drug interaction networks for possible repositioning of drugs against EGFR mutations: Clustering based DDI networks for EGFR mutations. *Comput. Biol. Chem.* **2018**, *75*, 24–31. [CrossRef] [PubMed]
15. Zainab, B.; Ayaz, Z.; Munir, A.; Hossam Mahmoud, A.; Soliman Elsheikh, M.; Mehmood, A.; Khan, S.; Rizwan, M.; Jahangir, K.; Mehmood Abbasi, A. Repositioning of strongly integrated drugs against achromatopsia (CNGB3). *J. King Saud Univ. Sci.* **2020**, *32*, 1793–1811. [CrossRef]
16. Cronin, M.T.D.T.D.D.; Jaworska, J.S.; Walker, J.D.; Comber, M.H.I.; Watts, C.D.; Worth, A.P.P.; Basketter, D.; Casati, S.; Gerberick, G.F.; Griem, P.; et al. Report of the EPAA-ECVAM workshop on the validation of Integrated Testing Strategies (ITS). *Altern. Lab. Anim.* **2008**, *27*, 258–284. [CrossRef]
17. Edelman, L.B.; Eddy, J.A.; Price, N.D. In silico models of cancer. *Wiley Interdiscip. Rev. Syst. Biol. Med.* **2010**, *2*, 438–459. [CrossRef] [PubMed]
18. Nath, O.; Singh, A.; Singh, I.K. In-Silico drug discovery approach targeting receptor tyrosine kinase-like orphan receptor 1 for cancer treatment. *Sci. Rep.* **2017**, *7*, 1–10. [CrossRef]
19. Cassidy, C.E.; Setzer, W.N. Cancer-relevant biochemical targets of cytotoxic Lonchocarpus flavonoids: A molecular docking analysis. *J. Mol. Model.* **2010**, *16*, 311–326. [CrossRef] [PubMed]
20. Majeed, A.I.; Ullah, A.; Jadoon, M.; Ahmad, W.; Riazuddin, S. Screening, diagnosis and genetic study of breast cancer patients in Pakistan. *Pakistan J. Med. Sci.* **2020**, *36*, 16–20. [CrossRef]
21. Safran, M.; Dalah, I.; Alexander, J.; Rosen, N.; Iny Stein, T.; Shmoish, M.; Nativ, N.; Bahir, I.; Doniger, T.; Krug, H.; et al. Gene Cards Version 3: The human gene integrator. *Database* **2010**, *2010*, 1–16. [CrossRef]
22. Rose, P.W.; Prlić, A.; Altunkaya, A.; Bi, C.; Bradley, A.R.; Christie, C.H.; Di Costanzo, L.; Duarte, J.M.; Dutta, S.; Feng, Z.; et al. The RCSB protein data bank: Integrative view of protein, gene and 3D structural information. *Nucleic Acids Res.* **2017**, *45*, D271–D281. [CrossRef] [PubMed]
23. Yang, H.; Lou, C.; Sun, L.; Li, J.; Cai, Y.; Wang, Z.; Li, W.; Liu, G.; Tang, Y. AdmetSAR 2.0: Web-service for prediction and optimization of chemical ADMET properties. *Bioinformatics* **2019**, *35*, 1067–1069. [CrossRef] [PubMed]
24. Wang, Y.; Bryant, S.H.; Cheng, T.; Wang, J.; Gindulyte, A.; Shoemaker, B.A.; Thiessen, P.A.; He, S.; Zhang, J. PubChem BioAssay: 2017 update. *Nucleic Acids Res.* **2017**, *45*, D955–D963. [CrossRef] [PubMed]
25. Wadood, A.; Ahmed, N.; Shah, L.; Ahmad, A.; Hassan, H.; Shams, S. In-silico drug design: An approach which revolutionarised the drug discovery process. *OA Drug Des. Deliv.* **2013**, *1*. [CrossRef]
26. Sapundzhi, F.I.; Dzimbova, T.A. Computer modelling of the CB1 receptor by Molecular Operating Environment. *Bulg. Chem. Commun.* **2018**, *50*, 15–19.
27. Hall, M.; Frank, E.; Holmes, G.; Pfahringer, B.; Reutemann, P.; Witten, I.H. The WEKA data mining software: An update. *ACM SIGKDD Explor. Newsl.* **2009**, *11*, 10–18. [CrossRef]
28. Bastian, M.; Heymann, S.; Jacomy, M. Gephi: An open source software for exploring and manipulating networks. BT-International AAAI Conference on Weblogs and Social. In Proceedings of the 3rd international AAAI conference on weblogs and social media, San Jose, CA, USA, 17–20 May 2009; pp. 361–362.
29. Réau, M.; Langenfeld, F.; Zagury, J.F.; Lagarde, N.; Montes, M. Decoys selection in benchmarking datasets: Overview and perspectives. *Front. Pharmacol.* **2018**, *9*. [CrossRef]
30. Prieto-Martínez, F.D.; Arciniega, M.; Medina-Franco, J.L. Acoplamiento molecular: Avances recientes y retos. *TIP Rev. Espec. Cienc. Químico Biológicas* **2018**, *21*, 65–87. [CrossRef]
31. Hathout, R.M.; Abdelhamid, S.G.; El-Housseiny, G.S.; Metwally, A.A. Comparing cefotaxime and ceftriaxone in combating meningitis through nose-to-brain delivery using bio/chemoinformatics tools. *Sci. Rep.* **2020**, *10*, 1–7. [CrossRef]
32. Belpomme, D.; Irigaray, P.; Hardell, L.; Clapp, R.; Montagnier, L.; Epstein, S.; Sasco, A.J. The multitude and diversity of environmental carcinogens. *Environ. Res.* **2007**, *105*, 414–429. [CrossRef]
33. Arrebola, J.P.; Fernández-Rodríguez, M.; Artacho-Cordón, F.; Garde, C.; Perez-Carrascosa, F.; Linares, I.; Tovar, I.; González-Alzaga, B.; Expósito, J.; Torne, P.; et al. Associations of persistent organic pollutants in serum and adipose tissue with breast cancer prognostic markers. *Sci. Total Environ.* **2016**, *566–567*, 41–49. [CrossRef] [PubMed]
34. Aubé, M.; Larochelle, C.; Ayotte, P. Differential effects of a complex organochlorine mixture on the proliferation of breast cancer cell lines. *Environ. Res.* **2011**, *111*, 337–347. [CrossRef] [PubMed]
35. Fitriah, A.; Holil, K.; Syarifah, U.; Fitriyah, F.; Utomo, D.H. In silico approach for revealing the anti-breast cancer and estrogen receptor alpha inhibitory activity of Artocarpus altilis. *AIP Conf. Proc.* **2018**, *2021*, 1–7. [CrossRef]
36. Science, E.; Ballschmiter, K. Persistent organic pollutants 274 review articles review articles: Persistent Organic Pollutants Man-made Chemicals Found in Remote Areas of the World. *Environ. Sci. Pollut. Res.* **2019**. [CrossRef]

37. Zheng, X.; Sun, Q.; Wang, S.; Li, X.; Liu, P.; Yan, Z.; Kong, X.; Fan, J. Advances in studies on toxic effects of short-chain chlorinated paraffins (SCCPs) and characterization of environmental pollution in china. *Arch. Environ. Contam. Toxicol.* **2020**, *78*, 501–512. [CrossRef]

38. UNEP. *Stockholm Convention on Persistent Organic Pollutants (POPs): Text and Annexesas Amended in 2009.Secretariat of the Stockholm Convention on Persistent Organic Pollutants*; United Nations Environment Programme (UNEP): Nairobi, Kenya; International Environment House: Geneva, Switzerland, 2009.

39. Fiedler, H. Short-chain chlorinated paraffins: Production, use and international regulations. In *Chlorinated Paraffins*; Springer: Berlin, Heidelberg, 2010; pp. 1–40. [CrossRef]

40. Gong, Y.; Zhang, H.; Geng, N.; Xing, L.; Fan, J.; Luo, Y.; Song, X.; Ren, X.; Wang, F.; Chen, J. Short-chain chlorinated paraffins (SCCPs) induced thyroid disruption by enhancement of hepatic thyroid hormone influx and degradation in male Sprague Dawley rats. *Sci. Total Environ.* **2018**, *625*, 657–666. [CrossRef] [PubMed]

41. Harno, E.; Ramamoorthy, T.G.; Coll, A.P.; White, A. POMC: The physiological power of hormone processing. *Physiol. Rev.* **2018**, *98*, 2381–2430. [CrossRef]

42. Jiang, Y.; Yang, S.; Liu, J.; Ren, T.; Zhang, Y.; Sun, X. Degradation of hexabromocyclododecane (HBCD) by nanoscale zero-valent aluminum (nZVAl). *Chemosphere* **2020**, *244*, 125536. [CrossRef]

43. Szabo, D.T. Hexabromocyclododecane. *Encycl. Toxicol. Third Ed.* **2014**, *2*, 864–868. [CrossRef]

44. Maras, M.; Vanparys, C.; Muylle, F.; Robbens, J.; Berger, U.; Barber, J.L.; Blust, R.; De Coen, W. Estrogen-like properties of fluorotelomer alcohols as revealed by MCF-7 breast cancer cell proliferation. *Environ. Health Perspect.* **2006**, *114*, 100–105. [CrossRef]

45. Zhou, J.; Ng, S.; Adesanya-Famuiya, O.; Anderson, K.; Bondy, C.A. Testosterone inhibits estrogen-induced mammary epithelial proliferation and suppresses estrogen receptor expression. *FASEB J.* **2000**, *14*, 1725–1730. [CrossRef]

46. Folkerd, E.; Dowsett, M. Sex hormones and breast cancer risk and prognosis. *Breast* **2013**, *22*, S38–S43. [CrossRef] [PubMed]

47. Somboonporn, W.; Davis, S.R. Testosterone effects on the breast: Implications for testosterone therapy for women. *Endocr. Rev.* **2004**, *25*, 374–388. [CrossRef] [PubMed]

48. Pasqualini, J.R.; Chetrite, G.S.; Pasqualini, J.R. Biological responses of progestogen metabolites in normal and cancerous human breast. *Horm. Mol. Biol. Clin. Investig.* **2010**, *3*, 427–435. [CrossRef] [PubMed]

49. Trabert, B.; Sherman, M.E.; Kannan, N.; Stanczyk, F.Z. Progesterone and breast cancer. *Endocr. Rev.* **2019**, *41*, 320–344. [CrossRef] [PubMed]

MDPI

St. Alban-Anlage 66

4052 Basel

Switzerland

Tel. +41 61 683 77 34

Fax +41 61 302 89 18

www.mdpi.com

Biology Editorial Office

E-mail: biology@mdpi.com

www.mdpi.com/journal/biology